PROGRESS IN

Molecular Biology and Translational Science

Volume 99

PROGRESS IN

Molecular Biology and Translational Science

Proteases in Health and Disease

edited by

Enrico Di Cera

Department of Biochemistry and Molecular Biology
Saint Louis University School of Medicine
St. Louis, Missouri, USA

Volume 99

ELSEVIER

AMSTERDAM • BOSTON • HEIDELBERG • LONDON
NEW YORK • OXFORD • PARIS • SAN DIEGO
SAN FRANCISCO • SINGAPORE • SYDNEY • TOKYO
Academic Press is an imprint of Elsevier

AP

Academic Press is an imprint of Elsevier
32 Jamestown Road, London, NW1 7BY, UK
Radarweg 29, PO Box 211, 1000 AE Amsterdam, The Netherlands
30 Corporate Drive, Suite 400, Burlington, MA 01803, USA
525 B Street, Suite 1900, San Diego, CA 92101-4495, USA

This book is printed on acid-free paper. ∞

Library of Congress Cataloging-in-Publication Data
A catalog record for this book is available from the Library of Congress

British Library Cataloguing in Publication Data
A catalogue record for this book is available from the British Library

ISBN: 978-0-12-385504-6
ISSN: 1877-1173

For information on all Academic Press publications
visit our website at elsevierdirect.com

Printed and bound by CPI Group (UK) Ltd, Croydon, CR0 4YY

Contents

Contributors ix
Preface xi

Membrane-Anchored Serine Proteases in Health and Disease 1

Toni M. Antalis, Thomas H. Bugge, and Qingyu Wu

I. Introduction 2
II. Structural Features 4
III. Regulation by Endogenous Inhibitors 7
IV. The Type I Transmembrane Serine Protease 7
V. The Type II Transmembrane Serine Proteases 8
VI. The GPI-Anchored Serine Proteases 30
VII. Perspectives 33
References 34

Complex Assemblies of Factors IX and X Regulate the Initiation, Maintenance, and Shutdown of Blood Coagulation 51

Thomas Zögg and Hans Brandstetter

I. Introduction 52
II. Proteolytic Activation of Trypsin-Like Serine Proteases 54
III. Low-Molecular Weight Cofactors 61
IV. Protein Cofactors 68
V. Membranes: Nonhumoral Cofactors of Coagulation Factors 79
VI. Protein Inhibitors of Coagulation Factors 81
VII. Pharmaceutical Opportunities and Challenges 90
References 92

Structure and Proteolytic Properties of ADAMTS13, A Metalloprotease Involved in the Pathogenesis of Thrombotic Microangiopathies 105

Stefano Lancellotti and Raimondo De Cristofaro

I. Introduction 106
II. Gene, Biosynthesis and Secretion of ADAMTS13 109
III. ADAMTS13 Structure and Function: General Aspects 109
IV. ADAMTS13: A Multidomain Enzyme 111
V. Regulation of ADAMTS13 Activity 118
VI. ADAMTS13 and Its Role in the Pathogenesis of Thrombotic Microangiopathy, A Pleiomorphic Clinical Setting 120
VII. Are ADAMTS13 Changes Always Associated with Thrombotic Microangiopathies? 132
VIII. Issues Concerning Laboratory Methods Used to Measure the ADAMTS13 Activity in Clinical Samples 133
IX. Future Directions 134
References 134

Thrombin as an Anticoagulant 145

Enrico Di Cera

I. Preamble 146
II. Thrombin Interactions 147
III. Thrombin Structure 148
IV. Thrombin is an Allosteric Enzyme 151
V. Structures of E*, E, and E:Na^+ 153
VI. Dissociating Procoagulant and Anticoagulant Activities 158
VII. WE: A Prototypic Anticoagulant Thrombin 163
VIII. Molecular Mechanism of Anticoagulant Activity of Thrombin Mutants. 165
IX. Beyond WE 170
X. Conclusions 175
References 175

Regulation of Proteases by Protein Inhibitors of the Serpin Superfamily 185

Steven T. Olson and Peter G.W. Gettins

I. Overview of the Serpin Superfamily 186
II. Mechanism of Action 197
III. Rates of Reaction, Specificity, and Regulation 211

IV. Engineering Serpins as Therapeutics 221
V. Concluding Remarks 230
References 231

Multiple Roles of Metalloproteinases in Neurological Disorders 241

Yi Yang, Jeff W. Hill, and Gary A. Rosenberg

I. Introduction 242
II. Biology of Matrix Metalloproteinases 243
III. MMPs in Neuroinflammation 247
IV. Tissue Inhibitors of Metalloproteinases and Apoptosis 248
V. The Neurovascular Unit, Tight Junction Proteins, and MMPs 250
VI. MMPs and Cell Death 251
VII. Neurological Disorders 253
VIII. Conclusions 257
References 259

Regulation of Caspases in the Nervous System: Implications for Functions in Health and Disease 265

Carol M. Troy, Nsikan Akpan, and Ying Y. Jean

I. The Caspase Family of Death Proteases 266
II. Caspases and Developmental Neuronal Death 275
III. Nonapoptotic Caspase Function in the Nervous System 276
IV. Caspase Function in Neurodegenerative Diseases 277
V. Future Directions 292
References 292

Index 307

Contributors

Numbers in parentheses indicate the pages on which the authors' contributions begin.

Nsikan Akpan, Department of Pathology & Cell Biology; and Department of Neurology, Taub Center for the Study of Alzheimer's Disease and the Aging Brain, Columbia University College of Physicians and Surgeons, New York, USA (265)

Toni M. Antalis, Center for Vascular and Inflammatory Diseases, University of Maryland School of Medicine, Baltimore, Maryland, USA (1)

Hans Brandstetter, Department of Molecular Biology, University of Salzburg, Salzburg, Austria (51)

Thomas H. Bugge, Proteases and Tissue Remodeling Section, Oral and Pharyngeal Cancer Branch, National Institute of Dental and Craniofacial Research, National Institutes of Health, Bethesda, Maryland, USA (1)

Raimondo De Cristofaro, Institute of Internal Medicine and Geriatrics, Physiopathology of Haemostasis Research Center, Catholic University School of Medicine, Rome, Italy (105)

Enrico Di Cera, Department of Biochemistry and Molecular Biology, Saint Louis University School of Medicine, St. Louis, Missouri, USA (145)

Peter G.W. Gettins, The Center for Structural Biology; and Department of Biochemistry and Molecular Genetics, University of Illinois at Chicago, Chicago, Illinois, USA (185)

Jeff W. Hill, Department of Neurology, University of New Mexico Health Sciences Center, Albuquerque, New Mexico, USA (241)

Ying Y. Jean, Department of Pathology & Cell Biology; and Department of Neurology, Taub Center for the Study of Alzheimer's Disease and the Aging Brain, Columbia University College of Physicians and Surgeons, New York, USA (265)

Stefano Lancellotti, Institute of Internal Medicine and Geriatrics, Physiopathology of Haemostasis Research Center, Catholic University School of Medicine, Rome, Italy (105)

Steven T. Olson, Center for Molecular Biology of Oral Diseases; and The Center for Structural Biology, University of Illinois at Chicago, Chicago, Illinois, USA (185)

Gary A. Rosenberg, Department of Neurology, University of New Mexico Health Sciences Center, Albuquerque, New Mexico, USA (241)

Carol M. Troy, Department of Pathology & Cell Biology; and Department of Neurology, Taub Center for the Study of Alzheimer's Disease and the Aging Brain, Columbia University College of Physicians and Surgeons, New York, USA (265)

Qingyu Wu, Molecular Cardiology/Nephrology & Hypertension, Lerner Research Institute/NB20, Cleveland Clinic, Cleveland, Ohio, USA (1)

Yi Yang, Department of Neurology, University of New Mexico Health Sciences Center, Albuquerque, New Mexico, USA (241)

Thomas Zögg, Department of Molecular Biology, University of Salzburg, Salzburg, Austria (51)

Preface

Proteases act as positive or negative effectors of numerous biological processes, either as nonspecific catalysts of protein degradation or highly selective agents controlling physiological events. Many biological pathways involving protease activity have been characterized, and a wealth of information is available. Five classes of proteolytic enzymes are recognized on the basis of their catalytic mechanism: aspartic, cysteine, metallo-, threonine, and serine peptidases. With the advent of whole genome sequencing, this classification system has expanded by the necessity to encompass the diverse catalytic repertoire found in nature. Much work remains to define the diversity of proteolytic events in biological systems and their spatial and temporal distribution in health and disease. This volume contains seven chapters contributed by leaders in their field and provides illustrative examples of the key role of proteases in biological processes, and how proteolytic function and regulation can be harnessed to define new strategies of therapeutic intervention. The reader will find these chapters both stimulating and informative.

In Chapter 1, Antalis, Bugge, and Wu discuss the exciting group of membrane-anchored serine proteases that have emerged as crucial contributors to processes related to development and maintenance of homeostasis.

In Chapter 2, Zogg and Brandstetter review the assembly of multiprotein complexes in blood coagulation, with emphasis on their structural organization, modes of action, and regulation, and outline therapeutic opportunities for the treatment of hemophilia and thrombosis.

In Chapter 3, Lancellotti and De Cristofaro detail the structural and functional properties of ADAMTS13, a disintegrin and metalloprotease involved in the pathogenesis of thrombotic microangiopathies via processing of matrix components.

In Chapter 4, Di Cera reviews the structure, function, and regulation of thrombin and how this knowledge has led to the rational engineering of thrombin variants with intriguing anticoagulant and antithrombotic properties, both *in vitro* and *in vivo*.

In Chapter 5, Olson and Gettins offer an extensive review of protein inhibitors of the serpin superfamily, detailing their structural and kinetic properties and how this knowledge may lead to engineering variants with desired specificity toward protease targets for therapeutic applications.

In Chapter 6, Yang, Hill, and Rosenberg discuss the multiple roles of metalloproteases in neurological disorders, especially their ability to open the blood-brain barrier and degrade the extracellular matrix around blood vessels, and outline the advantages of more specific inhibitors.

In Chapter 7, Troy, Akpan, and Jean deal with the regulation of caspases in the nervous system and discuss their function in the context of neurodegenerative diseases like Alzheimer's and cerebral ischemia.

I thank all the authors for their excellent contributions and their efforts to submit outstanding chapters by the stringent deadlines agreed with the publisher. Expert editorial assistance from Tracey Baird, Senior Administrative Assistant, is gratefully acknowledged and was key to the timely execution of this project.

September 1, 2010

Enrico Di Cera
St. Louis, Missouri

Membrane-Anchored Serine Proteases in Health and Disease

Toni M. Antalis,[*] Thomas H. Bugge,[†] and Qingyu Wu[‡]

[*]*Center for Vascular and Inflammatory Diseases, University of Maryland School of Medicine, Baltimore, Maryland, USA*

[†]*Proteases and Tissue Remodeling Section, Oral and Pharyngeal Cancer Branch, National Institute of Dental and Craniofacial Research, National Institutes of Health, Bethesda, Maryland, USA*

[‡]*Molecular Cardiology/Nephrology & Hypertension, Lerner Research Institute/ NB20, Cleveland Clinic, Cleveland, Ohio, USA*

I. Introduction 2
II. Structural Features 4
  A. Catalytic Domains 5
  B. Extracellular "Stem" Regions 6
  C. Membrane-Anchoring Domains 6
III. Regulation by Endogenous Inhibitors 7
IV. The Type I Transmembrane Serine Protease 7
  A. Tryptase γ1 7
V. The Type II Transmembrane Serine Proteases 8
  A. HEPSIN/TMPRSS Subfamily 8
  B. Matriptase Subfamily 17
  C. Corin Subfamily 24
  D. HAT/DESC Subfamily 28
VI. The GPI-Anchored Serine Proteases 30
  A. Prostasin 30
  B. Testisin 32
VII. Perspectives 33
  References 34

Serine proteases of the trypsin-like family have long been recognized to be critical effectors of biological processes as diverse as digestion, blood coagulation, fibrinolysis, and immunity. In recent years, a subgroup of these enzymes has been identified that are anchored directly to plasma membranes, either by a carboxy-terminal transmembrane domain (Type I), an amino-terminal transmembrane domain with a cytoplasmic extension (Type II or TTSP), or through

Progress in Molecular Biology
and Translational Science, Vol. 99
DOI: 10.1016/S1877-1173(11)99001-2

1877-1173/11 $35.00

a glycosylphosphatidylinositol (GPI) linkage. Recent biochemical, cellular, and *in vivo* analyses have now established that membrane-anchored serine proteases are key pericellular contributors to processes vital for development and the maintenance of homeostasis. This chapter reviews our current knowledge of the biological and physiological functions of these proteases, their molecular substrates, and their contributions to disease.

I. Introduction

Proteolytic enzymes comprise over 2% of the known proteome, and their participation in many essential biological processes is well established. The serine proteases constitute one of the largest families of proteolytic enzymes and are well recognized for their pivotal roles in physiological processes as diverse as development, digestion, coagulation, inflammation, and immunity. These enzymes share a common catalytic mechanism for selective cleavage of specific substrates and are frequently involved in consecutive proteolytic reactions or protease cascades, where one protease precursor or zymogen is the substrate for an active protease. This shared mechanism confers the advantage that a single signal may be specifically and irreversibly amplified every time a downstream zymogen is activated, providing the capacity for unleashing a burst of proteolytic potential.

Most of the well-characterized members of the S1 family of serine proteases are either secreted enzymes or exocytosed from secretory vesicles into the extracellular environment. Trypsin and chymotrypsin, the main intestinal digestive enzymes, are prototype members of the S1 family. Over the past 10 years, a structurally distinct group of S1 serine proteases, termed broadly as the membrane-anchored serine proteases, has emerged that are synthesized with amino- or carboxy-terminal extensions that serve to anchor their serine protease catalytic domains directly at the plasma membrane[1,2] (Fig. 1).

The largest group of membrane-anchored serine proteases is the Type II transmembrane serine proteases or TTSPs.[1] These proteases are synthesized with an amino-terminal signal anchor that is not removed during synthesis, but serves as a transmembrane domain that positions the protease in the plasma membrane with a cytoplasmic amino-terminal domain of variable length (20–160 amino acids) and the catalytic serine protease domain at the carboxy-terminus.[1] These serine proteases are synthesized as single-chain precursors or zymogens; activation produces a two-chain form with the chains linked by a disulfide bridge, so that the active enzyme remains membrane bound. Nineteen human TTSPs have been identified and may be categorized into four subfamilies: Hepsin/TMPRSS, Matriptase, HAT/DESC, and Corin[3,4] (Fig. 1).

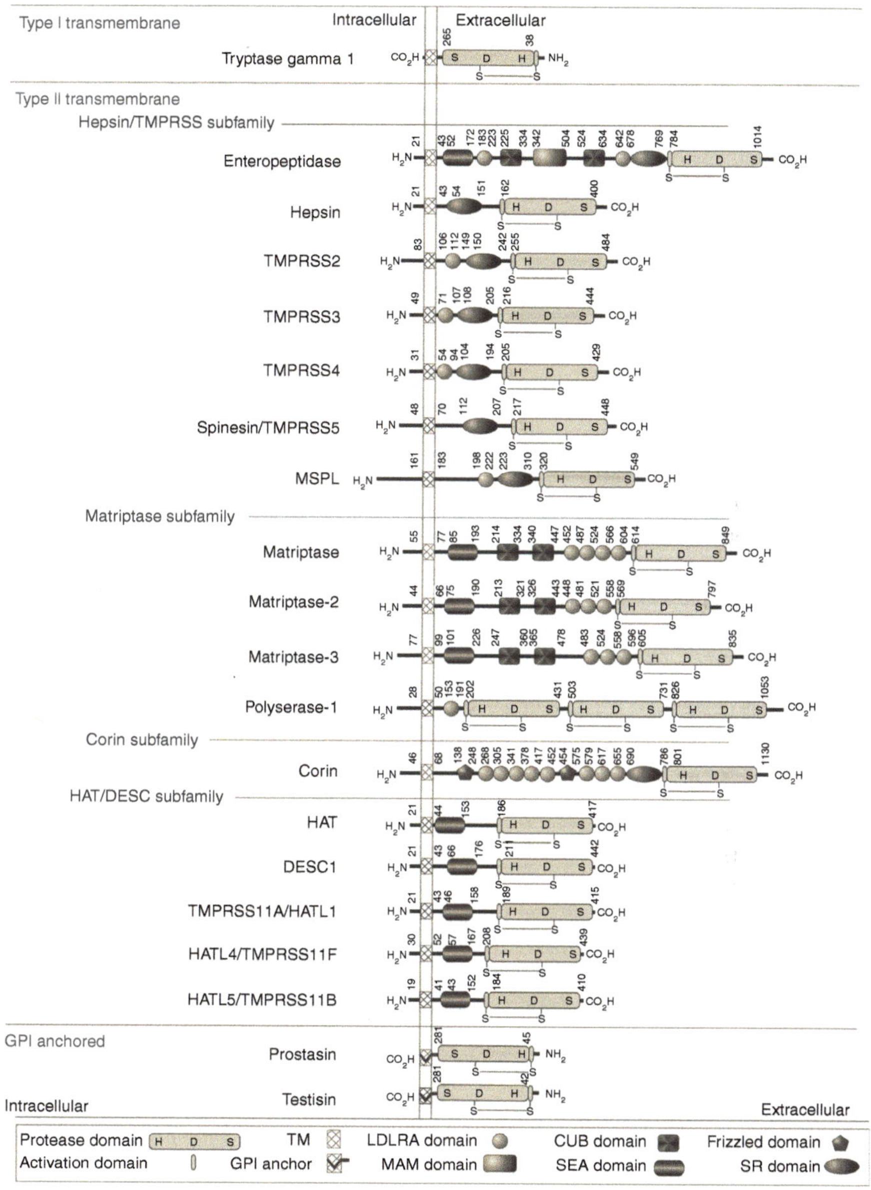

FIG. 1. *Domain structures of the human membrane-anchored serine proteases.* Structures are grouped according to similarity in domain structure to each other. Consensus domains are as indicated at the bottom of the figure. The location of each protein domain (amino acid numbering) is indicated above the domain. Amino and carboxy termini are as indicated. Protease domain: serine protease domain; activation domain: pro-domain; TM: transmembrane domain; GPI anchor: glycosylphosphatidylinositol linkage domain; LDLRA: LDL receptor class A domain; MAM: meprin, A5 antigen, and receptor protein phosphatase μ domain; CUB: Cls/Clr, urchin embryonic growth factor and bone morphogenetic protein-1; SEA: sea urchin sperm protein, enteropeptidase, agrin domain; Fz: frizzled domain; SR: Group A scavenger receptor domain. Modified from Ref. 2.

Mammalian orthologs, as well as different isoforms between humans and rodents, exist for many if not all of the TTSPs.[4–6] There also exist two nonmammalian TTSPs, *Drosophila* stubble-stubbloid (st-sb)[7] and corin.[8]

In contrast to the TTSPs, additional membrane-anchored serine proteases of the S1 family each possess an amino-terminal signal peptide and enter the secretory pathway. These enzymes are also synthesized as zymogens, with an amino-terminal extension that acts as a pro-peptide, requiring proteolytic cleavage to generate the active enzyme. The Type I transmembrane serine protease, tryptase $\gamma 1$, is the only human membrane-anchored serine protease synthesized with a carboxy-terminal hydrophobic extension that serves as a transmembrane domain.[9,10] The carboxy-terminal extensions of prostasin and testisin are modified posttranscriptionally with a glycosylphosphatidylinositol (GPI) linkage that anchors these proteases in the plasma membrane.[11–15]

The membrane-anchored serine proteases are proving to be key components of the cell machinery for activation of precursor molecules in the pericellular microenvironment, with several playing vital roles during development and the maintenance of homeostasis. There is also growing evidence for their participation in the pathogenesis of inflammatory and neoplastic diseases. Endogenous protein substrates targeted by membrane-anchored serine proteases include peptide hormones, growth and differentiation factors, receptors, enzymes, adhesion molecules, and viral coat proteins.[16] A number of insights into our understanding of the unique physiological functions of the membrane-anchored serine proteases and their involvement in human pathology have come from a combination of biochemical analyses, animal models, and human patient studies. However, our current understanding of the impact of the membrane-anchored serine proteases on many biological, physiological, and pathological processes is far from complete. This chapter provides a historical perspective on the discovery of these enzymes, current knowledge of their activities and their regulation, and the functional consequences of the activities of these proteases in mammalian physiology and disease. For the interested reader, several other reviews have focused on different aspects of their nomenclature, classifications into subgroups, gene structure and chromosomal localization, tissue- and cell-specific distribution, and biochemical properties.[1–4,16,17]

II. Structural Features

All of the membrane-anchored serine proteases have membrane-anchoring domains and structurally conserved serine protease catalytic domains. The TTSPs have additional extracellular stem regions that separate the catalytic domains from their transmembrane domains. The extracellular regions of the membrane-anchored serine proteases are believed to be essential to the biological and physiological functions ascribed to these enzymes.

A. Catalytic Domains

The zymogen forms of the membrane-anchored serine proteases are activated by proteolytic cleavage following an arginine or lysine amino acid present in a highly conserved activation motif separating the pro- and catalytic domains. The catalytic mechanism of the membrane serine proteases involves a catalytic triad of three amino acids, serine (nucleophile), aspartate (electrophile), and histidine (base), present in highly conserved sequence motifs. While the geometric orientations of these amino acid residues are similar, the protein folds are variable, which contribute to their selective substrate specificities. The catalytic reaction follows a two-step mechanism for hydrolysis of substrates in which a covalently linked enzyme-peptide intermediate is formed, with the loss of a peptide fragment.[18] This acylation step is followed by a deacylation step which occurs by a nucleophilic attack on the intermediate by water, resulting in hydrolysis of the peptide.

Some insights into the structural features that contribute to the unique catalytic and substrate specificities of the membrane-anchored serine proteases have been obtained through comparative analyses of amino acid sequences[1,2,19] combined with tertiary structural analyses.[20–26] Consistent with the family of S1 serine proteases, each catalytic domain possesses two adjacent, six-stranded β-barrel domains that are connected by three trans-domain segments. The catalytic triad amino acids are located along the junction between the two barrels, with the active site cleft running perpendicular to this junction.[27] The size, shape, and charge distribution within the formed binding pocket of the active enzyme are determinants of substrate specificity. These pockets are defined by differing substrate-binding subsites (e.g., S_4–S_2') and loop regions that surround the active site cleft.[23]

The specificity for cleavage of substrates with the positively charged amino acid residues, lysine or arginine, in the P_1 position (the position directly preceding the cleaved peptide bond) is conferred by the presence of a conserved aspartate residue at the bottom of the binding pocket of all of the activated enzymes.[2] The rate of cleavage is influenced by the amino acid residues surrounding the P_1 residue, numbered P_1 to P_n, counting outward from the amino-terminal side of the peptide bond that is cleaved during hydrolysis, and numbered P_1' through P_n' from the carboxy-terminal side.[28] In recent years, recombinant catalytic domains of several of the membrane-anchored serine proteases have been produced in various laboratories and peptide screening assays applied to quantitatively identify cleavage preferences for the P_1 and P_1' amino acids, and surrounding amino acid positions.[29–35]

Several of the biochemically purified TTSPs rapidly undergo autocatalytic activation *in vitro* (matriptase[29,36] matriptase-2,[37] hepsin,[38] TMPRSS2,[39] TMPRSS3,[40] TMPRSS4,[41] HAT-like 3/TMPRSS11C[42]). Thus, mutation of active site catalytic triad residues of several of these TTSPs prevents activation site cleavage.[3,36,42,43] It has been speculated that autoactivation of these TTSPs could contribute to the initiation of proteolytic cascades at the cell membrane, although the signals that induce internal cleavage or autoactivation *in vivo* have not yet been defined.

B. Extracellular "Stem" Regions

While the catalytic domains of the Type I and GPI-anchored serine proteases lie directly proximal to the membrane anchoring domain, the TTSPs are characterized by the presence of up to 11 extracellular structural domains present between their catalytic domains and the transmembrane domain (Fig. 1). These "stem" regions appear to serve as regulatory and/or binding interaction domains,[1] and are believed to variously contribute to the cellular localization, activation, inhibition, and/or substrate specificity of these enzymes.[44] The most common stem region structural domain is the low-density lipoprotein (LDL) receptor class A domain (LDLRA): corin contains eight, matriptase contains four, enteropeptidase two, and TMPRSS2 and TMPRSS4 one each of these domains (Fig. 1). In addition to the LDLRA domains, there are Group A scavenger receptor (SR) domains, frizzled domains, Cls/Clr, urchin embryonic growth factor and bone morphogenic protein 1 (CUB) domains, sea urchin sperm protein, enterokinase, agrin (SEA) domains, and meprin, A5 antigen, and receptor protein phosphatase μ (MAM) domains. The specific contributions of each stem domain to TTSP proteolytic activities have not yet been well defined, although for several of the TTSPs, the stem domain is required for efficient cleavage of their physiological substrates.[45,46]

Targeted release of the extracellular domain, or ectodomain shedding, occurs for several of the membrane-anchored serine proteases, enabling the release of the soluble catalytic domain into the extracellular space. Soluble shed protease forms have been reported for enteropeptidase, HAT, matriptase, matriptase-2, and prostasin.[11,47–49]

C. Membrane-Anchoring Domains

Surface localization studies demonstrate that membrane-anchored serine proteases normally localize to the cell surface and are differentially distributed on apical or basolateral surfaces of polarized cells in patterns unique for each protease.[3,4] Studies indicate that the TTSP transmembrane domains are dispensable for the catalytic activities of many of the TTSPs, including enteropeptidase, matriptase, hepsin, corin, and MSPL.[4] It appears that the function of the transmembrane domain in TTSPs is not to enhance their catalytic activities but to target the TTSPs to plasma membranes and localize their activities at specific sites. Similarly, the GPI-anchored serine proteases are found to associate with lipid raft microdomains at cell surfaces, but do not require these anchors for catalytic activity. Loss of membrane polarity, such as that in tumors, has been associated with redistribution and/or mislocalization of several of the membrane-anchored serine proteases.[50,51] In this regard, the significance of cell membrane association for the membrane-anchored serine proteases differs from that for blood-clotting proteases.

III. Regulation by Endogenous Inhibitors

As is the case with all serine proteases, enzyme activities must be tightly regulated in order to prevent inappropriate and frequently destructive proteolysis. The catalytic activities of the membrane-anchored serine proteases are regulated by endogenous protease inhibitors, specifically the Kunitz-type transmembrane serine protease inhibitors, hepatocyte growth factor (HGF) activator inhibitor (HAI)-1/SPINT1 and HAI-2/SPINT2, and several of the serine protease inhibitor family known as serpins. Kunitz-type inhibitors are a class of serine protease inhibitors present in all metazoa, whose prototype is the bovine pancreatic trypsin inhibitor (BPTI). They are competitive inhibitors acting in a substrate-like manner and form very stable complexes of 1:1 stoichiometry with their target enzymes, inhibiting their activity.[52] Matriptase,[53–56] hepsin,[57] and prostasin[31,48,57,58] are regulated by the Kunitz inhibitors, HAI-1 and/or HAI-2.

Serpin-type inhibitors utilize an irreversible conformational mechanism presenting a "pseudo-substrate" exposed binding loop to the protease which, upon cleavage, forms a covalent complex with the target protease.[59,60] The serpins α1-antitrypsin, α2-antiplasmin, antithrombin III, protein C inhibitor, PAI-1, and protease nexin I are effective inhibitors of various recombinant catalytic domains of the membrane-anchored serine proteases (reviewed in Ref. 16).

IV. The Type I Transmembrane Serine Protease

A. Tryptase γ1

Tryptase γ1 or transmembrane tryptase (TMT) was identified as a major granule constituent of numerous populations of human and mouse mast cells.[9,10,61,62] The gene that encodes tryptase γ1 lies on human chromosome 16p13.3, a large locus that contains 14 serine protease-like genes.[9,63] Expression of tryptase γ1 is restricted to cells of hematopoietic origin and has been studied in mast cells.[62] The Type I transmembrane domain at the carboxy-terminus of tryptase $\gamma 1$ causes its cellular retention when mast cells are induced to release the contents of their secretory granules, at which time it reaches the external face of the plasma membrane.[62]

The physiological functions of tryptase $\gamma 1$ remain unknown. Mast cells play beneficial immunosurveillance roles in host defense, particularly during parasitic infections, and are also implicated in asthma and other pathological conditions.[64,65] Administration of recombinant tryptase $\gamma 1$ into the trachea of mice promotes airway hyperresponsiveness and increased expression of interleukin-13 in bronchial alveolar lavage fluids of these animals. This effect was not obtained in STAT6- and IL-4Rα-null mice, indicating that tryptase $\gamma 1$ exerts its adverse effect in the lung in this animal model primarily by inducing the expression of IL-13, a central mediator of allergic asthma.

Tryptase $\gamma 1$ has been expressed as a recombinant catalytic domain and its substrate specificity characterized using combinatorial peptide substrate libraries.[33] Tryptase $\gamma 1$ prefers multibasic residues in the P_4–P_1 positions and has a strong preference for aromatic residues in the P_2 position.[33] It has also been shown to be rapidly inactivated by the human plasma serpin α_1-antitrypsin[62] and a submicromolar synthetic inhibitor,[33] *in vitro*.

V. The Type II Transmembrane Serine Proteases

A. HEPSIN/TMPRSS Subfamily

The HEPSIN/TMPRSS (transmembrane protease/serine) subfamily has seven members comprising hepsin, TMPRSS2, TMPRSS3, TMPRSS4, TMPRSS5/spinesin, MSPL (mosaic serine protease large-form), and enteropeptidase.[4] All of these enzymes possess a Group A SR domain in their stem region, preceded by a single LDLRA domain in TMPRSS2, TMPRSS3, TMPRSS4, and MSPL or by an array of SEA, LDLRA, CUB, and MAM domains in enteropeptidase (Fig. 1).

1. Enteropeptidase

Enteropeptidase is an intestinal protease discovered by Ivan Pavlov in Russia.[66] Using surgically modified dogs, Pavlov and colleagues studied the digestive system and found that pancreatic proteases were made as inactive forms. Upon entering the gut, latent pancreatic proteases were activated by another enzyme in the upper section of the intestine. Such an activation mechanism, which occurs outside of the pancreas, is important to prevent autoactivation of pancreatic proteases, thereby avoiding harmful tissue damage. Pavlov named the intestinal enzyme enterokinase for its transformative activation activity. Currently, the enzyme is called enteropeptidase to reflect its proteolytic, but not kinase, activity.

Initial biochemical studies on enteropeptidase were carried out in the early 1900s, notably by Ernst Waldschmidt-Leitz and colleagues at the Biochemical Institute of the German Technical College in Prague.[67] Porcine enteropeptidase was partially purified from duodenal fluids and shown to activate trypsinogen from pancreatic extracts. More kinetic studies were done by Moses Kunitz at the Rockefeller Institute for Medical Research in Princeton, New Jersey, demonstrating enzymatic conversion of crystalline trypsinogen to trypsin using purified porcine enteropeptidase.[68,69] By the 1970s, more advanced methods were used to purify enteropeptidase from porcine, bovine, and human intestinal mucosa.[70,71] The protein was found to have a heavy chain of approximately 120 kDa and a light chain of approximately 35 kDa connected by a disulfide bond. Such a two-chain structure was known for chymotrypsin and other serine proteases. Enteropeptidase cleaves trypsinogen at the activation site, DDDDK↓I. The activity is inhibited by Kunitz pancreatic trypsin inhibitor, BPTI, and small molecule serine protease inhibitors, such as diisopropyl fluorophosphate (DFP), *p*-aminobenzamidine, and benzamidine, but not by chicken ovomucoid, soybean trypsin inhibitor, chymostatin, pepstatin A, or bestatin.[66]

The partial amino-terminal sequence of the bovine enteropeptidase light chain was determined by Edman degradation, revealing its homology to other trypsin-like serine proteases.[72] This sequence information helped to clone human, bovine, and porcine enteropeptidase cDNAs.[73–76] The open reading frame of human enteropeptidase cDNA encodes a Type II transmembrane protein of 1019 amino acids with a calculated mass of 113 kDa and 17 potential N-linked glycosylation sites. The overall protein domain structure is shown in Fig. 1. Within the protease domain, the activation cleavage site, the catalytic triad, and the key cysteine residues are well conserved, indicating that enteropeptidase and other trypsin-like proteases have similar three-dimensional structures and comparable activation and catalytic mechanisms.

The human *PRSS7* gene, encoding enteropeptidase, is located on chromosome 21q21.[73] The gene has 25 exons and spans approximately 90-kb in length. Enteropeptidase mRNA is expressed mostly in the duodenum and, at lower levels, in the proximal segment of jejunum, consistent with Pavlov's finding that the enzyme was in the upper part of the small intestine. In these tissues, enteropeptidase mRNA is expressed in the enterocytes and goblet cells of the villus.[77] As a transmembrane protein, enteropeptidase is expected to remain at the site of expression, consistent with the previous observation that its activity was associated with the brush border of enterocytes.[78,79] Low levels of enteropeptidase mRNA were detected in other tissues, including stomach, colon, brain, and skin. The biological significance, if any, of its expression outside the digestive system remains unknown.

Enteropeptidase is anchored on the cell surface through its amino-terminal transmembrane domain. Other structural elements also play a role in its membrane targeting. The mucin-like repeats in the SEA domain and *N*-glycans in the protease domain have been found necessary for apical sorting.[80,81] Like many membrane proteins, enteropeptidase is shed from the cell surface. Hadorn *et al.* showed that enteropeptidase was released into the small intestine lumen when brush-border membranes were treated with bile-acids or cholescystokinin–pancreozymin.[82] To date, enzymes responsible for enteropeptidase shedding have not been characterized. It remains to be determined whether the shedding represents a physiological mechanism regulating enteropeptidase activity in the gut.

The mechanism responsible for the activation of the enteropeptidase zymogen is unclear. In transfected cells, enteropeptidase was expressed as a single-chain molecule and no autoactivation was observed, indicating that other proteases are required to activate enteropeptidase.[80] Trypsin activates enteropeptidase efficiently, suggesting a possible mechanism of reciprocal activation between enteropeptidase and trypsin.[66] For such an activation cycle to start, however, at least a fraction of enteropeptidase needs to be activated before trypsinogen enters the gut. Zamolodchikova *et al.* identified a novel serine protease, duodenase, from bovine duodenal mucosa.[83] The enzyme, which is approximately 30 kDa and consists of 226 amino acids, is expressed in the secretory epithelial cells of Brunner's glands in the proximal segment of the duodenum.[84] Duodenase has a preferred cleavage sequence with Lys at P_1 and Pro at P_2. It activated recombinant enteropeptidase, although the rate of the activation was approximately 100 times slower than that of trypsin.[85] It remains to be firmly established whether duodenase is a physiological enteropeptidase activator. More recently, other proteases have been shown to modulate enteropeptidase activity. For example, beta-site APP-cleaving enzyme1 (BACE1) was detected in pancreatic islets and shown to inactivate enteropeptidase by cleavage of its light chain.[86] This inactivation may represent a defense mechanism to inhibit enteropeptidase that enters the pancreatic duct accidentally, thereby preventing trypsinogen activation in the pancreas.

One of the remarkable features of enteropeptidase is its unique substrate specificity, which recognizes Lys at P_1 and a cluster of four Asp residues at P_2–P_5. Within this recognition sequence, a Lys or an Arg residue at P_1 and Lys residues at P_2 and P_3 appear to be most important for efficient cleavage.[87] The structural determinants for enteropeptidase substrate specificity have been localized in its protease domain. There is a group of four conserved basic residues, R/KRRK at positions 96–99, which were suspected to interact with the acidic P_2–P_5 residues in the trypsinogen activation site.[74] In a crystal structure of bovine enteropeptidase light chain, Lys99 residue was found to have extensive contacts with the P_2 and P_4 Asp residues.[22] Lys99 residue is conserved in enteropeptidase from many species. Substitution of Lys99 with Ala by mutagenesis prevented enteropeptidase from activating trypsinogen. In contrast, substituting Lys96, Arg97, and Arg98 residues for bovine enteropeptidase activity had less significant effects.[22] The unique enteropeptidase substrate specificity has been exploited in protein engineering. The DDDDK↓I sequence is frequently used in recombinant proteins that require specific cleavage.

Enteropeptidase deficiency impairs food digestion and absorption.[88] To date, a number of patients who had low or undetectable enteropeptidase activity in intestinal biopsies or duodenal fluid samples have been identified. These patients suffer from diarrhea, vomiting, edema, anemia, and hypoproteinemia. Consequently, the patients fail to gain weight in early infancy. Genetic studies have shown deletion or nonsense mutations in the *PRSS7* gene in enteropeptidase-deficient patients.[66] For these patients, treatment with pancreatic extracts, which contain active trypsin, helps to initiate proteolytic reactions in the gut. Usually, the therapy is effective for the patients to absorb nutrients and gain weight. Apparently, the digestive enzyme reactions, once started, are self-sustained. Pancreatic extracts can be discontinued over time without causing gastrointestinal symptoms in these patients.[89]

2. Hepsin

Hepsin was originally cloned in Earl Davie's Laboratory at the University of Washington. Leytus *et al.* screened human liver cDNA libraries, which were rich in serine proteases.[90] A novel cDNA encoding a trypsin-like serine protease was identified by degenerate oligonucleotides. The protease was named hepsin for its hepatic expression. Rat and mouse hepsin cDNAs were isolated subsequently, which share high sequence homology with human hepsin.[91,92] The full-length human hepsin cDNA, approximately 1.8-kb in length, encodes a polypeptide of 417 amino acids with a calculated mass of 45 kDa. Its protein domain structure is shown in Fig. 1. There is one potential N-linked glycosylation site at residue 112 in the SR-like domain.

The human *HPN* gene, encoding hepsin, is on chromosome 19 at q11-13.2 and has 14 exons. Hepsin mRNA expression is abundant in the liver.[93] Other tissues such as the kidney, pancreas, stomach, prostate, and thyroid express low levels of hepsin mRNA. The promoter sequence of the mouse *Hpn* gene has been characterized.[92] In a 274-bp sequence upstream of the transcription initiation site, several potential binding sites for transcription factors such as

SP1, AP2, C/EBP, LF-A1, and E-box have been identified. This sequence showed strong promoter activities when tested in human HepG2 cells, indicating that it retains the essential elements required for hepatic expression.

In hepatocytes, hepsin is synthesized as a single-chain zymogen. Human hepsin purified from hepatocytes had an apparent mass of approximately 51 kDa on Western blots.[93] A similar molecular mass was found for native rat and recombinant human hepsin. During protein purification, human and rat hepsin were readily activated to become a two-chain molecule. It is unclear whether the activation was mediated by autocleavage or other proteases in hepatocytes. The membrane association and topology of hepsin have been confirmed by immunostaining and proteolytic digestion.

In peptide substrate-based assays, hepsin favors basic residues at the P_1 position. Thr/Leu/Asn, Gln/Lys, and Pro/Lys residues are favored at the P_2, P_3, and P_4 positions, respectively.[31] This substrate profile does not appear to be unique. Somoza *et al.* have solved the crystal structure of a soluble human hepsin that included the SR-like and protease domains.[21] As expected, hepsin protease domain has an architecture of two six-stranded β barrels. There are several structurally distinct loops, including an especially large one between residues 241 and 256, which may interact with its substrates. The activity of hepsin is inhibited by nonspecific inhibitors such as leupeptin, aprotinin, antipain, 4-amidiophenylmethylsulfonyl fluoride, N^{α}-tosyl-L-lysine chloromethyl ketone, soybean trypsin inhibitor, and antithrombin III.[94] No inhibition was detected with EDTA, indicating that divalent ions are not required for hepsin activity toward peptide substrates. More recently, HAI-1 and HAI-2 have been reported to be potent hepsin inhibitors.[31,95]

While many proteins are activated or cleaved by hepsin *in vitro*,[96–99] the physiological function of hepsin remains unclear. These proteins include blood clotting factors VII, IX, and XII; pro-urokinase; pro-HGF; liver microsomal glutathione transferase; matriptase; prostasin; epidermal growth factor receptor; and laminin. It is unknown how many of them are hepsin substrates *in vivo*. Studies have suggested that hepsin may play a role in blood coagulation, hepatocyte proliferation/growth, and embryonic development.[43,100,101] Such functions, however, have not been validated *in vivo*, as hepsin null mice are fertile, grow normally, and exhibit no spontaneous bleeding.[102,103] Unexpectedly, however, hepsin appears to be important in normal hearing. Guipponi *et al.* detected hepsin mRNA expression in the inner ear.[104] Hepsin null mice have been found to have abnormal cochlea, reduced myelin protein expression in the auditory nerve, and severely impaired hearing. The biochemical basis for hepsin function in cochlear development has not been determined. Apparently, hepsin null mice have low levels of plasma thyroxine, a hormone important for cochlear development.[104] It is unknown whether impaired hearing in hepsin null mice is a result of thyroid hormone deficiency.

Multiple lines of evidence suggest a potential role of hepsin in human prostate cancer growth and progression. Microarray studies detected high levels of hepsin mRNA in human prostate cancers.[105–108] Elevated hepsin mRNA expression appears to correlate with the disease severity. Single nucleotide polymorphisms (SNPs) in or near the human *HPN* gene are associated with prostate cancer susceptibility and tumor aggressiveness.[109] In functional studies, Xuan *et al.* showed that antibodies neutralizing hepsin protease activity did not block prostate or ovarian cancer cell growth in culture but inhibited their invasion in a Matrigel basement membrane.[110] In other studies, however, prostate or other cancer cell lines stably transfected with hepsin cDNA were less invasive in culture or nude mice. Because stably transfected cells often undergo phenotypic changes during selection, it is difficult to determine whether observed differences were related to different levels of hepsin expression. To circumvent this problem, Klezovitch *et al.* created transgenic mice with high levels of hepsin in prostate epithelium, mimicking hepsin overexpression in prostate cancer patients.[111] The results show that hepsin overexpression did not affect cell proliferation but altered basement membrane structures in prostates in these mice. When the mice were crossed with a nonmetastasizing prostate cancer model, prostate cancer cells became more invasive, causing metastasis in the liver, lung, and bone. Thus, hepsin promotes prostate cancer progression and metastasis *in vivo*, suggesting that hepsin inhibitors may be developed to treat prostate cancer in patients.

3. TMPRSS2

TMPRSS2 cDNA was originally cloned by exon trapping in a gene mapping study of human chromosome 21.[112] The *TMPRSS2* gene, located at human chromosome 21q22.3,[112] is approximately 44 kb in length and includes 14 exons.[113] The full-length TMPRSS2 cDNA encodes a polypeptide of 492 amino acids, with a domain structure shown in Fig. 1. Mouse TMPRSS2, also called epitheliasin, was isolated from a kidney cDNA library.[114] Human TMPRSS2 mRNA is expressed in many tissues, including the prostate, breast, bile duct, kidney, colon, small intestine, pancreas, ovary, salivary gland, stomach, and lung.[114,115] In these tissues, TMPRSS2 protein and mRNA are mostly in the epithelial cells. In human prostate cancers, TMPRSS2 protein is localized on the apical membrane of secretory epithelia and in the lumen of the glands.[39] Similar luminal staining patterns are found in human colon cancers. Many human prostate- and colon-derived cells, including APC-4, LNCaP, LoVo, T84, and Colo-205, also express TMPRSS2 mRNA.[39]

In vitro-translated TMPRSS2 protein is detected as a zymogen of approximately 54 kDa,[39] whereas native and recombinant human TMPRSS2 proteins have a higher molecular mass (~ 60–70 kDa) due to N-linked glycosylation.[39,116]

In human prostate cancer tissues or cells, approximately 32-kDa TMPRSS2 protease domain may also be detected, suggesting that the protease is partially activated. This cleavage appears to be due to TMPRSS2 autoactivation, as active site mutation prevented the cleavage of TMPRSS2 in transfected HEK 293T cells.[39] To date, TMPRSS2 substrate specificity and catalytic properties have not been well characterized.

The physiological roles of TMPRSS2 are currently unknown. In a *Xenopus* oocyte expression system, TMPRSS2 expression reduced epithelial sodium channel (ENaC) current and protein levels.[117] In addition, TMPRSS2 activates influenza virus by cleaving hemagglutinin, suggesting that the enzyme may contribute to virus invasion of human airways.[118,119] Secreted forms of TMPRSS2 have recently been found in human seminal prostasomes, suggesting its potential role in regulating sperm function.[116] In prostate cancer cells, TMPRSS2 was shown to activate PAR-2.[120] It is unclear whether these activities are critical *in vivo*. TMPRSS2 null mice are viable and fertile with no reported abnormalities.[121]

The human *TMPRSS2* gene promoter has a 15-bp androgen response element at position -148 relative to the putative transcription start site. Consistently, TMPRSS2 mRNA expression is elevated in androgen-stimulated prostate cancer (LNCaP) cells.[115] The upregulation of TMPRSS2 mRNA by androgen appears to be mediated by the androgen receptor.[39] Androgen treatment is also reported to increase TMPRSS2 zymogen activation in cell culture and in a mouse xenograft model, suggesting that TMPRSS2 may contribute to prostate cancer development and progression in an androgen-dependent manner.

Frequent gene fusions between *TMPRSS2* and members of the E26 transformation specific (ETS) transcription factor family have been identified in many prostate cancer patients.[122] In these patients, the 5′-untranslated region of *TMPRSS2* is fused with the transcription factor *ERG* or *ETV* genes. As a result, the androgen-responsive promoter elements of *TMPRSS2* drive the expression of the ETS family transcription factors to promote prostate cancer progression and invasion. Among these chromosomal rearrangements, *TMPRSS2–ERG* gene fusions are most frequent, accounting for approximately 50% of prostate cancers. Molecular studies show that abnormal ERG expression disrupts normal androgen receptor signaling and activates epigenetic programs, thereby inducing tumorigenesis.[123] Although ERG enhances the expression of many proteases, including urokinase-type plasminogen activator (uPA), matrix metalloproteinase (MMP)-3, and MMP9, TMPRSS2 protease activity is unlikely to have a direct role in ERG downstream events, because the *TMPRSS2* protease domain coding sequence is never included in the fusion genes.[124] Studies have shown that in prostate cancers, ETS transcription factor genes are also fused to other genes that contain androgen-responsive elements.[125] These rearrangements have been attributed recently to androgen-induced co-recruitment of the DNA-processing enzyme topoisomerase 2B to sites of genomic breakpoints.[126]

4. TMPRSS3

TMPRSS3 was identified as a novel serine protease overexpressed in ovarian cancers and was originally named TADG-12 (tumor-associated differentially expressed gene-12).[127] Subsequent reports showed that this protease was also overexpressed in pancreatic and breast cancers. Independently, the gene encoding TMPRSS3 was identified as a candidate gene located within the disease locus (*DFNB10*) for autosomal recessive deafness,[128] previously mapped by linkage analysis to chromosome 21q22.[129] The human *TMPRSS3* is approximately 24 kb in length and contains 13 exons.[130] The approximately 2.4-kb TMPRSS3 cDNA encodes a polypeptide of 454 amino acids, with an overall domain structure similar to TMPRSS2 (Fig. 1). In fact, these two proteins share approximately 63% sequence similarities. TMPRSS3 mRNA is expressed in a variety of tissues, including the kidney, lung, colon, thymus, stomach, and cochlea.[130]

TMPRSS3 mutations have been found in patients with congenital and childhood onset autosomal recessive deafness. Scott *et al.* first identified an 8-bp deletion and insertion of 18 β-satellite tandem repeats in the *TMPRSS3* gene in a Palestinian family with congenital deafness.[128] In the same study, a second splice site mutation in the *TMPRSS3* gene causing a frameshift was found in a Pakistani family with childhood-onset deafness. To date, deletion, insertion, frameshift, and nonsense *TMPRSS3* gene mutations have been reported in over 20 patient families with nonsyndromic autosomal recessive deafness.[130] Most mutations occur in exons coding for the LDL receptor repeat, SR repeat, or protease domain, resulting in complete or near complete loss of the protease activity. These studies demonstrate that defects in the *TMPRSS3* gene are responsible for hearing loss in these patient families.

Biochemical mechanisms by which TMPRSS3 regulates normal hearing remain unknown. Analogous to hepsin, TMPRSS3 mRNA is expressed in the cochlea.[131] Considering that hepsin null mice have severely impaired hearing, it is possible that the TTSPs participate in a proteolytic pathway required for normal cochlear structure and/or hearing function. ENaC is known to regulate Na^+ levels in the perilymph of the inner ear. Guipponi *et al.* have shown that TMPRSS3 activates ENaC *in vitro* and that most naturally occurring mutations prevented TMPRSS3 activation in cell-based experiments.[40] It is unclear, however, whether ENaC is a physiological TMPRSS3 substrate in the inner ear. It has been reported that pseudohypoaldosteronism type 1 patients, who are homozygous for mutations in the ENaCα subunit, do not suffer hearing loss,[132] suggesting that TMPRSS3 may act on other substrates *in vivo*.

5. TMPRSS4

TMPRSS4, initially called TMPRSS3 and also known as CAP2, was originally identified in pancreatic cancers.[133] The gene encoding TMPRSS4 is located on human chromosome 11q23.3. Human TMPRSS4 protein consists of 437 amino acids and a domain structure, as shown in Fig. 1. TMPRSS4 mRNA is expressed in the pancreas, lung, stomach, colon, kidney, skin, and eye.[131,133] The biological function of TMPRSS4 in these tissues is unknown. Like TMPRSS2, TMPRSS4 is able to activate ENaC when coexpressed in Xenopus oocytes[40,134], and TMPRSS4 was recently shown to cleave in the inhibitory ENaCγ subunit.[135] TMPRSS4 activates influenza virus hemagglutinin, suggesting that it may facilitate viral infection in lung tissues.[136]

TMPRSS4 mRNA overexpression has been documented in many cancers, including pancreatic, gastric, colorectal, lung, and thyroid.[133,137] In lung and colon cancer cells, knockdown of TMPRSS4 expression by siRNA reduced cell proliferation and invasion, whereas TMPRSS4 overexpression promoted epithelial–mesenchymal-like transition in culture and metastasis in nude mice.[138,139] The effects of TMPRSS4 in cancer cells appeared to be mediated by upregulating integrin α5 and its downstream molecules.[140] These data suggest a role of TMPRSS4 in cancer progression and metastasis, although its proteolytic targets remain unknown.

6. Spinesin

Spinesin, also known as TMPRSS5, was cloned from a human spinal cord library,[141] and named spinesin for spinal cord-enriched trypsin-like protease. The human *TMPRSS5* gene is located on human chromosome 11q23. The spinesin protein domain structure is shown in Fig. 1. On Western blots, human spinesin appeared as a major band of approximately 52 kDa, representing its zymogen form.[141] In cerebrospinal fluids, a smaller spinesin band of approximately 50 kDa was detected, suggesting that spinesin may be shed from the cell surface.[142] Recombinant soluble spinesin is active toward peptide substrates with an optimal pH of approximately 10.[142] The functions of spinesin are unknown, although its predominant expression in the brain and spinal cord may suggest that its primary function will be in the central nervous system. More recently, spinesin mRNA was detected in rat inner ear tissues, indicating a possible role in the auditory system.[131]

7. MSPL

Mosaic serine protease long-form (MSPL), also known as TMPRSS13, was isolated from a human lung cDNA library.[143] The human *TMPRSS13* gene has 12 exons and is located on chromosome 11q23.2. Several alternatively spliced forms from this gene have been identified.[34] MSPL mRNA is expressed in the lung, placenta, pancreas, and prostate, and encodes a protease of 581 amino acids (Fig. 1). The recombinant human MSPL migrated at approximately 60 kDa and prefers synthetic peptide substrates with Arg at the P_1 position. MSPL activity is inhibited by aprotinin, benzamidine, and soybean trypsin inhibitor.[34] The physiological functions of MSPL remain unknown. Similar to other TTSPs, MSPL was recently shown to induce avian influenza virus replication by activating the viral hemagglutinin, suggesting a potential role in human airway viral infections.[144]

B. Matriptase Subfamily

The matriptase subfamily consists of the three highly homologous proteases: matriptase, matriptase-2, and matriptase-3, and a protein with an atypical mosaic structure, polyserase-1 (Fig. 1). All matriptases exhibit a similar stem region with one SEA, two CUB, and three (matriptase-2 and matriptase-3) or four (matriptase) LDLRA domains. In polyserase-1, the transmembrane domain is followed by a single LDLRA domain and a tandem repeat of three serine protease catalytic domains, referred to as serase-1, -2, and -3. Although structurally similar, the members of this TTSP subfamily appear to have quite divergent biological functions.

1. Matriptase

In 1993, Matriptase (also known as MT-SP1, TADG-15, CAP3, epithin, and ST14) was identified as a new gelatinolytic activity in conditioned medium from cultured breast cancer cells.[145] It was molecularly cloned by several groups at the turn of the millennium.[53,146,147] The matriptase gene, *ST14*, is located on human chromosome 11q24-25 and encodes a polypeptide of 855 amino acids. Orthologs of matriptase are present in all vertebrate genomes examined to date, indicating conserved evolutionary functions.[3] Matriptase is the most widely expressed member of the matriptase subfamily and is found in the epithelial compartments of most embryonic and adult tissues.[148–153] A number of studies in different laboratories have revealed insights into the catalytic properties and unique physiological functions of matriptase.

Matriptase is an 80- to 90-kDa cell surface glycoprotein, with the modular structure illustrated in Fig. 1. Matriptase is synthesized as an inactive, single-chain zymogen and its activation is extraordinarily complex (reviewed in[44]). Matriptase activation requires two sequential endoproteolytic cleavages, the first in the amino-terminal SEA domain and the second within the highly conserved activation cleavage site R↓VVGG, in the serine protease domain. Proteolytic autoactivation of matriptase appears to be controlled by the stem region, posttranslational modifications, and the cellular localization of the protease[44]; however, the specific mechanisms that trigger the activation of matriptase are incompletely understood.[154] The inhibition of activated matriptase by HAI-1 was first documented by the identification of matriptase/HAI-1 complexes in human milk, in conditioned medium of cultured mammary epithelial cells, and in a number of cancer cell lines.[50]

Matriptase is widely expressed in various epithelial cells during mouse development. Matriptase can first be detected on embryonic day (E)10 in the epithelial lining of several tissues of the embryo proper, such as the forming olfactory placode, oral cavity and foregut, intestine, inner ear, and apical ectodermal regions of the limbs.[149] However, matriptase may be expressed even earlier in development, possibly already at the pre- or peri-implantation stage.[155] At E14.5 and E16.5, respectively, matriptase becomes expressed in the developing hair follicles and the interfollicular epidermis.[148] Starting on E8.5, matriptase is also present in chorionic trophoblasts of the mouse placenta, and a similar pattern of expression is also observed in the human placenta.[149,151] During postnatal life, the highest levels of matriptase expression are found in the epithelium of the gastrointestinal tract, matrix cells of the hair follicles, and the distal and collecting tubules of the kidney, while lower levels of matriptase are present in the granular/transitional layer of the epidermis and in the epithelial compartments of developing teeth, vomeronasal cavity, trachea, bronchioles, thymus, inner ear, gall bladder, urinary bladder, ureter, prostate, seminal vesicle, epididymis, uterus, and oviduct.[148,152,153] In addition to epithelial tissues, matriptase is also expressed by certain leukocyte populations in humans, including monocytoid cells and mast cells.[152,156,157]

Matriptase has emerged to play a critical role in skin formation, in epidermal differentiation and skin function (Fig. 2). Initial analyses of matriptase null mice uncovered a critical function for matriptase in the development of epidermal tissues.[158,159] Matriptase-ablated mice die shortly after birth, due to a severe dehydration that results from an impaired epidermal barrier function. Follicular structures are also affected by matriptase loss, as shown by the absence of whiskers and generalized hypoplasia of pelage hair follicles of null mice. These phenotypes were linked to defects in the initiation of caspase-14-calpain I, bleomycin hydrolase-mediated processing of the epidermal polyprotein, profilaggrin, into filaggrin monomer units and subsequently, into free hygroscopic amino acids, which partake in the formation of the cornified envelope of the uppermost layers of the epidermis and serve as a source of water binding free amino acids that contribute to skin hydration.[160] Matriptase-deficient epidermis also exhibits a defect in the formation of lamellar granules, specialized secretory vesicles that contain lipid material required for the formation of extracellular lipid lamellae within the cornified layer, and display impaired formation of epidermal tight junctions within the granular layer.[158,161]

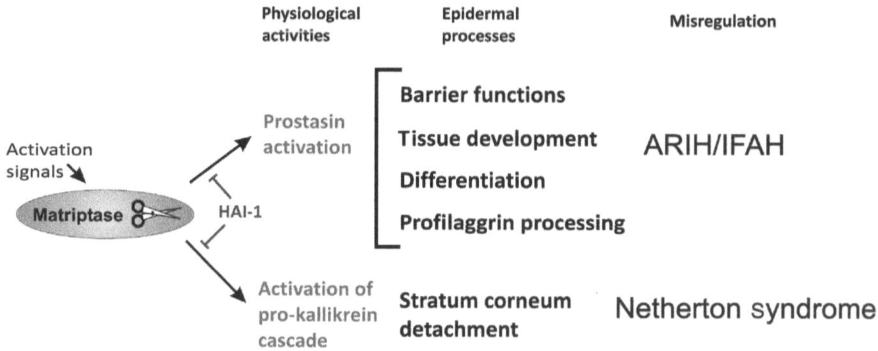

FIG. 2. *Matriptase is critical for the development and functions of the epidermis.* Matriptase activation, triggered by unknown mechanisms in skin, activates the GPI-anchored prostasin zymogen. Prostasin activation is required for epidermal tight junction formation, epidermal lipid synthesis, and induction of caspase-14-calpain I, and bleomysin hydrolase-mediated processing of the epidermal polyprotein, profilaggrin, into filaggrin monomer units, and subsequently, into free hygroscopic amino acids, which participate in the formation of the cornified envelope and contribute to skin hydration. Matriptase deficiency is linked to a rare form of skin disease, referred to as autosomal recessive ichthyosis with hypotrichosis (ARIH), or ichthyosis, follicular atrophoderma, and hypotrichosis (IFAH). Matriptase is also capable of activating proinflammatory pro-kallikrein-related peptidases that are associated with stratum corneum detachment and are responsible for the runaway kallikrein proteolytic cascade associated with LEKTI-deficiency/Netherton syndrome. HAI-1 regulates matriptase in the epidermis.

More recent studies suggest that matriptase is part of an epidermal proteolytic cascade and enables epidermal differentiation through the activation of the GPI-anchored membrane serine protease, prostasin. This is evidenced in part by the identical phenotype of matriptase- and prostasin-deficient mouse skin, and by the absence of active prostasin in matriptase-deficient epidermis.[150,162] Matriptase and prostasin are found together in a variety of other simple, stratified, and pseudo-stratified epithelia,[163] suggesting that matriptase may be a candidate activator of the prostasin zymogen in additional physiological settings.

Several recent reports have linked a rare form of skin disease, referred to, respectively, as autosomal recessive ichthyosis with hypotrichosis (ARIH) or ichthyosis, follicular atrophoderma, and hypotrichosis (IFAH), to homozygosity or compound heterozygosity for an assortment of mutations in the human *ST14* gene, encoding matriptase.[164–166] ARIH/IFAH patients present with mild to moderate ichthyosis, indicative of impaired barrier function, and hair follicle hypoplasia associated with fragile, brittle, dry, and slow-growing scalp hair. All these symptoms resemble transplanted skin from matriptase null mice or skin from matriptase hypomorphic mice.[158,167] Importantly, prostasin activation is impaired in ARIH/IFAH epidermis,[165] further supporting the belief that activation of the prostasin zymogen may be a principal function for matriptase during epidermal differentiation.

Matriptase is also linked to the initiation of the Netherton syndrome in a LEKTI-deficient mouse model through the premature activation of an epidermal pro-kallikrein protease cascade.[168] Deficiency in the serine protease inhibitor, LEKTI, is the etiological origin of the Netherton syndrome, and causes detachment of the stratum corneum and chronic inflammation due to excessive kallikrein-related protease activity. Matriptase was demonstrated to initiate activation of proinflammatory pro-kallikrein-related peptidases that are associated with stratum corneum detachment. Moreover, ablation of matriptase from LEKTI-deficient mice dampened inflammation, eliminated aberrant protease activity, prevented detachment of the stratum corneum, and improved the barrier function of the epidermis, demonstrating that the runaway kallikrein proteolytic cascade associated with LEKTI-deficiency was dependent on matriptase activity.

The wide expression of matriptase in epithelial tissues, including simple epithelia of most organs, suggested that matriptase could have functions in epithelial biology beyond the epidermis, but attempts to investigate this notion were initially hampered by the perinatal lethality of matriptase null mice. The subsequent generation of viable matriptase hypomorphic mice[167] and matriptase conditional knockout mice[161] afforded the opportunity to explore the function of matriptase in global epithelial biology. Tissue-specific embryonic and postnatal ablation of matriptase from a variety of epithelial tissues, including orofacial epithelium, salivary gland epithelium, lacrimal gland epithelium, and the epithelium of stomach, and small and large intestine, in all cases were associated with severe epithelial dysfunction, with two distinct phenotypes emerging: (a) loss of epithelial function, such as barrier and secretory capacity, but preservation of epithelial anatomy (salivary glands, tear glands, stomach, small intestine) and (b) loss of epithelial function, followed by loss of anatomical integrity (skin, orofacial, and colonic epithelium). In all epithelia, the absence of matriptase uniformly was associated with the rapid loss of epithelial tight junction function, compatible with the localization of matriptase on the lateral membrane of polarized epithelial cells.[161,169] In polarized monolayers of epithelial cells cultured *ex vivo*, down-regulation of matriptase decreased transepithelial electrical resistance and increased paracellular diffusion of macromolecules. Decreased matriptase was associated with enhanced expression and incorporation of the permeability-associated, "leaky" tight junction protein, claudin-2, at intercellular junctions, suggesting that the reduced barrier integrity was caused, at least in part, by an inability to regulate claudin-2 expression and incorporation into tight junctions.[169]

Matriptase is nearly ubiquitously coexpressed with the Kunitz-type transmembrane serine protease inhibitors, HAI-1 and HAI-2, in adult and embryonic tissues[56,149,151,155,170–172], and both HAI-1 and HAI-2 display potent matriptase inhibitory activity in purified systems.[36,56,170,173,174] Knockout studies have revealed that HAI-1 and HAI-2 are both required for the completion of embryonic development in mice.[149,151,155,175] However, both HAI-1/matriptase double-deficient and HAI-2/matriptase double-deficient mice complete embryonic development, thereby revealing matriptase as a principal inhibitory target for both protease inhibitors.[149,155] Loss of HAI-1 inhibition of matriptase in mice leads to failure of the chorionic epithelium to differentiate into the placental labyrinth, which prevents embryonic development beyond mid-gestation,[149] whereas other developmental processes proceed normally. HAI-2 inhibition of matriptase, however, is essential at three distinct developmental processes: (a) early (< E8.5) embryonic development, which proceeds normally in HAI-2 null embryos in the absence, but not in the presence of matriptase; (b) placental development, where the failure of the placental labyrinth to undergo branching morphogenesis in HAI-2 null embryos can be prevented by loss of matriptase; and (c) neural tube closure, in which the development of exencephaly and spina bifida is partially rescued by the matriptase deficiency.[155] The functional interactions between matriptase, HAI-1, and HAI-2 during development appear to be uniquely complex for a protease–protease inhibitor system. Thus, not only the complete loss of HAI-1 or HAI-2 but also the combined haploinsufficiency for both inhibitors causes embryonic lethality (nonallelic noncomplementation), which can be prevented by haploinsufficiency for matriptase, suggesting that the two inhibitors functionally cooperate in regulating the activity of matriptase.[155]

The requirement for matriptase regulation by HAI-1 is not restricted to development. Thus, the loss of HAI-1 from adult tissues is associated with fatal epithelial dysfunction in mice,[176] but HAI-1-deficient mice in which the level of matriptase is genetically reduced by 85–99% are fertile, healthy, have normal long-term survival, and possess histologically unremarkable epithelia. Similar findings were obtained from studies in zebrafish, where the combined loss of function of the two HAI-1 encoding genes compromises the integrity of embryonic epidermal keratinocytes and causes skin inflammation and embryonic death 18–26 h postfertilization. All of these defects can be fully rescued by a simultaneous downregulation of matriptase.[177]

Matriptase has received significant attention in the field of cancer biology due to its remarkably consistent expression in tumors of epithelial origin. Following the first description of matriptase as a major gelatinolytic activity in breast carcinoma,[145,178] the protease was reported to be expressed in a wide variety of other benign and malignant tumors of epithelial origin, including prostate, ovarian, cervical, gastric, colon, renal cell, esophageal, oral squamous cell carcinoma (SCC), and malignant pleural mesothelioma (reviewed in Ref. 179). In most carcinomas, malignant progression is associated with a significant increase in matriptase mRNA or protein expression. For example, in ductal mammary carcinoma, elevated levels of matriptase correlate with both tumor and nodal staging.[180] A clinical study in node-negative breast cancer patients showed a tight correlation between the expression of matriptase, cMet, and HAI-1, and poor patient outcome.[181] Likewise, more than fivefold overexpression of matriptase was detected in a study of ovarian cancer, compared to normal ovarian tissues, and this increased expression of matriptase correlated with tumor aggressiveness.[182,183] Furthermore, although matriptase was more frequently expressed in stage I/II tumors than in more advanced-stage III/IV tumors, an increased matriptase/HAI-1 ratio was indicative of the poor clinical outcome of advanced-stage tumors, suggesting that loss of protease inhibition may play a role in the late stages of the disease.[184] In prostate and cervical carcinoma, a several-fold increase in matriptase expression was reported for malignant versus nonmalignant tissues, and this correlated with Gleason score and histopathological grade, respectively, of the tumors.[185,186] Other common human cancers with significantly elevated matriptase levels include carcinomas of the pancreas, lung, kidney, and liver.[179] In contrast, significant downregulation of matriptase and HAI-1 has been detected in the samples of gastric and colorectal carcinoma.[172] However, an analysis of the clinicopathological parameters in colorectal adenoma and carcinoma tumors showed an increased matriptase/HAI-1 ratio in the tumors compared to the corresponding tissue from control individuals.[187] This finding may suggest that in tissues of the gastrointestinal tract, which exhibit high expression of matriptase and HAI-1,[152,188,189] the tumorigenic process does not involve further elevation in the expression of the two proteins, but rather may be associated with altered levels of an HAI-1-free, uninhibited form of matriptase. SCC of the skin represents a second example of matriptase being dysregulated, not by overexpression, but rather by the translocation of the expression of the protease from a differentiated postmitotic compartment in the epidermis to a basal compartment with high tumorigenic potential.[148]

Matriptase is unique among pericellular proteases in that it possesses a strong oncogenic potential when misexpressed in an epithelial tissue.[173] Even low-level matriptase expression in basal keratinocytes of transgenic mice suffices to induce severe hyperproliferation of epidermal keratinocytes, which gradually progresses to invasive SCC. Both the epidermal hyperproliferation and the formation of matriptase-induced skin tumors are completely abolished by coexpression of HAI-1, demonstrating that the proteolytic activity of matriptase is critical to its oncogenic potential. The molecular mechanisms of matriptase-induced epithelial carcinogenesis remain to be determined. In cell-free or cell-based assays, matriptase can activate several proteins that have been previously associated with malignant progression, including pro-hepatocyte growth factor/scatter factor (pro-HGF),[29,190] pro-macrophage stimulating protein-1 (pro-MSP-1),[191] pro-urokinase plasminogen activator (pro-uPA),[29,190] protease activated receptor-2 (PAR_2),[29,192] and the src-associated transmembrane protein SIMA135/CDCP1.[193,194]

2. Matriptase-2

Matriptase-2, also called TMPRSS6, was independently cloned from liver tissues by two groups in 2002 and 2003 and shown to express a membrane serine protease with homology to matriptase that displayed proteolytic activity toward various macromolecular substrates.[37,195] Matriptase-2 is encoded by the *TMPRSS6* gene located on human chromosome 22q12.3. Matriptase-2 is approximately 90-kDa cell surface glycoprotein with a modular structure (Fig. 1) and is synthesized as an inactive, single-chain zymogen. Cell surface matriptase-2 is efficiently shed into the conditioned medium of transfected cells in an active two-chain form by proteolytic cleavage within the second CUB domain of the noncatalytic stem region.[196] Whereas matriptase is expressed in a large number of embryonic and adult epithelia, matriptase-2 expression is largely confined to adult and fetal liver in humans and mice, with minor expression in the kidney, uterus, and nasal cavity.[37,195]

A breakthrough in the understanding of the physiological function of matriptase-2 was enabled by the generation of matriptase-2 knockout mice and by the identification of loss of function mutations in the *TMPRSS6* gene as a cause of the human autosomal recessive disorder, iron-refractory iron deficiency anemia (IRIDA). Both mice and humans with matriptase-2 deficiency suffer from very low iron levels and severe microcytic anemia.[197–199] Matriptase-2 expressed by liver cells functions as a suppressor of the hepatic hormone, hepcidin, which in turn internalizes the iron export protein, ferroportin, on enterocytes and macrophages to reduce iron uptake. Thus, matriptase-2 is a key regulator of systemic iron hemostasis. Hepcidin suppression by matriptase-2 appears to occur at the transcriptional level, as hepcidin mRNA levels are elevated in both matriptase-2-deficient humans and mice.[197–199] This suppression of hepcidin gene transcription has been linked to matriptase-2-mediated degradation of hemojuvelin, a cofactor for bone morphogenetic protein, and a key regulator of hepcidin gene activation.[200,201]

3. Matriptase-3

Matriptase-3 was identified by bioinformatic analysis in 2005 as a protease with high homology to matriptase.[202] The *TMPRSS7* gene encoding matriptase-3 is located on human chromosome 3q13.2 and encodes an approximately 90-kDa glycosylated protein detected on the cell surface.[202] Orthologs of the matriptase-3 gene are present in all vertebrates analysed to date, including fish, birds, rodents, canines, and primates. The pattern of expression of matriptase-3 is conserved in mice and humans, with the highest levels of mRNA expression in the brain, skin, eye, salivary gland, and the reproductive tissues, including prostate, testis, epididymis, ovary, and uterus. The recombinant matriptase-3 catalytic domain can hydrolyse synthetic peptide substrates with a strong preference for arginine in the P_1 position, and shows proteolytic activity toward several macromolecular substrates, including gelatin, casein, and albumin.[202] The activated matriptase-3 catalytic domain forms stable inhibitor complexes with an array of serpins, including PAI-1, PCI, α1-proteinase inhibitor, α2-antiplasmin, and antithrombin III *in vitro*. Matriptase-3 loss of function studies in animals or humans with matriptase-3 deficiency has not been reported, and the physiological function of the protease remains to be determined.[4]

4. Polyserase-1/Serase1B

Polyserase was cloned from a human liver cDNA as a unique type II transmembrane serine protease with three serine protease domains, two of which display catalytic activity.[203] mRNA capable of encoding full-length human polyserase-1 is detected predominantly in the skeletal muscle, heart, kidney, liver, placenta, and the brain. In addition, a shorter splice variant, termed serase-1B, which contains only the first of the three serine protease domains of polyserase-1, has been described in mice and humans, with its highest expression detected in the liver, small intestine, pancreas, testes, and the peripheral blood CD14+ and CD8+ cells.[35,203] Recombinant serase-1B shows proteolytic activity toward synthetic peptide substrates, converted pro-uPA into active uPA, and could be inhibited by the serpins, aprotinin, α2-antiplasmin, and PAI-1.[35]

The human polyserase gene, called *TMPRSS9*, is located on human chromosome 19p13.3. The 5′ promoter region of the mouse polyserase-1 gene has been characterized and contains a GATA motif, a glucocorticoid responsible element, and an E-box sequence required for maximal promoter activity.[204] The physiological function of polyserase has not been elucidated.

C. Corin Subfamily

Corin, the single member of the corin subfamily, is characterized by a complex stem region composed of two frizzled domains, eight LDLRA domains, and one Group A SR domain (Fig. 1).

1. Corin

Human corin cDNA was originally cloned from the heart. In a search for novel serine proteases in the cardiovascular system, Yan *et al.* identified a partially expressed sequence in a genomic database and subsequently cloned the full-length cDNA encoding corin.[205] The protease was named corin for its abundant cardiac expression. Independently, Hooper *et al.* isolated human corin cDNA from HeLa-derived cells that were resistant to tumor necrosis factor-α-induced apoptosis.[206] Corin was also cloned from a mouse heart library as a novel protein containing LDL receptor-like repeats, and was named LRP4 (LDL receptor-related protein 4).[207]

The human *CORIN* gene is present on chromosome 4p12-13, consists of 22 exons, and spans approximately 200 kb.[208] Corin is highly expressed in fetal and adult cardiomyocytes. Low levels of corin mRNA are detected in the kidney, bone, brain, skin, and pregnant uterus.[109] In both human and mouse corin genes, there is a 5′-flanking region that contains conserved binding sites for TBX5, GATA, NKX2.5, NF-AT, and Kruppel-like transcription factors.[208] Among them, a conserved GATA-binding site is critical for interacting with GATA-4 transcription factor that is important for heart-specific gene expression. GATA-4 protein is known to act downstream of the calcineurin/NF-ATc3 signaling pathway in cardiomyocytes. Most likely, this signaling pathway is involved in regulating corin expression in the heart.

The full-length human corin cDNA encodes a polypeptide of 1042 amino acids with a calculated molecular mass of approximately 116 kDa and a domain structure shown in Fig. 1. The presence of two frizzled-like cysteine-rich domains in the corin extracellular region is unique among the TTSPs.[4] Native and recombinant human, rat, and mouse corin proteins appear as bands of approximately 150–200 kDa on immunoblots, largely due to N-glycosylation which has been confirmed by glycosidase digestion.[99] Human corin has 19 predicted N-linked glycosylation sites in its extracellular region, whereas rat and mouse corin has 17 N-glycosylation sites. Corin does not contain detectable amounts of *O*-glycans or sialic acids.[209,210]

Corin is synthesized as a zymogen with no detectable activity in functional assays, indicating that cleavage at Arg^{801} within the conserved activation site, R↓ILGG is necessary for its activity.[211] A small fraction of activated corin is detected in transfected HEK 293 cells and cardiomyocytes[210]; however, the enzyme(s) responsible for corin activation in these cells have not been determined. Recombinant corin with an engineered activation site has a trypsin-like catalytic activity, favoring peptide substrates with Arg/Lys residues at P_1, Pro/Phe/Gly at P_2, and small neutral amino acids at the P_3 position.[211] The activity is inhibited by leupeptin, aprotinin, and soybean trypsin inhibitor and, unlike many serine proteases, corin activity is not inhibited in the presence of human plasma.[211]

The topology of corin on the cell membrane has been confirmed by cell surface protein labeling and protease digestion.[99,209] In addition to the transmembrane domain, *N*-glycans on corin also play a role in its cell surface targeting and activation.[210,212] Inhibition of N-glycosylation or removal of certain N-glycosylation sites by site-directed mutagenesis impaired corin cell surface expression and zymogen activation. The transmembrane domain is not necessary for corin enzyme activity, since soluble corin without the transmembrane domain exhibited similar activities toward small peptide and biological substrates when compared to that of wild-type corin.[211]

Wu *et al.* identified corin as the physiological pro-atrial natriuretic peptide (pro-ANP) convertase.[213] ANP, also called atrial natriuretic factor (ANF), is a cardiac hormone that regulates blood pressure and cardiac function by promoting natriuresis, diuresis, and vasodilation. The function of ANP is mediated by its receptor, NPR-A, which promotes intracellular cGMP production. In mice, deficiency in ANP or its receptor causes spontaneous hypertension, demonstrating the importance of the ANP pathway in regulating blood pressure.[213] In humans, SNPs and mutations in the *NPPA* gene encoding ANP have been reported in patients with hypertension and heart disease.[214]

In cardiomyocytes, ANP is made as a precursor molecule, pro-ANP. Upon secretion under high blood volume or pressure, pro-ANP is converted to active ANP by proteolytic cleavage. This activation mechanism was recognized for many years, but the responsible enzyme remained unidentified. In cell-based studies, corin activated pro-ANP in a sequence-specific manner.[215] This activity was confirmed in cardiomyocytes and with purified recombinant corin. Chan *et al.* made corin null mice, which were viable and fertile.[216] In these mice, lack of corin prevents pro-ANP processing in the heart and causes salt-sensitive hypertension. Corin null mice develop cardiac hypertrophy and exhibit impaired cardiac function. This hypertrophic heart phenotype has been confirmed in a naturally occurring mutant mouse strain, $Kit^{W\text{-}sh}$, in which the *corin* gene is disrupted by genetic inversion.[217] These data show that corin-mediated pro-ANP processing is critical for regulating blood pressure and cardiac function (Fig. 3). Curiously, corin null mice appear to have a lighter coat color, and this phenotype depends on the *agouti* gene.[218] As corin mRNA and protein are expressed in the dermal papilla in mice, corin may act in the skin to regulate hair color formation by an agouti pathway-dependent mechanism.

In addition to ANP, BNP (B-type natriuretic peptide) and CNP (C-type natriuretic peptide) are additional members of the human natriuretic peptide family, which share similar protein sequences and structures. The function of BNP is to promote natriuresis and diuresis, whereas CNP is involved in endothelial cell growth and chondrocyte differentiation. In cell-based studies, corin also cleaved pro-BNP, although the reaction was less efficient than that for pro-ANP.[215] Recent studies indicate that both furin and corin convert pro-BNP to BNP.[219] In contrast, pro-CNP is processed by furin but not corin.[220] It is interesting that these natriuretic peptides have evolved from an ancestor gene, but their activation processing requires different enzymes.

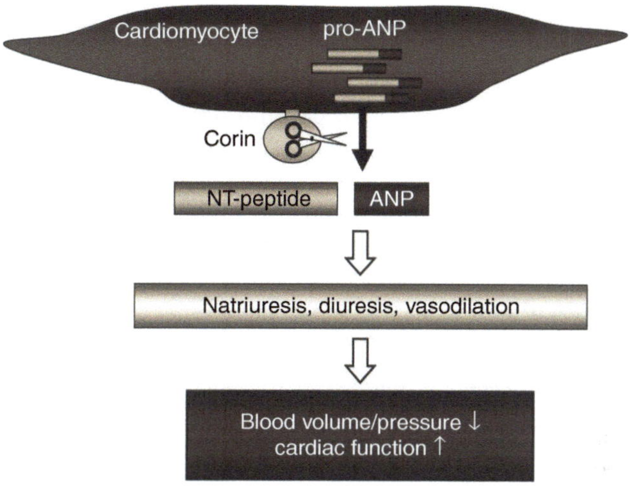

FIG. 3. *Corin-mediated pro-ANP processing in the heart.* In cardiomyocytes, ANP is made as a precursor, pro-ANP. Upon secretion, pro-ANP is cleaved by corin, generating an inactive N-terminal (NT) peptide and an active ANP. ANP promotes natriuresis, diuresis, and vasodilation, thereby reducing blood volume/pressure and improving cardiac function.

Corin variants exist in hypertensive patients. Dries *et al.* have identified two SNPs (T555I/Q568P) in the human *CORIN* gene.[221] These SNPs are located on a minor corin gene allele that is present primarily in African-Americans with hypertension. It is known that African-Americans have a high prevalence of cardiovascular disease, but the underlying mechanisms remain unclear. Rame *et al.* have shown that individuals with these corin SNPs have an enhanced cardiac response to high blood pressure, which leads to severe left ventricular hypertrophy.[222] Under pathological conditions, cardiac hypertrophy impairs heart functions. In patients with heart failure, the presence of this corin variant allele is associated with worse clinical outcomes such as hospitalization and death.[223] Changes in the amino acids, T555I/Q568P, encoded by the SNPs occur within the second frizzled-like domain, an extracellular region demonstrated to be functionally important.[46] Examination of the effect of these amino acid substitutions on corin function in cell-based studies showed that the corin variant, T555I/Q568P, exhibited markedly reduced activity in processing pro-ANP and pro-BNP.[224] Apparently, the reduced activity was caused by impaired corin zymogen activation but not cell surface expression.[224] The data indicate that these polymorphisms alter corin structure and function, which may contribute to elevated blood pressure and cardiac hypertrophy in patients. Thus, corin defects may present an important mechanism underlying cardiovascular disease.

Soluble corin has been detected in human plasma, suggesting that corin is shed from the cell surface and subsequently circulates in the blood.[225] Interestingly, plasma corin levels were significantly lower in patients with heart failure and the reduction correlated with the disease severity.[226] In contrast, patients with acute myocardial infarction had similar levels of plasma corin compared to that of normal controls, suggesting that low plasma corin levels are associated more closely with pathological changes in heart failure than that in acute ischemic cardiac injury. The protease(s) responsible for corin shedding have not been determined. Low levels of plasma soluble corin in patients with heart failure may reflect reduced corin production, accelerated clearance of plasma corin, and/or downregulation of corin shedding in failing hearts. Additional studies will be required to elucidate corin shedding mechanisms under physiological and pathological conditions, and whether decreased plasma corin could be used as a biomarker for the diagnosis of heart failure.

D. HAT/DESC Subfamily

The HAT/DESC (human airway trypsin-like protease/differentially expressed in squamous cell carcinoma) subfamily comprises five members in humans, HAT, DESC1, TMPRSS11A, HAT-like 4/TMPRSS11F, and HAT-like 5, and two additional members in rodents, HAT-like 2/DESC4 and HAT-like 3.[3,4] All genes encoding members of this subfamily are located within a single gene cluster on chromosome 4q in humans (5 E1 in mice), suggesting that they originated by gene duplication from a common ancestor.[227] Of all TTSPs, members of the HAT/DESC subfamily exhibit the simplest modular structure of the stem region, which consists of a single SEA domain located adjacent to the cell surface. SEA domains are structurally homologous protein modules that have been identified in many mucin-like proteins. These domains undergo spontaneous posttranslational confirmation-driven hydrolysis of the glycine–serine peptide bond within a conserved GSVVV sequence, although the two peptides remain tightly associated.[228] Little is known about most of the members of the HAT/DESC subfamily, with HAT, DESC1, and TMPRSS11A being the best characterized.

1. HAT

HAT was originally characterized as a novel 28-kDa protease from the sputum of patients with chronic airway diseases.[229] The cDNA was subsequently cloned and the protease was named HAT for human airway trypsin-like protease.[230] HAT cDNA encodes an integral membrane protein of 418 amino acids, with a predicted molecular mass of approximately 46-kDa, which is released in a soluble form from the cell surface.[230] Hence, significant HAT-related activity can be detected in the extracellular airway fluids of patients with chronic airway inflammatory diseases, particularly in patients with asthma.[229,231,232] Within human tissues, the airways and the skin have been shown to express HAT.[233,234] In the airways, HAT is expressed in the apical cilia layer of tracheobronchial epithelial cells.[234,235]

There is little information regarding the physiological functions and candidate proteolytic targets of HAT; however, HAT upregulation is found to be associated with inflammatory environments.[233] The available evidence, obtained from the study of *in vitro* cleavage reactions and cell-based assays, supports a potential role for HAT in respiratory homeostasis during inflammatory responses through a capacity to increase mucin gene expression, control fibrin deposition, and stimulate bronchial fibroblast proliferation in airway epithelial cells *in vitro*.[231,235,236] HAT has been demonstrated to have the capacity to degrade fibrinogen, to activate pro-uPA,[232,235] to activate membrane receptors such as PAR-2,[232,235,236], and to cleave the D1–D2 linker sequence of human uPAR (h-uPAR), which prevents interaction with the extracellular matrix component, vitronectin.[237] The physiological relevance of these cleavages, if any, remains untested.

HAT is one of the several TTSPs found to cleave the surface glycoprotein hemaglutinin (HA) of the influenza virus.[118,238] HA is responsible for initiating influenza virus replication by mediating binding to sialic acid-containing cell surface receptors and fusion of the viral envelope with the endosomal membrane. HA precursor proteins require cleavage at a single arginine amino acid to trigger membrane fusion. Membrane-bound HATwas shown to cleave newly synthesized HA before or during the release of progeny virions as well as HA of incoming viruses prior to endocytosis at the cell surface, supporting its contribution to the facilitation of virus activation and spread.

2. DESC1

Differentially expressed in squamous cell carcinoma (DESC1) was originally identified by representational difference comparative analysis of RNA from SCC of the head and neck compared with matched normal tissue.[239] DESC1 is expressed in normal epithelial cells of prostate, skin, testes, head, and neck, whereas it is downregulated in 11/12 SCCs,[239] suggesting that it may serve as a tumor marker. The human *DESC1* gene encodes a polypeptide of 442 amino acids. The mouse ortholog of DESC1 shares 72% identity with human DESC1.[227] Both proteases are expressed in similar anatomical locations and are predicted to have common functions in the development and maintenance of oral epidermal tissues and the male reproductive tract.[227]

The physiological functions of DESC1 are not known. DESC1 is upregulated during the induction of terminal keratinocyte differentiation.[240] Madin-Darby canine kidney (MDCK) cells expressing exogenous human DESC1 display enhanced motility and an increase of tubular forms in a three-dimensional collagen lattice following HGF treatment.[241] Kinetic studies using internally quenched peptides[227,242] and structural analyses of the DESC1 catalytic domain[23] reveal DESC1 substrate preference for large hydrophobic residues in P_4/P_3, for small residues in P_2, Arg or Lys in P_1, and hydrophobic residues in P_1' and P_3'. Mouse DESC1 forms stable inhibitory complexes with PAI-1 and PCI, suggesting that these serpins might be regulators of DESC1 proteolytic activities in DESC1 expressing tissues.[227]

3. TMPRSS11A

Most of the remaining members of the HAT/DESC family have not yet been well characterized. HATL1, also known as TMPRSS11A, is expressed in the upper respiratory tract (pharynx and trachea) and has been shown to have the capability to cleave recombinant, native, full-length S-protein trimer (triSpike) of the severe acute respiratory syndrome, coronavirus (SARS-CoV) *in vitro.*[243]

VI. The GPI-Anchored Serine Proteases

A. Prostasin

Prostasin, also known as PRSS8 and channel-activating protease (CAP)-1, was originally identified and characterized from human seminal fluid.[11] Subsequent studies using a *Xenopus* oocyte expression system and mammalian epithelial cell cultures independently showed that prostasin expression increased amiloride-sensitive ENaC current.[134,244–246] As a result, prostasin became the first membrane-anchored serine protease to be implicated in the modulation of fluid and electrolyte regulation via proteolysis of ENaCs.

The *PRSS8* gene encoding prostasin belongs to the family of genes located on the syntenic regions of human chromosome 16p13.3 and mouse chromosome 17[14,247–249], and this family includes genes encoding both testisin and tryptase γ1.[9] The mouse frizzy (fr) and rat "hairless" (fr(CR)) mutations are natural variants of the murine prostasin gene (*Prss8*). The deduced amino acid sequence of *PRSS8* predicts a preproenzyme consisting of 343 amino acids, which shows 34–42% identity to human acrosin, plasma kallikrein, and hepsin. Prostasin is posttranscriptionally modified with GPI-anchor[11–13] and associates with lipid rafts.[13] In polarized epithelia, prostasin localizes to apical membranes.[12,250,251] Soluble forms of prostasin are found in human urine[252,253] and are elevated in hypertensive patients.[254] Prostasin has been found to be released from the cell surface by an endogenous GPI-specific phospholipase D1[13] or, alternatively, may be shed via a tryptic-like proteolytic cleavage in its hydrophobic C-terminal domain.[11,255]

Tertiary structures of prostasin[24–26] and analyses for substrate preference using positional scanning combinatorial substrate libraries[32] have revealed insights into prostasin's unique catalytic activity. Prostasin was found to have a substrate preference for polybasic substrates: in position P_4, preference was for arginine or lysine; in P_3, for histidine, lysine, or arginine; in P_2, for basic or large hydrophobic amino acids; and in P_1, for arginine and lysine.[32] Prostasin showed no activity with substrates containing isoleucine in position P_1', providing an explanation for the inability of prostasin to autoactivate. Prostasin activity was also highly influenced by mono- and divalent metal ions, which were potent inhibitors and substrate specific modulators of enzymatic activity. Interestingly, structural analyses showed that the S_1 subsite loop of prostasin exhibits a large degree of conformational variation and directly binds the divalent cation, Ca^{2+}, being able to move to block or to expose the S_1 subsite.[24]

The prostasin zymogen is proteolytically cleaved to an active enzyme by matriptase *in vitro* and *in vivo* in skin[150] and when coexpressed with hepsin in cultured cells.[96] Low levels of plasmin found in urine from patients with nephritic syndrome may also activate prostasin in the context of ENaC activation.[256] Intriguingly, the zymogen form of matriptase was also able to be converted to an active protease by the addition of active prostasin, suggesting that, in certain cellular contexts, prostasin might function both upstream and downstream of matriptase.[192] Comparison of the skin phenotypes of conditional prostasin-deficient mice with matriptase-deficient mice supports the participation of both serine proteases in the same protease-signaling cascade in the skin.[162] Antithrombin III and protease nexin I are inhibitors of cell surface-associated activities of prostasin[257], and protease nexin I is an effective inhibitor of prostasin-mediated activation of ENaC.[257] Prostasin is also inhibited by the Kunitz-type inhibitors, HAI-1/SPINT1[31,48,57,58] and HAI-2/SPINT2,[32,258] similar to matriptase.

The data suggest that the "open probability" of ENaCs is increased by prostasin, resulting in an increased cellular uptake of sodium (Na^+) which can, in turn, regulate the homeostasis of extracellular fluid volume, blood pressure, and Na^+ reabsorption.[259] Prostasin cleaves the ENaCγ subunit in the extracellular loop at a site (K186) distal to a furin cleavage site, and releases a 43-amino acid inhibitory peptide that results in increased open probability of the channel and full activation.[260,261] Recent studies have localized this inhibitory peptide to a key 11-mer tract, R158-F168 (RFLNLIPLLVF), which is capable of inhibiting wild-type ENaC expressed in *Xenopus* oocytes and endogenous channels in airway epithelial cells.[262] The channel-activating activity of prostasin is likely to be of pathophysiological significance. In the lung, adequate levels of ENaC activity are essential for controlling airway surface liquid volumes. Mice with targeted disruption of the prostasin gene in lung epithelium show impaired ENaC-mediated alveolar fluid clearance.[263] In addition, increased ENaC activation in the airway of cystic fibrosis patients has been linked to excessive prostasin activity on the surface of lung epithelium.[264,265] Prostasin has been identified as a candidate gene for the development of hypertension in youths,[266] which may also be associated with increased ENaC activation.[267] Increased levels of urinary prostasin are found in hypertensive patients,[252] and inhibition of prostasin in a rat model of hypertension significantly reduced blood pressure.[268] Inhibition of prostasin-dependent activation of ENaCs is considered a potential therapeutic strategy for modulating surface liquid volume in cystic fibrosis and salt-sensitive hypertension.

Prostasin also exhibits pleiotropic activities associated with the activation of growth factors, G protein-coupled receptors, and the activation of proteolytic cascades.[269–271] Less well-characterized potential targets of prostasin include PAR_2 and the epidermal growth factor receptor.[269] Soluble recombinant prostasin has also been reported to activate PAR_2 indirectly, through a matriptase-dependent mechanism.[192] Prostasin activity is further implicated in the regulation of epithelial barrier function,[13] which is believed to be unrelated to ENaC processing activity, suggesting that prostasin may also target as yet unidentified substrates involved in the regulation of epithelial barrier integrity and permeability. Downregulation of prostasin is associated with tumor progression[269,272,273] through as yet unknown mechanisms.

B. Testisin

Testisin was originally cloned from human eosinophils and from HeLa cervical carcinoma cells.[15,274] Human testisin has been known as eosinophil serine protease 1 (esp-1)[274] and murine testisin[247] has been called tryptase 4[248] and TESP5.[14] The testisin gene (*PRSS21*) belongs to the family of genes on the syntenic regions of human chromosome 16p13.3 and mouse chromosome 17.[9,14,247–249] Several isoforms of human testisin have been identified, which are believed to be generated by alternate intron–exon junctional sliding.[249,275] Testisin is abundantly expressed by male germ cells and sperm[15,247,276], and is also present in capillary endothelial cells[277] and in eosinophils.[248,274]

Analyses of the promoter sequences of the human and mouse genes encoding testisin have revealed that the region surrounding the transcription initiation site lacks a TATA consensus sequence, but contains a CCAAT sequence and includes a 385-bp 5′-CpG island.[247,249] There is a strong correlation with DNA hypermethylation of the 5′-CpG rich island and silencing of testisin gene expression.[278] Furthermore, the 5′-flanking region of the testisin gene contains several consensus response elements, including Sp1, AP1, and several testis-specific elements.[247,249,275]

Testisin is posttranscriptionally modified by the addition of a carboxy-terminal GPI-membrane anchor and has been demonstrated to be compartmentalized at plasma membranes within the dynamic microenvironment of cholesterol-rich membrane microdomains or lipid rafts.[14] Naturally occurring and recombinant testisin proteins are cell-associated,[14,15,248] but may be released from cell membranes by exposure to exogenous bacterial phosphatidylinositol-specific phospholipase C (PI-PLC).[14]

Developmental evidence links testisin to the maturation and function of mammalian sperm. Testisin is detected on spermatogenic cells throughout spermatogenesis and on mature sperm in humans,[276] mice,[14,247,276] and rats,[279] which suggests a similar function among these mammalian species. Mice deficient in testisin display several functional abnormalities associated with epididymal sperm maturation and fertilizing ability.[276,280] Epididymal sperm lacking testisin show an increased tendency toward decapitation, heterogeneity in sperm form and angulated flagella, decreased numbers of motile sperm, and abnormal sperm volume regulation. Although testisin-deficient mice exhibit normal fertility when bred by continuous mating, these functional sperm defects result in a decreased ability to fertilize oocytes in short-term mating studies.[276] The data suggest an essential requirement for testisin during sperm cell maturation processes which occur during epididymal transit and are required for normal fertilizing ability. Functional overlap between testisin and the related fertilization serine protease, acrosin, may compensate for respective functional fertilizing abilities, since acrosin-deficient mice exhibit normal fertility, but acrosin/testisin double knockout mice are subfertile.[281] It has been proposed that aberrant regulation of testisin may contribute to certain secondary male infertility syndromes such as "easily decapitated" spermatozoa in humans.[276]

Absence of testisin expression is associated with testicular tumor formation and progression[15], and overexpression of testisin promotes advanced-stage disease in ovarian carcinomas.[282] Epigenetic gene silencing may contribute to the downregulation of both prostasin and testisin expression in cancers.[272,278,282–284]

VII. Perspectives

Over the past few decades, membrane-anchored serine proteases have emerged to play key roles in many diverse aspects of mammalian physiology, including food digestion, fluid balance, blood pressure regulation, hearing and inner ear development, epithelial barrier function, sperm maturation, and regulation of iron homeostasis (Table I). The associations of mutant alleles of the membrane-anchored serine proteases with disease, the misregulation of these enzymes in malignant growth, and the hijacking of their activities by viruses to facilitate infection and propagation, all highlight the pathological impact of misregulation of their activities. Despite remarkable progress in the past few years, many questions remain unanswered. How these proteases

TABLE I
MEMBRANE-ANCHORED SERINE PROTEASE PHYSIOLOGICAL FUNCTIONS IN MAMMALIAN DEVELOPMENT AND TISSUE HOMEOSTASIS

Enteropeptidase	Proteolytic activation cascade in intestine
Hepsin	Cochlear development and hearing (mice)
TMPRSS3	Hearing
TMPRSS5	Hearing
Matriptase	Epidermal and hair follicle development and epithelial barrier functions
Matriptase-2	Regulates iron hemostasis; deficiency or mutation increases hepcidin levels, leading to iron-refractory iron deficiency anemia (IRIDA)
Corin	Processes critical cardiac pronatriatic peptides involved in the regulation of blood pressure
Prostasin	Epidermal and hair follicle development and epithelial barrier functions; regulation of ENaC-mediated alveolar sodium and water transport (mice)
Testisin	Subfertility and aberrant sperm morphology (mice)

intersect and modulate the protease cascades in the pericellular microenvironment, the signals that trigger their activation, the repertoire of endogenous substrates, and the inhibitors that modulate their activities, are important issues to address. As suitable molecular tools, such as antibodies, activity-based probes, animal models, and human mutational analyses, are refined and developed, we will look forward to gaining a better understanding of the critical roles of these enzymes both in mammalian physiology and in disease.

ACKNOWLEDGMENTS

This work was supported by grants from the National Institutes of Health (CA098369, HL084387, DK081376) (T. M. A.), (HL089298 and HL089298-S1) (Q. W.), the NIH Intramural Program (T. H. B.), and the Mary Kay Ash Charitable Foundation 075-07 (T. M. A.).

REFERENCES

1. Hooper JD, Clements JA, Quigley JP, Antalis TM. Type II transmembrane serine proteases. Insights into an emerging class of cell surface proteolytic enzymes. *J Biol Chem* 2001;**276**:857–60.
2. Netzel-Arnett S, Hooper JD, Szabo R, Madison EL, Quigley JP, Bugge TH, et al. Membrane anchored serine proteases, a rapidly expanding group of cell surface proteolytic enzymes with potential roles in cancer. *Cancer Metastasis Rev* 2003;**22**:237–58.
3. Szabo R, Bugge TH. Type II transmembrane serine proteases in development and disease. *Int J Biochem Cell Biol* 2008;**40**:1297–316.
4. Bugge TH, Antalis TM, Wu Q. Type II transmembrane serine proteases. *J Biol Chem* 2009;**284**:23177–81.

5. Hansen IA, Fassnacht M, Hahner S, Hammer F, Schammann M, Meyer SR, et al. The adrenal secretory serine protease AsP is a short secretory isoform of the transmembrane airway trypsin-like protease. *Endocrinology* 2004;**145**:1898–905.
6. Bhagwandin VJ, Hau LW, Mallen-St Clair J, Wolters PJ, Caughey GH. Structure and activity of human pancreasin, a novel tryptic serine peptidase expressed primarily by the pancreas. *J Biol Chem* 2003;**278**:3363–71.
7. Appel LF, Prout M, Abu-Shumays R, Hammonds A, Garbe JC, Fristrom D, et al. The Drosophila Stubble-stubbloid gene encodes an apparent transmembrane serine protease required for epithelial morphogenesis. *Proc Natl Acad Sci USA* 1993;**90**:4937–41.
8. Irving P, Troxler L, Heuer TS, Belvin M, Kopczynski C, Reichhart JM, et al. A genome-wide analysis of immune responses in Drosophila. *Proc Natl Acad Sci USA* 2001;**98**:15119–24.
9. Caughey GH, Raymond WW, Blount JL, Hau LW, Pallaoro M, Wolters PJ, et al. Characterization of human gamma-tryptases, novel members of the chromosome 16p mast cell tryptase and prostasin gene families. *J Immunol* 2000;**164**:6566–75.
10. Wong GW, Tang Y, Feyfant E, Sali A, Li L, Li Y, et al. Identification of a new member of the tryptase family of mouse and human mast cell proteases which possesses a novel COOH-terminal hydrophobic extension. *J Biol Chem* 1999;**274**:30784–93.
11. Yu JX, Chao L, Chao J. Prostasin is a novel human serine proteinase from seminal fluid. Purification, tissue distribution, and localization in prostate gland. *J Biol Chem* 1994;**269**:18843–8.
12. Chen LM, Skinner ML, Kauffman SW, Chao J, Chao L, Thaler CD, et al. Prostasin is a glycosylphosphatidylinositol-anchored active serine protease. *J Biol Chem* 2001;**276**:21434–42.
13. Verghese GM, Gutknecht MF, Caughey GH. Prostasin regulates epithelial monolayer function, cell-specific Gpld1-mediated secretion and functional role for GPI anchor. *Am J Physiol Cell Physiol* 2006;**291**:C1258–70.
14. Honda A, Yamagata K, Sugiura S, Watanabe K, Baba T. A mouse serine protease TESP5 is selectively included into lipid rafts of sperm membrane presumably as a glycosylphosphatidylinositol-anchored protein. *J Biol Chem* 2002;**277**:16976–84.
15. Hooper JD, Nicol DL, Dickinson JL, Eyre HJ, Scarman AL, Normyle JF, et al. Testisin, a new human serine proteinase expressed by premeiotic testicular germ cells and lost in testicular germ cell tumors. *Cancer Res* 1999;**59**:3199–205.
16. Antalis TM, Buzza MS, Hodge KM, Hooper JD, Netzel-Arnett S. The cutting edge, membrane-anchored serine protease activities in the pericellular microenvironment. *Biochem J* 2010;**428**:325–46.
17. Puente XS, Sanchez LM, Overall CM, Lopez-Otin C. Human and mouse proteases, a comparative genomic approach. *Nat Rev Genet* 2003;**4**:544–58.
18. Fastrez J, Fersht AR. Demonstration of the acyl-enzyme mechanism for the hydrolysis of peptides and anilides by chymotrypsin. *Biochemistry* 1973;**12**:2025–34.
19. Szabo R, Wu Q, Dickson RB, Netzel-Arnett S, Antalis TM, Bugge TH. Type II transmembrane serine proteases. *Thromb Haemost* 2003;**90**:185–93.
20. Friedrich R, Fuentes-Prior P, Ong E, Coombs G, Hunter M, Oehler R, et al. Catalytic domain structures of MT-SP1/matriptase, a matrix-degrading transmembrane serine proteinase. *J Biol Chem* 2002;**277**:2160–8.
21. Somoza JR, Ho JD, Luong C, Ghate M, Sprengeler PA, Mortara K, et al. The structure of the extracellular region of human hepsin reveals a serine protease domain and a novel scavenger receptor cysteine-rich (SRCR) domain. *Structure* 2003;**11**:1123–31.
22. Lu D, Futterer K, Korolev S, Zheng X, Tan K, Waksman G, et al. Crystal structure of enteropeptidase light chain complexed with an analog of the trypsinogen activation peptide. *J Mol Biol* 1999;**292**:361–73.

23. Kyrieleis OJ, Huber R, Ong E, Oehler R, Hunter M, Madison EL, et al. Crystal structure of the catalytic domain of DESC1, a new member of the type II transmembrane serine proteinase family. *FEBS J* 2007;**274**:2148–60.
24. Spraggon G, Hornsby M, Shipway A, Tully DC, Bursulaya B, Danahay H, et al. Active site conformational changes of prostasin provide a new mechanism of protease regulation by divalent cations. *Protein Sci* 2009;**18**:1081–94.
25. Tully DC, Vidal A, Chatterjee AK, Williams JA, Roberts MJ, Petrassi HM, et al. Discovery of inhibitors of the channel-activating protease prostasin (CAP1/PRSS8) utilizing structure-based design. *Bioorg Med Chem Lett* 2008;**18**:5895–9.
26. Rickert KW, Kelley P, Byrne NJ, Diehl RE, Hall DL, Montalvo AM, et al. Structure of human prostasin, a target for the regulation of hypertension. *J Biol Chem* 2008;**283**:34864–72.
27. Rawlings ND, Barrett AJ. Introduction, serine peptidases and their clans. In: Barrett AJ, Rawlings ND, Woessner JF, editors. *Handbook of proteolytic enzymes*. 2nd ed. London, UK: Elsevier Ltd.; 2004. p. 1417–39.
28. Schechter I, Berger A. On the size of the active site in proteases. I. Papain. *Biochem Biophys Res Commun* 1967;**27**:157–62.
29. Takeuchi T, Harris JL, Huang W, Yan KW, Coughlin SR, Craik CS. Cellular localization of membrane-type serine protease 1 and identification of protease-activated receptor-2 and single-chain urokinase-type plasminogen activator as substrates. *J Biol Chem* 2000;**275**:26333–42.
30. Boulware KT, Daugherty PS. Protease specificity determination by using cellular libraries of peptide substrates (CLiPS). *Proc Natl Acad Sci USA* 2006;**103**:7583–8.
31. Herter S, Piper DE, Aaron W, Gabriele T, Cutler G, Cao P, et al. Hepatocyte growth factor is a preferred in vitro substrate for human hepsin, a membrane-anchored serine protease implicated in prostate and ovarian cancers. *Biochem J* 2005;**390**:125–36.
32. Shipway A, Danahay H, Williams JA, Tully DC, Backes BJ, Harris JL. Biochemical characterization of prostasin, a channel activating protease. *Biochem Biophys Res Commun* 2004;**324**:953–63.
33. Yuan J, Beltman J, Gjerstad E, Nguyen MT, Sampang J, Chan H, et al. Expression and characterization of recombinant gamma-tryptase. *Protein Expr Purif* 2006;**49**:47–54.
34. Kido H, Okumura Y. MSPL/TMPRSS13. *Front Biosci* 2008;**13**:754–8.
35. Okumura Y, Hayama M, Takahashi E, Fujiuchi M, Shimabukuro A, Yano M, et al. Serase-1B, a new splice variant of polyserase-1/TMPRSS9, activates urokinase-type plasminogen activator and the proteolytic activation is negatively regulated by glycosaminoglycans. *Biochem J* 2006;**400**:551–61.
36. Oberst MD, Williams CA, Dickson RB, Johnson MD, Lin CY. The activation of matriptase requires its noncatalytic domains, serine protease domain, and its cognate inhibitor. *J Biol Chem* 2003;**278**:26773–9.
37. Velasco G, Cal S, Quesada V, Sanchez LM, Lopez-Otin C. Matriptase-2, a membrane-bound mosaic serine proteinase predominantly expressed in human liver and showing degrading activity against extracellular matrix proteins. *J Biol Chem* 2002;**277**:37637–46.
38. Qiu D, Owen K, Gray K, Bass R, Ellis V. Roles and regulation of membrane-associated serine proteases. *Biochem Soc Trans* 2007;**35**:583–7.
39. Afar DE, Vivanco I, Hubert RS, Kuo J, Chen E, Saffran DC, et al. Catalytic cleavage of the androgen-regulated TMPRSS2 protease results in its secretion by prostate and prostate cancer epithelia. *Cancer Res* 2001;**61**:1686–92.
40. Guipponi M, Vuagniaux G, Wattenhofer M, Shibuya K, Vazquez M, Dougherty L, et al. The transmembrane serine protease (TMPRSS3) mutated in deafness DFNB8/10 activates the epithelial sodium channel (ENaC) in vitro. *Hum Mol Genet* 2002;**11**:2829–36.

41. Andreasen D, Vuagniaux G, Fowler-Jaeger N, Hummler E, Rossier BC. Activation of epithelial sodium channels by mouse channel activating proteases (mCAP) expressed in Xenopus oocytes requires catalytic activity of mCAP3 and mCAP2 but not mCAP1. *J Am Soc Nephrol* 2006;**17**:968–76.
42. Stallmach R, Gloor SM. Neurobin/TMPRSS11c, a novel type II transmembrane serine protease that cleaves fibroblast growth factor-2 *in vitro*. *Biochem J* 2008;**412**:81–91.
43. Vu TK, Liu RW, Haaksma CJ, Tomasek JJ, Howard EW. Identification and cloning of the membrane-associated serine protease, hepsin, from mouse preimplantation embryos. *J Biol Chem* 1997;**272**:31315–20.
44. List K, Bugge TH, Szabo R. Matriptase, potent proteolysis on the cell surface. *Mol Med* 2006;**12**:1–7.
45. Lu D, Yuan X, Zheng X, Sadler JE. Bovine proenteropeptidase is activated by trypsin, and the specificity of enteropeptidase depends on the heavy chain. *J Biol Chem* 1997;**272**:31293–300.
46. Knappe S, Wu F, Madlansacay MR, Wu Q. Identification of domain structures in the propeptide of corin essential for the processing of proatrial natriuretic peptide. *J Biol Chem* 2004;**279**:34464–71.
47. Desilets A, Beliveau F, Vandal G, McDuff FO, Lavigne P, Leduc R. Mutation G827R in matriptase causing autosomal recessive ichthyosis with hypotrichosis yields an inactive protease. *J Biol Chem* 2008;**283**:10535–42.
48. Lin CY, Tseng IC, Chou FP, Su SF, Chen YW, Johnson MD, et al. Zymogen activation, inhibition, and ectodomain shedding of matriptase. *Front Biosci* 2008;**13**:621–35.
49. Jin X, Hirosaki T, Lin CY, Dickson RB, Higashi S, Kitamura H, et al. Production of soluble matriptase by human cancer cell lines and cell surface activation of its zymogen by trypsin. *J Cell Biochem* 2005;**95**:632–47.
50. Benaud CM, Oberst M, Dickson RB, Lin CY. Deregulated activation of matriptase in breast cancer cells. *Clin Exp Metastasis* 2002;**19**:639–49.
51. Lucas JM, True L, Hawley S, Matsumura M, Morrissey C, Vessella R, et al. The androgen-regulated type II serine protease TMPRSS2 is differentially expressed and mislocalized in prostate adenocarcinoma. *J Pathol* 2008;**215**:118–25.
52. Laskowski Jr. M, Kato I. Protein inhibitors of proteinases. *Annu Rev Biochem* 1980;**49**:593–626.
53. Lin CY, Anders J, Johnson M, Sang QA, Dickson RB. Molecular cloning of cDNA for matriptase, a matrix-degrading serine protease with trypsin-like activity. *J Biol Chem* 1999;**274**:18231–6.
54. Oberst MD, Chen LY, Kiyomiya K, Williams CA, Lee MS, Johnson MD, et al. HAI-1 regulates activation and expression of matriptase, a membrane-bound serine protease. *Am J Physiol Cell Physiol* 2005;**289**:C462–70.
55. Kojima K, Tsuzuki S, Fushiki T, Inouye K. Roles of functional and structural domains of hepatocyte growth factor activator inhibitor type 1 in the inhibition of matriptase. *J Biol Chem* 2008;**283**:2478–87.
56. Szabo R, Hobson JP, List K, Molinolo A, Lin CY, Bugge TH. Potent inhibition and global co-localization implicate the transmembrane Kunitz-type serine protease inhibitor hepatocyte growth factor activator inhibitor-2 in the regulation of epithelial matriptase activity. *J Biol Chem* 2008;**283**:29495–504.
57. Kirchhofer D, Peek M, Lipari MT, Billeci K, Fan B, Moran P. Hepsin activates pro-hepatocyte growth factor and is inhibited by hepatocyte growth factor activator inhibitor-1B (HAI-1B) and HAI-2. *FEBS Lett* 2005;**579**:1945–50.
58. Fan B, Wu TD, Li W, Kirchhofer D. Identification of hepatocyte growth factor activator inhibitor-1B as a potential physiological inhibitor of prostasin. *J Biol Chem* 2005;**280**:34513–20.

59. Silverman GA, Bird PI, Carrell RW, Church FC, Coughlin PB, Gettins PG, et al. The serpins are an expanding superfamily of structurally similar but functionally diverse proteins. Evolution, mechanism of inhibition, novel functions, and a revised nomenclature. *J Biol Chem* 2001;**276**:33293–6.
60. Gettins PG. Serpin structure, mechanism, and function. *Chem Rev* 2002;**102**:4751–804.
61. Wong GW, Tang Y, Stevens RL. Cloning of the human homolog of mouse transmembrane tryptase. *Int Arch Allergy Immunol* 1999;**118**:419–21.
62. Wong GW, Foster PS, Yasuda S, Qi JC, Mahalingam S, Mellor EA, et al. Biochemical and functional characterization of human transmembrane tryptase (TMT)/tryptase gamma. TMT is an exocytosed mast cell protease that induces airway hyperresponsiveness in vivo via an interleukin-13/interleukin-4 receptor alpha/signal transducer and activator of transcription (STAT) 6-dependent pathway. *J Biol Chem* 2002;**277**:41906–15.
63. Reimer JM, Samollow PB, Hellman L. High degree of conservation of the multigene tryptase locus over the past 150-200 million years of mammalian evolution. *Immunogenetics* 2010;**62**:369–82.
64. Metcalfe DD, Baram D, Mekori YA. Mast cells. *Physiol Rev* 1997;**77**:1033–79.
65. Caughey GH. Mast cell tryptases and chymases in inflammation and host defense. *Immunol Rev* 2007;**217**:141–54.
66. Zheng XL, Kitamoto Y, Sadler JE. Enteropeptidase, a type II transmembrane serine protease. *Front Biosci (Elite Ed)* 2009;**1**:242–9.
67. Waldschmidt-Leitz E. The chemical nature of enzymes. *Science* 1933;**78**:189–90.
68. Kunitz M. Formation of trypsin from crystalline trypsinogen by means of enterokinase. *J Gen Physiol* 1939;**22**:429–46.
69. Kunitz M. Purification and concentration of enterokinase. *J Gen Physiol* 1939;**22**:447–50.
70. Light A, Janska H. Enterokinase (enteropeptidase), comparative aspects. *Trends Biochem Sci* 1989;**14**:110–2.
71. Mann NS, Mann SK. Enterokinase. *Proc Soc Exp Biol Med* 1994;**206**:114–8.
72. Light A, Janska H. The amino-terminal sequence of the catalytic subunit of bovine enterokinase. *J Protein Chem* 1991;**10**:475–80.
73. Kitamoto Y, Veile RA, Donis-Keller H, Sadler JE. cDNA sequence and chromosomal localization of human enterokinase, the proteolytic activator of trypsinogen. *Biochemistry* 1995;**34**:4562–8.
74. Kitamoto Y, Yuan X, Wu Q, McCourt DW, Sadler JE. Enterokinase, the initiator of intestinal digestion, is a mosaic protease composed of a distinctive assortment of domains. *Proc Natl Acad Sci USA* 1994;**91**:7588–92.
75. LaVallie ER, Rehemtulla A, Racie LA, DiBlasio EA, Ferenz C, Grant KL, et al. Cloning and functional expression of a cDNA encoding the catalytic subunit of bovine enterokinase. *J Biol Chem* 1993;**268**:23311–7.
76. Matsushima M, Ichinose M, Yahagi N, Kakei N, Tsukada S, Miki K, et al. Structural characterization of porcine enteropeptidase. *J Biol Chem* 1994;**269**:19976–82.
77. Yuan X, Zheng X, Lu D, Rubin DC, Pung CY, Sadler JE. Structure of murine enterokinase (enteropeptidase) and expression in small intestine during development. *Am J Physiol* 1998;**274**:G342–9.
78. Hermon-Taylor J, Perrin J, Grant DA, Appleyard A, Bubel M, Magee AI. Immunofluorescent localisation of enterokinase in human small intestine. *Gut* 1977;**18**:259–65.
79. Lojda Z, Gossrau R. Histochemical demonstration of enteropeptidase activity. New method with a synthetic substrate and its comparison with the trypsinogen procedure. *Histochemistry* 1983;**78**:251–70.

80. Zheng X, Lu D, Sadler JE. Apical sorting of bovine enteropeptidase does not involve detergent-resistant association with sphingolipid-cholesterol rafts. *J Biol Chem* 1999;**274**:1596–605.

81. Zheng X, Sadler JE. Mucin-like domain of enteropeptidase directs apical targeting in Madin-Darby canine kidney cells. *J Biol Chem* 2002;**277**:6858–63.

82. Hadorn B, Steiner N, Sumida C, Peters TJ. Intestinal enterokinase. Mechanisms of tts "secretion" into the lumen of the small intestine. *Lancet* 1971;**1**:165–6.

83. Zamolodchikova TS, Vorotyntseva TI, Antonov VK. Duodenase, a new serine protease of unusual specificity from bovine duodenal mucosa. Purification and properties. *Eur J Biochem* 1995;**227**:866–72.

84. Zamolodchikova TS, Sokolova EA, Alexandrov SL, Mikhaleva II, Prudchenko IA, Morozov IA, et al. Subcellular localization, substrate specificity and crystallization of duodenase, a potential activator of enteropeptidase. *Eur J Biochem* 1997;**249**:612–21.

85. Zamolodchikova TS, Sokolova EA, Lu D, Sadler JE. Activation of recombinant proenteropeptidase by duodenase. *FEBS Lett* 2000;**466**:295–9.

86. Hoffmeister A, Dietz G, Zeitschel U, Mossner J, Rossner S, Stahl T. BACE1 is a newly discovered protein secreted by the pancreas which cleaves enteropeptidase in vitro. *JOP* 2009;**10**:501–6.

87. Kim YT, Nishii W, Matsushima M, Inoue H, Ito H, Park SJ, et al. Substrate specificities of porcine and bovine enteropeptidases toward the peptide Val-(Asp)4-Lys-Ile-Val-Gly and its analogs. *Biosci Biotechnol Biochem* 2008;**72**:905–8.

88. Hadorn B, Tarlow MJ, Lloyd JK, Wolff OH. Intestinal enterokinase deficiency. *Lancet* 1969;**1**:812–3.

89. Hadorn B, Haworth JC, Gourley B, Prasad A, Troesch V. Intestinal enterokinase deficiency. Occurrence in two sibs and age dependency of clinical expression. *Arch Dis Child* 1975;**50**:277–82.

90. Leytus SP, Loeb KR, Hagen FS, Kurachi K, Davie EW. A novel trypsin-like serine protease (hepsin) with a putative transmembrane domain expressed by human liver and hepatoma cells. *Biochemistry* 1988;**27**:1067–74.

91. Farley D, Reymond F, Nick H. Cloning and sequence analysis of rat hepsin, a cell surface serine proteinase. *Biochim Biophys Acta* 1993;**1173**:350–2.

92. Kawamura S, Kurachi S, Deyashiki Y, Kurachi K. Complete nucleotide sequence, origin of isoform and functional characterization of the mouse hepsin gene. *Eur J Biochem* 1999;**262**:755–64.

93. Kurachi K, Torres-Rosado A, Tsuji A. Hepsin. *Methods Enzymol* 1994;**244**:100–14.

94. Wu Q. Gene targeting in hemostasis. Hepsin. *Front Biosci* 2001;**6**:D192–200.

95. Li W, Wang BE, Moran P, Lipari T, Ganesan R, Corpuz R, et al. Pegylated kunitz domain inhibitor suppresses hepsin-mediated invasive tumor growth and metastasis. *Cancer Res* 2009;**69**:8395–402.

96. Chen M, Chen LM, Lin CY, Chai KX. Hepsin activates prostasin and cleaves the extracellular domain of the epidermal growth factor receptor. *Mol Cell Biochem* 2010;**337**:259–66.

97. Nakama S, Oshiro N, Aniya Y. Activation of rat liver microsomal glutathione transferase by hepsin. *Biol Pharm Bull* 2010;**33**:561–7.

98. Tripathi M, Nandana S, Yamashita H, Ganesan R, Kirchhofer D, Quaranta V. Laminin-332 is a substrate for hepsin, a protease associated with prostate cancer progression. *J Biol Chem* 2008;**283**:30576–84.

99. Wu Q. The serine protease corin in cardiovascular biology and disease. *Front Biosci* 2007;**12**:4179–90.

100. Kazama Y, Hamamoto T, Foster DC, Kisiel W. Hepsin, a putative membrane-associated serine protease, activates human factor VII and initiates a pathway of blood coagulation on the cell surface leading to thrombin formation. *J Biol Chem* 1995;**270**:66–72.
101. Torres-Rosado A, O'Shea KS, Tsuji A, Chou SH, Kurachi K. Hepsin, a putative cell-surface serine protease, is required for mammalian cell growth. *Proc Natl Acad Sci USA* 1993;**90**:7181–5.
102. Wu Q, Yu D, Post J, Halks-Miller M, Sadler JE, Morser J. Generation and characterization of mice deficient in hepsin, a hepatic transmembrane serine protease. *J Clin Invest* 1998;**101**:321–6.
103. Yu IS, Chen HJ, Lee YS, Huang PH, Lin SR, Tsai TW, et al. Mice deficient in hepsin, a serine protease, exhibit normal embryogenesis and unchanged hepatocyte regeneration ability. *Thromb Haemost* 2000;**84**:865–70.
104. Guipponi M, Tan J, Cannon PZ, Donley L, Crewther P, Clarke M, et al. Mice deficient for the type II transmembrane serine protease, TMPRSS1/hepsin, exhibit profound hearing loss. *Am J Pathol* 2007;**171**:608–16.
105. Dhanasekaran SM, Barrette TR, Ghosh D, Shah R, Varambally S, Kurachi K, et al. Delineation of prognostic biomarkers in prostate cancer. *Nature* 2001;**412**:822–6.
106. Magee JA, Araki T, Patil S, Ehrig T, True L, Humphrey PA, et al. Expression profiling reveals hepsin overexpression in prostate cancer. *Cancer Res* 2001;**61**:5692–6.
107. Stamey TA, Warrington JA, Caldwell MC, Chen Z, Fan Z, Mahadevappa M, et al. Molecular genetic profiling of Gleason grade 4/5 prostate cancers compared to benign prostatic hyperplasia. *J Urol* 2001;**166**:2171–7.
108. Welsh JB, Sapinoso LM, Su AI, Kern SG, Wang-Rodriguez J, Moskaluk CA, et al. Analysis of gene expression identifies candidate markers and pharmacological targets in prostate cancer. *Cancer Res* 2001;**61**:5974–8.
109. Wu Q, Parry G. Hepsin and prostate cancer. *Front Biosci* 2007;**12**:5052–9.
110. Xuan JA, Schneider D, Toy P, Lin R, Newton A, Zhu Y, et al. Antibodies neutralizing hepsin protease activity do not impact cell growth but inhibit invasion of prostate and ovarian tumor cells in culture. *Cancer Res* 2006;**66**:3611–9.
111. Klezovitch O, Chevillet J, Mirosevich J, Roberts RL, Matusik RJ, Vasioukhin V. Hepsin promotes prostate cancer progression and metastasis. *Cancer Cell* 2004;**6**:185–95.
112. Paoloni-Giacobino A, Chen H, Peitsch MC, Rossier C, Antonarakis SE. Cloning of the TMPRSS2 gene, which encodes a novel serine protease with transmembrane, LDLRA, and SRCR domains and maps to 21q22.3. *Genomics* 1997;**44**:309–20.
113. Jacquinet E, Rao NV, Rao GV, Zhengming W, Albertine KH, Hoidal JR. Cloning and characterization of the cDNA and gene for human epitheliasin. *Eur J Biochem* 2001;**268**:2687–99.
114. Jacquinet E, Rao NV, Rao GV, Hoidal JR. Cloning, genomic organization, chromosomal assignment and expression of a novel mosaic serine proteinase, epitheliasin. *FEBS Lett* 2000;**468**:93–100.
115. Lin B, Ferguson C, White JT, Wang S, Vessella R, True LD, et al. Prostate-localized and androgen-regulated expression of the membrane-bound serine protease TMPRSS2. *Cancer Res* 1999;**59**:4180–4.
116. Chen YW, Lee MS, Lucht A, Chou FP, Huang W, Havighurst TC, et al. TMPRSS2, a serine protease expressed in the prostate on the apical surface of luminal epithelial cells and released into semen in prostasomes, is misregulated in prostate cancer cells. *Am J Pathol* 2010;**176**:2986–96.
117. Donaldson SH, Hirsh A, Li DC, Holloway G, Chao J, Boucher RC, et al. Regulation of the epithelial sodium channel by serine proteases in human airways. *J Biol Chem* 2002;**277**:8338–45.

118. Bottcher-Friebertshauser E, Freuer C, Sielaff F, Schmidt S, Eickmann M, Uhlendorff J, et al. Cleavage of influenza virus hemagglutinin by airway proteases TMPRSS2 and HAT differs in subcellular localization and susceptibility to protease inhibitors. *J Virol* 2010;**84**:5605–14.
119. Bottcher E, Matrosovich T, Beyerle M, Klenk HD, Garten W, Matrosovich M. Proteolytic activation of influenza viruses by serine proteases TMPRSS2 and HAT from human airway epithelium. *J Virol* 2006;**80**:9896–8.
120. Wilson S, Greer B, Hooper J, Zijlstra A, Walker B, Quigley J, et al. The membrane-anchored serine protease, TMPRSS2, activates PAR-2 in prostate cancer cells. *Biochem J* 2005;**388**:967–72.
121. Kim TS, Heinlein C, Hackman RC, Nelson PS. Phenotypic analysis of mice lacking the Tmprss2-encoded protease. *Mol Cell Biol* 2006;**26**:965–75.
122. Tomlins SA, Rhodes DR, Perner S, Dhanasekaran SM, Mehra R, Sun XW, et al. Recurrent fusion of TMPRSS2 and ETS transcription factor genes in prostate cancer. *Science* 2005;**310**:644–8.
123. Yu J, Yu J, Mani RS, Cao Q, Brenner CJ, Cao X, et al. An integrated network of androgen receptor, polycomb, and TMPRSS2-ERG gene fusions in prostate cancer progression. *Cancer Cell* 2010;**17**:443–54.
124. Tomlins SA, Laxman B, Varambally S, Cao X, Yu J, Helgeson BE, et al. Role of the TMPRSS2-ERG gene fusion in prostate cancer. *Neoplasia* 2008;**10**:177–88.
125. Tomlins SA, Laxman B, Dhanasekaran SM, Helgeson BE, Cao X, Morris DS, et al. Distinct classes of chromosomal rearrangements create oncogenic ETS gene fusions in prostate cancer. *Nature* 2007;**448**:595–9.
126. Haffner MC, Aryee MJ, Toubaji A, Esopi DM, Albadine R, Gurel B, et al. Androgen-induced TOP2B-mediated double-strand breaks and prostate cancer gene rearrangements. *Nat Genet* 2010;**42**:668–75.
127. Underwood LJ, Shigemasa K, Tanimoto H, Beard JB, Schneider EN, Wang Y, et al. Ovarian tumor cells express a novel multi-domain cell surface serine protease. *Biochim Biophys Acta* 2000;**1502**:337–50.
128. Scott HS, Kudoh J, Wattenhofer M, Shibuya K, Berry A, Chrast R, et al. Insertion of beta-satellite repeats identifies a transmembrane protease causing both congenital and childhood onset autosomal recessive deafness. *Nat Genet* 2001;**27**:59–63.
129. Bonne-Tamir B, DeStefano AL, Briggs CE, Adair R, Franklyn B, Weiss S, et al. Linkage of congenital recessive deafness (gene DFNB10) to chromosome 21q22.3. *Am J Hum Genet* 1996;**58**:1254–9.
130. Guipponi M, Antonarakis SE, Scott HS. TMPRSS3, a type II transmembrane serine protease mutated in non-syndromic autosomal recessive deafness. *Front Biosci* 2008;**13**:1557–67.
131. Guipponi M, Toh MY, Tan J, Park D, Hanson K, Ballana E, et al. An integrated genetic and functional analysis of the role of type II transmembrane serine proteases (TMPRSSs) in hearing loss. *Hum Mutat* 2008;**29**:130–41.
132. Peters TA, Levtchenko E, Cremers CW, Curfs JH, Monnens LA. No evidence of hearing loss in pseudohypoaldosteronism type 1 patients. *Acta Otolaryngol* 2006;**126**:237–9.
133. Wallrapp C, Hahnel S, Muller-Pillasch F, Burghardt B, Iwamura T, Ruthenburger M, et al. A novel transmembrane serine protease (TMPRSS3) overexpressed in pancreatic cancer. *Cancer Res* 2000;**60**:2602–6.
134. Vuagniaux G, Vallet V, Jaeger NF, Hummler E, Rossier BC. Synergistic activation of ENaC by three membrane-bound channel-activating serine proteases (mCAP1, mCAP2, and mCAP3) and S. *J Gen Physiol* 2002;**120**:191–201.
135. Garcia-Caballero A, Dang Y, He H, Stutts MJ. ENaC proteolytic regulation by channel-activating protease 2. *J Gen Physiol* 2008;**132**:521–35.

136. Chaipan C, Kobasa D, Bertram S, Glowacka I, Steffen I, Tsegaye TS, et al. Proteolytic activation of the 1918 influenza virus hemagglutinin. *J Virol* 2009;**83**:3200–11.
137. Kebebew E, Peng M, Reiff E, Duh QY, Clark OH, McMillan A. ECM1 and TMPRSS4 are diagnostic markers of malignant thyroid neoplasms and improve the accuracy of fine needle aspiration biopsy. *Ann Surg* 2005;**242**:353–61.
138. Cheng H, Fukushima T, Takahashi N, Tanaka H, Kataoka H. Hepatocyte growth factor activator inhibitor type 1 regulates epithelial to mesenchymal transition through membrane-bound serine proteinases. *Cancer Res* 2009;**69**:1828–35.
139. Jung H, Lee KP, Park SJ, Park JH, Jang YS, Choi SY, et al. TMPRSS4 promotes invasion, migration and metastasis of human tumor cells by facilitating an epithelial-mesenchymal transition. *Oncogene* 2008;**27**:2635–47.
140. Kim S, Kang HY, Nam EH, Choi MS, Zhao XF, Hong CS, et al. TMPRSS4 induces invasion and epithelial-mesenchymal transition through upregulation of integrin alpha5 and its signaling pathways. *Carcinogenesis* 2010;**31**:597–606.
141. Yamaguchi N, Okui A, Yamada T, Nakazato H, Mitsui S. Spinesin/TMPRSS5, a novel transmembrane serine protease, cloned from human spinal cord. *J Biol Chem* 2002;**277**:6806–12.
142. Mitsui S, Watanabe Y, Yamaguchi T, Yamaguchi N. Mosaic serine proteases in the mammalian central nervous system. *Front Biosci* 2008;**13**:1991–2000.
143. Kim DR, Sharmin S, Inoue M, Kido H. Cloning and expression of novel mosaic serine proteases with and without a transmembrane domain from human lung. *Biochim Biophys Acta* 2001;**1518**:204–9.
144. Okumura Y, Takahashi E, Yano M, Ohuchi M, Daidoji T, Nakaya T, et al. Novel type II transmembrane serine proteases, MSPL and TMPRSS13, proteolytically activate membrane fusion activity of the hemagglutinin of highly pathogenic avian influenza viruses and induce their multicycle replication. *J Virol* 2010;**84**:5089–96.
145. Shi YE, Torri J, Yieh L, Wellstein A, Lippman ME, Dickson RB. Identification and characterization of a novel matrix-degrading protease from hormone-dependent human breast cancer cells. *Cancer Res* 1993;**53**:1409–15.
146. Tanimoto H, Shigemasa K, Tian X, Gu L, Beard JB, Sawasaki T, et al. Transmembrane serine protease TADG-15 (ST14/Matriptase/MT-SP1), expression and prognostic value in ovarian cancer. *Br J Cancer* 2005;**92**:278–83.
147. Kim MG, Chen C, Lyu MS, Cho EG, Park D, Kozak C, et al. Cloning and chromosomal mapping of a gene isolated from thymic stromal cells encoding a new mouse type II membrane serine protease, epithin, containing four LDL receptor modules and two CUB domains. *Immunogenetics* 1999;**49**:420–8.
148. List K, Szabo R, Molinolo A, Nielsen BS, Bugge TH. Delineation of matriptase protein expression by enzymatic gene trapping suggests diverging roles in barrier function, hair formation, and squamous cell carcinogenesis. *Am J Pathol* 2006;**168**:1513–25.
149. Szabo R, Molinolo A, List K, Bugge TH. Matriptase inhibition by hepatocyte growth factor activator inhibitor-1 is essential for placental development. *Oncogene* 2007;**26**:1546–56.
150. Netzel-Arnett S, Currie BM, Szabo R, Lin CY, Chen LM, Chai KX, et al. Evidence for a matriptase-prostasin proteolytic cascade regulating terminal epidermal differentiation. *J Biol Chem* 2006;**281**:32941–5.
151. Fan B, Brennan J, Grant D, Peale F, Rangell L, Kirchhofer D. Hepatocyte growth factor activator inhibitor-1 (HAI-1) is essential for the integrity of basement membranes in the developing placental labyrinth. *Dev Biol* 2007;**303**:222–30.
152. Oberst MD, Singh B, Ozdemirli M, Dickson RB, Johnson MD, Lin CY. Characterization of matriptase expression in normal human tissues. *J Histochem Cytochem* 2003;**51**:1017–25.

153. Takeuchi T, Shuman MA, Craik CS. Reverse biochemistry, use of macromolecular protease inhibitors to dissect complex biological processes and identify a membrane-type serine protease in epithelial cancer and normal tissue. *Proc Natl Acad Sci USA* 1999;**96**:11054–61.

154. Benaud C, Oberst M, Hobson JP, Spiegel S, Dickson RB, Lin CY. Sphingosine 1-phosphate, present in serum-derived lipoproteins, activates matriptase. *J Biol Chem* 2002;**277**:10539–46.

155. Szabo R, Hobson JP, Christoph K, Kosa P, List K, Bugge TH. Regulation of cell surface protease matriptase by HAI2 is essential for placental development, neural tube closure and embryonic survival in mice. *Development* 2009;**136**:2653–63.

156. Kilpatrick LM, Harris RL, Owen KA, Bass R, Ghorayeb C, Bar-Or A, et al. Initiation of plasminogen activation on the surface of monocytes expressing the type II transmembrane serine protease matriptase. *Blood* 2006;**108**:2616–23.

157. Cheng MF, Jin JS, Wu HW, Chiang PC, Sheu LF, Lee HS. Matriptase expression in the normal and neoplastic mast cells. *Eur J Dermatol* 2007;**17**:375–80.

158. List K, Szabo R, Wertz PW, Segre J, Haudenschild CC, Kim SY, et al. Loss of proteolytically processed filaggrin caused by epidermal deletion of Matriptase/MT-SP1. *J Cell Biol* 2003;**163**:901–10.

159. List K, Haudenschild CC, Szabo R, Chen W, Wahl SM, Swaim W, et al. Matriptase/MT-SP1 is required for postnatal survival, epidermal barrier function, hair follicle development, and thymic homeostasis. *Oncogene* 2002;**21**:3765–79.

160. Kamata Y, Taniguchi A, Yamamoto M, Nomura J, Ishihara K, Takahara H, et al. Neutral cysteine protease bleomycin hydrolase is essential for the breakdown of deiminated filaggrin into amino acids. *J Biol Chem* 2009;**284**:12829–36.

161. List K, Kosa P, Szabo R, Bey AL, Wang CB, Molinolo A, et al. Epithelial integrity is maintained by a matriptase-dependent proteolytic pathway. *Am J Pathol* 2009;**175**:1453–63.

162. Leyvraz C, Charles RP, Rubera I, Guitard M, Rotman S, Breiden B, et al. The epidermal barrier function is dependent on the serine protease CAP1/Prss8. *J Cell Biol* 2005;**170**:487–96.

163. List K, Hobson JP, Molinolo A, Bugge TH. Co-localization of the channel activating protease prostasin/(CAP1/PRSS8) with its candidate activator, matriptase. *J Cell Physiol* 2007;**213**:237–45.

164. Basel-Vanagaite L, Attia R, Ishida-Yamamoto A, Rainshtein L, Ben Amitai D, Lurie R, et al. Autosomal recessive ichthyosis with hypotrichosis caused by a mutation in ST14, encoding type II transmembrane serine protease matriptase. *Am J Hum Genet* 2007;**80**:467–77.

165. Alef T, Torres S, Hausser I, Metze D, Tursen U, Lestringant GG, et al. Ichthyosis, follicular atrophoderma, and hypotrichosis caused by mutations in ST14 is associated with impaired profilaggrin processing. *J Invest Dermatol* 2009;**129**:862–9.

166. Avrahami L, Maas S, Pasmanik-Chor M, Rainshtein L, Magal N, Smitt J, et al. Autosomal recessive ichthyosis with hypotrichosis syndrome, further delineation of the phenotype. *Clin Genet* 2008;**74**:47–53.

167. List K, Currie B, Scharschmidt TC, Szabo R, Shireman J, Molinolo A, et al. Autosomal ichthyosis with hypotrichosis syndrome displays low matriptase proteolytic activity and is phenocopied in ST14 hypomorphic mice. *J Biol Chem* 2007;**282**:36714–23.

168. Sales KU, Masedunskas A, Bey AL, Rasmussen AL, Weigert R, List K, et al. Matriptase initiates activation of epidermal pro-kallikrein and disease onset in a mouse model of Netherton syndrome. *Nat Genet* 2010;**42**:676–83.

169. Buzza MS, Netzel-Arnett S, Shea-Donohue T, Zhao A, Lin CY, List K, et al. Membrane-anchored serine protease matriptase regulates epithelial barrier formation and permeability in the intestine. *Proc Natl Acad Sci USA* 2010;**107**:4200–5.

170. Lin CY, Anders J, Johnson M, Dickson RB. Purification and characterization of a complex containing matriptase and a Kunitz-type serine protease inhibitor from human milk. *J Biol Chem* 1999;**274**:18237–42.
171. Oberst M, Anders J, Xie B, Singh B, Ossandon M, Johnson M, et al. Matriptase and HAI-1 are expressed by normal and malignant epithelial cells *in vitro* and *in vivo*. *Am J Pathol* 2001;**158**:1301–11.
172. Zeng L, Cao J, Zhang X. Expression of serine protease SNC19/matriptase and its inhibitor hepatocyte growth factor activator inhibitor type 1 in normal and malignant tissues of gastrointestinal tract. *World J Gastroenterol* 2005;**11**:6202–7.
173. List K, Szabo R, Molinolo A, Sriuranpong V, Redeye V, Murdock T, et al. Deregulated matriptase causes ras-independent multistage carcinogenesis and promotes ras-mediated malignant transformation. *Genes Dev* 2005;**19**:1934–50.
174. Kirchhofer D, Peek M, Li W, Stamos J, Eigenbrot C, Kadkhodayan S, et al. Tissue expression, protease specificity, and Kunitz domain functions of hepatocyte growth factor activator inhibitor-1B (HAI-1B), a new splice variant of HAI-1. *J Biol Chem* 2003;**278**:36341–9.
175. Tanaka H, Nagaike K, Takeda N, Itoh H, Kohama K, Fukushima T, et al. Hepatocyte growth factor activator inhibitor type 1 (HAI-1) is required for branching morphogenesis in the chorioallantoic placenta. *Mol Cell Biol* 2005;**25**:5687–98.
176. Nagaike K, Kawaguchi M, Takeda N, Fukushima T, Sawaguchi A, Kohama K, et al. Defect of hepatocyte growth factor activator inhibitor type 1/serine protease inhibitor, Kunitz type 1 (Hai-1/Spint1) leads to ichthyosis-like condition and abnormal hair development in mice. *Am J Pathol* 2008;**173**:1464–75.
177. Carney TJ, von der Hardt S, Sonntag C, Amsterdam A, Topczewski J, Hopkins N, et al. Inactivation of serine protease Matriptase1a by its inhibitor Hai1 is required for epithelial integrity of the zebrafish epidermis. *Development* 2007;**134**:3461–71.
178. Lin CY, Wang JK, Torri J, Dou L, Sang QA, Dickson RB. Characterization of a novel, membrane-bound, 80-kDa matrix-degrading protease from human breast cancer cells. Monoclonal antibody production, isolation, and localization. *J Biol Chem* 1997;**272**:9147–52.
179. List K. Matriptase, a culprit in cancer? *Future Oncol* 2009;**5**:97–104.
180. Jin JS, Cheng TF, Tsai WC, Sheu LF, Chiang H, Yu CP. Expression of the serine protease, matriptase, in breast ductal carcinoma of Chinese women, correlation with clinicopathological parameters. *Histol Histopathol* 2007;**22**:305–9.
181. Kang JY, Dolled-Filhart M, Ocal IT, Singh B, Lin CY, Dickson RB, et al. Tissue microarray analysis of hepatocyte growth factor/Met pathway components reveals a role for Met, matriptase, and hepatocyte growth factor activator inhibitor 1 in the progression of node-negative breast cancer. *Cancer Res* 2003;**63**:1101–5.
182. Jin JS, Hsieh DS, Loh SH, Chen A, Yao CW, Yen CY. Increasing expression of serine protease matriptase in ovarian tumors, tissue microarray analysis of immunostaining score with clinicopathological parameters. *Mod Pathol* 2006;**19**:447–52.
183. Santin AD, Cane' S, Bellone S, Bignotti E, Palmieri M, Las Casas LE, et al. The novel serine protease tumor-associated differentially expressed gene-15 (matriptase/MT-SP1) is highly overexpressed in cervical carcinoma. *Cancer* 2003;**98**:1898–904.
184. Oberst MD, Johnson MD, Dickson RB, Lin CY, Singh B, Stewart M, et al. Expression of the serine protease matriptase and its inhibitor HAI-1 in epithelial ovarian cancer, correlation with clinical outcome and tumor clinicopathological parameters. *Clin Cancer Res* 2002;**8**:1101–7.
185. Saleem M, Adhami VM, Zhong W, Longley BJ, Lin CY, Dickson RB, et al. A novel biomarker for staging human prostate adenocarcinoma, overexpression of matriptase with concomitant loss of its inhibitor, hepatocyte growth factor activator inhibitor-1. *Cancer Epidemiol Biomarkers Prev* 2006;**15**:217–27.

186. Lee JW, Yong SongS, Choi JJ, Lee SJ, Kim BG, Park CS, et al. Increased expression of matriptase is associated with histopathologic grades of cervical neoplasia. *Hum Pathol* 2005;**36**:626–33.

187. Vogel LK, Saebo M, Skjelbred CF, Abell K, Pedersen ED, Vogel U, et al. The ratio of Matriptase/HAI-1 mRNA is higher in colorectal cancer adenomas and carcinomas than corresponding tissue from control individuals. *BMC Cancer* 2006;**6**:176.

188. Kishi K, Yamazaki K, Yasuda I, Yahagi N, Ichinose M, Tsuchiya Y, et al. Characterization of a membrane-bound arginine-specific serine protease from rat intestinal mucosa. *J Biochem* 2001;**130**:425–30.

189. Satomi S, Yamasaki Y, Tsuzuki S, Hitomi Y, Iwanaga T, Fushiki T. A role for membrane-type serine protease (MT-SP1) in intestinal epithelial turnover. *Biochem Biophys Res Commun* 2001;**287**:995–1002.

190. Lee SL, Dickson RB, Lin CY. Activation of hepatocyte growth factor and urokinase/plasminogen activator by matriptase, an epithelial membrane serine protease. *J Biol Chem* 2000;**275**:36720–5.

191. Bhatt AS, Welm A, Farady CJ, Vasquez M, Wilson K, Craik CS. Coordinate expression and functional profiling identify an extracellular proteolytic signaling pathway. *Proc Natl Acad Sci USA* 2007;**104**:5771–6.

192. Camerer E, Barker A, Duong DN, Ganesan R, Kataoka H, Cornelissen I, et al. Local protease signaling contributes to neural tube closure in the mouse embryo. *Dev Cell* 2010;**18**:25–38.

193. Bhatt AS, Erdjument-Bromage H, Tempst P, Craik CS, Moasser MM. Adhesion signaling by a novel mitotic substrate of src kinases. *Oncogene* 2005;**24**:5333–43.

194. He Y, Wortmann A, Burke LJ, Reid JC, Adams MN, Abdul-Jabbar I, et al. Proteolysis induced amino terminal ectodomain shedding of the integral membrane glycoprotein CUB domain containing protein 1 (CDCP1) is accompanied by tyrosine phosphorylation of its carboxy terminal domain and recruitment of Src and PKC{delta}. *J Biol Chem* 2010;**285**:26162–73.

195. Hooper JD, Campagnolo L, Goodarzi G, Truong TN, Stuhlmann H, Quigley JP. Mouse matriptase-2, identification, characterization and comparative mRNA expression analysis with mouse hepsin in adult and embryonic tissues. *Biochem J* 2003;**373**:689–702.

196. Stirnberg M, Maurer E, Horstmeyer A, Kolp S, Frank S, Bald T, et al. Proteolytic processing of the serine protease matriptase-2, identification of the cleavage sites required for its autocatalytic release from the cell surface. *Biochem J* 2010;**430**:87–95.

197. Folgueras AR, de Lara FM, Pendas AM, Garabaya C, Rodriguez F, Astudillo A, et al. Membrane-bound serine protease matriptase-2 (Tmprss6) is an essential regulator of iron homeostasis. *Blood* 2008;**112**:2539–45.

198. Du X, She E, Gelbart T, Truksa J, Lee P, Xia Y, et al. The serine protease TMPRSS6 is required to sense iron deficiency. *Science* 2008;**320**:1088–92.

199. Finberg KE, Heeney MM, Campagna DR, Aydinok Y, Pearson HA, Hartman KR, et al. Mutations in TMPRSS6 cause iron-refractory iron deficiency anemia (IRIDA). *Nat Genet* 2008;**40**:569–71.

200. Silvestri L, Pagani A, Nai A, De Domenico I, Kaplan J, Camaschella C. The serine protease matriptase-2 (TMPRSS6) inhibits hepcidin activation by cleaving membrane hemojuvelin. *Cell Metab* 2008;**8**:502–11.

201. Silvestri L, Guillem F, Pagani A, Nai A, Oudin C, Silva M, et al. Molecular mechanisms of the defective hepcidin inhibition in TMPRSS6 mutations associated with iron-refractory iron deficiency anemia. *Blood* 2009;**113**:5605–8.

202. Szabo R, Netzel-Arnett S, Hobson JP, Antalis TM, Bugge TH. Matriptase-3 is a novel phylogenetically preserved membrane-anchored serine protease with broad serpin reactivity. *Biochem J* 2005;**390**:231–42.

203. Cal S, Quesada V, Garabaya C, Lopez-Otin C. Polyserase-I, a human polyprotease with the ability to generate independent serine protease domains from a single translation product. *Proc Natl Acad Sci USA* 2003;**100**:9185–90.

204. Hayama M, Okumura Y, Takahashi E, Shimabukuro A, Tamura M, Takeda N, et al. Identification and analysis of the promoter region of the type II transmembrane serine protease polyserase-1 and its transcript variants. *Biol Chem* 2007;**388**:853–8.

205. Yan W, Sheng N, Seto M, Morser J, Wu Q. Corin, a mosaic transmembrane serine protease encoded by a novel cDNA from human heart. *J Biol Chem* 1999;**274**:14926–35.

206. Hooper JD, Scarman AL, Clarke BE, Normyle JF, Antalis TM. Localization of the mosaic transmembrane serine protease corin to heart myocytes. *Eur J Biochem* 2000;**267**:6931–7.

207. Tomita Y, Kim DH, Magoori K, Fujino T, Yamamoto TT. A novel low-density lipoprotein receptor-related protein with type II membrane protein-like structure is abundant in heart. *J Biochem* 1998;**124**:784–9.

208. Pan J, Hinzmann B, Yan W, Wu F, Morser J, Wu Q. Genomic structures of the human and murine corin genes and functional GATA elements in their promoters. *J Biol Chem* 2002;**277**:38390–8.

209. Gladysheva IP, Robinson BR, Houng AK, Kovats T, King SM. Corin is co-expressed with pro-ANP and localized on the cardiomyocyte surface in both zymogen and catalytically active forms. *J Mol Cell Cardiol* 2008;**44**:131–42.

210. Liao X, Wang W, Chen S, Wu Q. Role of glycosylation in corin zymogen activation. *J Biol Chem* 2007;**282**:27728–35.

211. Knappe S, Wu F, Masikat MR, Morser J, Wu Q. Functional analysis of the transmembrane domain and activation cleavage of human corin, design and characterization of a soluble corin. *J Biol Chem* 2003;**278**:52363–70.

212. Kohn EC, Liotta LA. Molecular insights into cancer invasion, strategies for prevention and intervention. *Cancer Res* 1995;**55**:1856–62.

213. Wu Q, Xu-Cai YO, Chen S, Wang W. Corin, new insights into the natriuretic peptide system. *Kidney Int* 2009;**75**:142–6.

214. Hodgson-Zingman DM, Karst ML, Zingman LV, Heublein DM, Darbar D, Herron KJ, et al. Atrial natriuretic peptide frameshift mutation in familial atrial fibrillation. *N Engl J Med* 2008;**359**:158–65.

215. Yan W, Wu F, Morser J, Wu Q. Corin, a transmembrane cardiac serine protease, acts as a pro-atrial natriuretic peptide-converting enzyme. *Proc Natl Acad Sci USA* 2000;**97**:8525–9.

216. Chan JC, Knudson O, Wu F, Morser J, Dole WP, Wu Q. Hypertension in mice lacking the proatrial natriuretic peptide convertase corin. *Proc Natl Acad Sci USA* 2005;**102**:785–90.

217. Nigrovic PA, Gray DH, Jones T, Hallgren J, Kuo FC, Chaletzky B, et al. Genetic inversion in mast cell-deficient (W(sh)) mice interrupts corin and manifests as hematopoietic and cardiac aberrancy. *Am J Pathol* 2008;**173**:1693–701.

218. Enshell-Seijffers D, Lindon C, Morgan BA. The serine protease Corin is a novel modifier of the Agouti pathway. *Development* 2008;**135**:217–25.

219. Semenov AG, Tamm NN, Seferian KR, Postnikov AB, Karpova NS, Serebryanaya DV, et al. Processing of pro-B-type natriuretic peptide, furin and corin as candidate convertases. *Clin Chem* 2010;**56**:1166–76.

220. Wu C, Wu F, Pan J, Morser J, Wu Q. Furin-mediated processing of Pro-C-type natriuretic peptide. *J Biol Chem* 2003;**278**:25847–52.

221. Dries DL, Victor RG, Rame JE, Cooper RS, Wu X, Zhu X, et al. Corin gene minor allele defined by 2 missense mutations is common in blacks and associated with high blood pressure and hypertension. *Circulation* 2005;**112**:2403–10.

222. Rame JE, Drazner MH, Post W, Peshock R, Lima J, Cooper RS, et al. Corin I555(P568) allele is associated with enhanced cardiac hypertrophic response to increased systemic afterload. *Hypertension* 2007;**49**:857–64.
223. Rame JE, Tam SW, McNamara D, Worcel M, Sabolinski ML, Wu AH, et al. Dysfunctional corin I555(P568) allele is associated with impaired brain natriuretic peptide processing and adverse outcomes in blacks with systolic heart failure, results from the genetic risk assessment in heart failure substudy. *Circ Heart Fail* 2009;**2**:541–8.
224. Wang W, Liao X, Fukuda K, Knappe S, Wu F, Dries DL, et al. Corin variant associated with hypertension and cardiac hypertrophy exhibits impaired zymogen activation and natriuretic peptide processing activity. *Circ Res* 2008;**103**:502–8.
225. Peleg A, Jaffe AS, Hasin Y. Enzyme-linked immunoabsorbent assay for detection of human serine protease corin in blood. *Clin Chim Acta* 2009;**409**:85–9.
226. Dong N, Chen S, Yang J, He L, Liu P, Zheng D, et al. Plasma soluble corin in patients with heart failure. *Circ Heart Fail* 2010;**3**:207–11.
227. Hobson JP, Netzel-Arnett S, Szabo R, Rehault SM, Church FC, Strickland DK, et al. Mouse DESC1 is located within a cluster of seven DESC1-like genes and encodes a type II transmembrane serine protease that forms serpin inhibitory complexes. *J Biol Chem* 2004;**279**:46981–94.
228. Macao B, Johansson DG, Hansson GC, Hard T. Autoproteolysis coupled to protein folding in the SEA domain of the membrane-bound MUC1 mucin. *Nat Struct Mol Biol* 2006;**13**:71–6.
229. Yasuoka S, Ohnishi T, Kawano S, Tsuchihashi S, Ogawara M, Masuda K, et al. Purification, characterization, and localization of a novel trypsin-like protease found in the human airway. *Am J Respir Cell Mol Biol* 1997;**16**:300–8.
230. Yamaoka K, Masuda K, Ogawa H, Takagi K, Umemoto N, Yasuoka S. Cloning and characterization of the cDNA for human airway trypsin-like protease. *J Biol Chem* 1998;**273**:11895–901.
231. Matsushima R, Takahashi A, Nakaya Y, Maezawa H, Miki M, Nakamura Y, et al. Human airway trypsin-like protease stimulates human bronchial fibroblast proliferation in a protease-activated receptor-2-dependent pathway. *Am J Physiol Lung Cell Mol Physiol* 2006;**290**: L385–95.
232. Yoshinaga S, Nakahori Y, Yasuoka S. Fibrinogenolytic activity of a novel trypsin-like enzyme found in human airway. *J Med Invest* 1998;**45**:77–86.
233. Iwakiri K, Ghazizadeh M, Jin E, Fujiwara M, Takemura T, Takezaki S, et al. Human airway trypsin-like protease induces PAR-2-mediated IL-8 release in psoriasis vulgaris. *J Invest Dermatol* 2004;**122**:937–44.
234. Takahashi M, Sano T, Yamaoka K, Kamimura T, Umemoto N, Nishitani H, et al. Localization of human airway trypsin-like protease in the airway, an immunohistochemical study. *Histochem Cell Biol* 2001;**115**:181–7.
235. Miki M, Nakamura Y, Takahashi A, Nakaya Y, Eguchi H, Masegi T, et al. Effect of human airway trypsin-like protease on intracellular free Ca^{2+} concentration in human bronchial epithelial cells. *J Med Invest* 2003;**50**:95–107.
236. Chokki M, Eguchi H, Hamamura I, Mitsuhashi H, Kamimura T. Human airway trypsin-like protease induces amphiregulin release through a mechanism involving protease-activated receptor-2-mediated ERK activation and TNF alpha-converting enzyme activity in airway epithelial cells. *FEBS J* 2005;**272**:6387–99.
237. Beaufort N, Leduc D, Eguchi H, Mengele K, Hellmann D, Masegi T, et al. The human airway trypsin-like protease modulates the urokinase receptor (uPAR, CD87) structure and functions. *Am J Physiol Lung Cell Mol Physiol* 2007;**292**:L1263–72.
238. Bertram S, Glowacka I, Steffen I, Kuhl A, Pohlmann S. Novel insights into proteolytic cleavage of influenza virus hemagglutinin. *Rev Med Virol.* 2010;**20**(5):298–310.

239. Lang JC, Schuller DE. Differential expression of a novel serine protease homologue in squamous cell carcinoma of the head and neck. *Br J Cancer* 2001;**84**:237–43.

240. Sedghizadeh PP, Mallery SR, Thompson SJ, Kresty L, Beck FM, Parkinson EK, et al. Expression of the serine protease DESC1 correlates directly with normal keratinocyte differentiation and inversely with head and neck squamous cell carcinoma progression. *Head Neck* 2006;**28**:432–40.

241. Viloria CG, Peinado JR, Astudillo A, Garcia-Suarez O, Gonzalez MV, Suarez C, et al. Human DESC1 serine protease confers tumorigenic properties to MDCK cells and it is upregulated in tumours of different origin. *Br J Cancer* 2007;**97**:201–9.

242. Beliveau F, Desilets A, Leduc R. Probing the substrate specificities of matriptase, matriptase-2, hepsin and DESC1 with internally quenched fluorescent peptides. *FEBS J* 2009;**276**:2213–26.

243. Kam YW, Okumura Y, Kido H, Ng LF, Bruzzone R, Altmeyer R. Cleavage of the SARS coronavirus spike glycoprotein by airway proteases enhances virus entry into human bronchial epithelial cells in vitro. *PLoS One* 2009;**4**:e7870.

244. Adachi M, Kitamura K, Miyoshi T, Narikiyo T, Iwashita K, Shiraishi N, et al. Activation of epithelial sodium channels by prostasin in Xenopus oocytes. *J Am Soc Nephrol* 2001;**12**:1114–21.

245. Vuagniaux G, Vallet V, Jaeger NF, Pfister C, Bens M, Farman N, et al. Activation of the amiloride-sensitive epithelial sodium channel by the serine protease mCAP1 expressed in a mouse cortical collecting duct cell line. *J Am Soc Nephrol* 2000;**11**:828–34.

246. Tong Z, Illek B, Bhagwandin VJ, Verghese GM, Caughey GH. Prostasin, a membrane-anchored serine peptidase, regulates sodium currents in JME/CF15 cells, a cystic fibrosis airway epithelial cell line. *Am J Physiol Lung Cell Mol Physiol* 2004;**287**:L928–35.

247. Scarman AL, Hooper JD, Boucaut KJ, Sit ML, Webb GC, Normyle JF, et al. Organization and chromosomal localization of the murine Testisin gene encoding a serine protease temporally expressed during spermatogenesis. *Eur J Biochem* 2001;**268**:1250–8.

248. Wong GW, Li L, Madhusudhan MS, Krilis SA, Gurish MF, Rothenberg ME, et al. Tryptase 4, a new member of the chromosome 17 family of mouse serine proteases. *J Biol Chem* 2001;**276**:20648–58.

249. Hooper JD, Bowen N, Marshall H, Cullen LM, Sood R, Daniels R, et al. Localization, expression and genomic structure of the gene encoding the human serine protease testisin. *Biochim Biophys Acta* 2000;**1492**:63–71.

250. Vallet V, Chraibi A, Gaeggeler HP, Horisberger JD, Rossier BC. An epithelial serine protease activates the amiloride-sensitive sodium channel. *Nature* 1997;**389**:607–10.

251. Selzer-Plon J, Bornholdt J, Friis S, Bisgaard HC, Lothe IM, Tveit KM, et al. Expression of prostasin and its inhibitors during colorectal cancer carcinogenesis. *BMC Cancer* 2009;**9**:201.

252. Koda A, Wakida N, Toriyama K, Yamamoto K, Iijima H, Tomita K, et al. Urinary prostasin in humans, relationships among prostasin, aldosterone and epithelial sodium channel activity. *Hypertens Res* 2009;**32**:276–81.

253. Zhu H, Chao J, Guo D, Li K, Huang Y, Hawkins K, et al. Urinary prostasin, a possible biomarker for renal pressure natriuresis in black adolescents. *Pediatr Res* 2009;**65**:443–6.

254. Olivieri O, Castagna A, Guarini P, Chiecchi L, Sabaini G, Pizzolo F, et al. Urinary prostasin, a candidate marker of epithelial sodium channel activation in humans. *Hypertension* 2005;**46**:683–8.

255. Iwashita K, Kitamura K, Narikiyo T, Adachi M, Shiraishi N, Miyoshi T, et al. Inhibition of prostasin secretion by serine protease inhibitors in the kidney. *J Am Soc Nephrol* 2003;**14**:11–6.

256. Svenningsen P, Uhrenholt TR, Palarasah Y, Skjoedt K, Jensen BL, Skott O. Prostasin-dependent activation of epithelial Na^+ channels by low plasmin concentrations. *Am J Physiol Regul Integr Comp Physiol* 2009;**297**:R1733–41.
257. Wakida N, Kitamura K, Tuyen DG, Maekawa A, Miyoshi T, Adachi M, et al. Inhibition of prostasin-induced ENaC activities by PN-1 and regulation of PN-1 expression by TGF-beta1 and aldosterone. *Kidney Int* 2006;**70**:1432–8.
258. Coote K, Atherton-Watson HC, Sugar R, Young A, MacKenzie-Beevor A, Gosling M, et al. Camostat attenuates airway epithelial sodium channel function in vivo through the inhibition of a channel-activating protease. *J Pharmacol Exp Ther* 2009;**329**:764–74.
259. Planes C, Caughey GH. Regulation of the epithelial Na^+ channel by peptidases. *Curr Top Dev Biol* 2007;**78**:23–46.
260. Bruns JB, Carattino MD, Sheng S, Maarouf AB, Weisz OA, Pilewski JM, et al. Epithelial Na^+ channels are fully activated by furin- and prostasin-dependent release of an inhibitory peptide from the gamma-subunit. *J Biol Chem* 2007;**282**:6153–60.
261. Carattino MD, Hughey RP, Kleyman TR. Proteolytic processing of the epithelial sodium channel gamma subunit has a dominant role in channel activation. *J Biol Chem* 2008;**283**:25290–5.
262. Passero CJ, Carattino MD, Kashlan OB, Myerburg MM, Hughey RP, Kleyman TR. Defining an inhibitory domain in the gamma subunit of the epithelial sodium channel. *Am J Physiol Renal Physiol.* 2010;**299**(4):F854–61.
263. Planes C, Randrianarison NH, Charles RP, Frateschi S, Cluzeaud F, Vuagniaux G, et al. ENaC-mediated alveolar fluid clearance and lung fluid balance depend on the channel-activating protease 1. *EMBO Mol Med* 2010;**2**:26–37.
264. Donaldson SH, Boucher RC. Sodium channels and cystic fibrosis. *Chest* 2007;**132**:1631–6.
265. Myerburg MM, McKenna EE, Luke CJ, Frizzell RA, Kleyman TR, Pilewski JM. Prostasin expression is regulated by airway surface liquid volume and is increased in cystic fibrosis. *Am J Physiol Lung Cell Mol Physiol* 2008;**294**:L932–41.
266. Zhu H, Guo D, Li K, Yan W, Tan Y, Wang X, et al. Prostasin, a possible candidate gene for human hypertension. *Am J Hypertens* 2008;**21**:1028–33.
267. Hummler E. Implication of ENaC in salt-sensitive hypertension. *J Steroid Biochem Mol Biol* 1999;**69**:385–90.
268. Maekawa A, Kakizoe Y, Miyoshi T, Wakida N, Ko T, Shiraishi N, et al. Camostat mesilate inhibits prostasin activity and reduces blood pressure and renal injury in salt-sensitive hypertension. *J Hypertens* 2009;**27**:181–9.
269. Chen LM, Hatfield ML, Fu YY, Chai KX. Prostasin regulates iNOS and cyclin D1 expression by modulating protease-activated receptor-2 signaling in prostate epithelial cells. *Prostate* 2009;**69**:1790–801.
270. Chen M, Fu YY, Lin CY, Chen LM, Chai KX. Prostasin induces protease-dependent and independent molecular changes in the human prostate carcinoma cell line PC-3. *Biochim Biophys Acta* 2007;**1773**:1133–40.
271. Fu YY, Gao WL, Chen M, Chai KX, Wang YL, Chen LM. Prostasin regulates human placental trophoblast cell proliferation via the epidermal growth factor receptor signaling pathway. *Hum Reprod* 2010;**25**:623–32.
272. Sakashita K, Mimori K, Tanaka F, Tahara K, Inoue H, Sawada T, et al. Clinical significance of low expression of Prostasin mRNA in human gastric cancer. *J Surg Oncol* 2008;**98**:559–64.
273. Ma XJ, Fu YY, Li YX, Chen LM, Chai K, Wang YL. Prostasin inhibits cell invasion in human choriocarcinomal JEG-3 cells. *Histochem Cell Biol* 2009;**132**:639–46.
274. Inoue M, Kanbe N, Kurosawa M, Kido H. Cloning and tissue distribution of a novel serine protease esp-1 from human eosinophils. *Biochem Biophys Res Commun* 1998;**252**:307–12.

275. Inoue M, Isobe M, Itoyama T, Kido H. Structural analysis of esp-1 gene (PRSS 21). *Biochem Biophys Res Commun* 1999;**266**:564–8.
276. Netzel-Arnett S, Bugge TH, Hess RA, Carnes K, Stringer BW, Scarman AL, et al. The glycosylphosphatidylinositol-anchored serine protease PRSS21 (testisin) imparts murine epididymal sperm cell maturation and fertilizing ability. *Biol Reprod* 2009;**81**:921–32.
277. Aimes RT, Zijlstra A, Hooper JD, Ogbourne SM, Sit ML, Fuchs S, et al. Endothelial cell serine proteases expressed during vascular morphogenesis and angiogenesis. *Thromb Haemost* 2003;**89**:561–72.
278. Manton KJ, Douglas ML, Netzel-Arnett S, Fitzpatrick DR, Nicol DL, Boyd AW, et al. Hypermethylation of the 5′ CpG island of the gene encoding the serine protease Testisin promotes its loss in testicular tumorigenesis. *Br J Cancer* 2005;**92**:760–9.
279. Nakamura Y, Inoue M, Okumura Y, Shiota M, Nishikawa M, Arase S, et al. Cloning, expression analysis, and tissue distribution of esp-1/testisin, a membrane-type serine protease from the rat. *J Med Invest* 2003;**50**:78–86.
280. Yamashita M, Honda A, Ogura A, Kashiwabara S, Fukami K, Baba T. Reduced fertility of mouse epididymal sperm lacking Prss21/Tesp5 is rescued by sperm exposure to uterine microenvironment. *Genes Cells* 2008;**13**:1001–13.
281. Kawano N, Kang W, Yamashita M, Koga Y, Yamazaki T, Hata T. Mice lacking two sperm serine proteases, ACR and PRSS21, are subfertile, but the mutant sperm are infertile in vitro. *Biol Reprod* 2010;**83**(3):359–69.
282. Shigemasa K, Underwood LJ, Beard J, Tanimoto H, Ohama K, Parmley TH, et al. Overexpression of testisin, a serine protease expressed by testicular germ cells, in epithelial ovarian tumor cells. *J Soc Gynecol Investig* 2000;**7**:358–62.
283. Tang T, Kmet M, Corral L, Vartanian S, Tobler A, Papkoff J. Testisin, a glycosyl-phosphatidylinositol-linked serine protease, promotes malignant transformation in vitro and in vivo. *Cancer Res* 2005;**65**:868–78.
284. Kempkensteffen C, Christoph F, Weikert S, Krause H, Kollermann J, Schostak M, et al. Epigenetic silencing of the putative tumor suppressor gene testisin in testicular germ cell tumors. *J Cancer Res Clin Oncol* 2006;**132**:765–70.

Complex Assemblies of Factors IX and X Regulate the Initiation, Maintenance, and Shutdown of Blood Coagulation

THOMAS ZÖGG AND HANS BRANDSTETTER

Department of Molecular Biology, University of Salzburg, Salzburg, Austria

I. Introduction 52
A. Role of Coagulation Factors IX and X in Hemostasis 52
B. Aims of This Chapter 52
II. Proteolytic Activation of Trypsin-Like Serine Proteases 54
A. Activation Requires Correct N-Terminus Formation 54
B. Release of the Activation Peptide 60
III. Low-Molecular Weight Cofactors 61
A. Ethylene Glycol 61
B. Heparin 63
C. Metal Ions: Calcium and Sodium 64
D. Long Distance-Communication Line Connects Ca^{2+} Exosite with the Active Site 66
IV. Protein Cofactors 68
A. Factor VII: Tissue Factor 69
B. Factor VIII 70
C. Substrate's Role in Xase Function 74
D. An Atomic Model of Xase-Like Activity Enhancement 75
E. Factor V 78
V. Membranes: Nonhumoral Cofactors of Coagulation Factors 79
A. The Injured Vasculature: Endothelial Cells 79
B. Activated Platelets and Phospholipid Membranes 80
VI. Protein Inhibitors of Coagulation Factors 81
A. Endogenous Inhibitors 82
B. Exogenous Inhibitors 88
VII. Pharmaceutical Opportunities and Challenges 90
A. Antihemophilic Drug Discovery 90
B. Antithrombotic Drugs 91
References 92

Blood hemostasis is accomplished by a complex network of (anti-)coagulatory and fibrinolytic processes. These physiological processes are implemented by the assembly of multiprotein complexes involving both humoral and cellular components. Coagulation factor X, and particularly, factor IX, exemplify the

Progress in Molecular Biology
and Translational Science, Vol. 99
DOI: 10.1016/S1877-1173(11)99002-4

dramatic enhancement that is obtained by the synergistic interaction of cell surface, inorganic and protein cofactors, protease, and substrate. With a focus on structure–function relationship, we review the current knowledge of activity modulation principles in the coagulation proteases factors IX and X and indicate future challenges for hemostasis research. This chapter is organized by describing the principles of hierarchical activation of blood coagulation proteases, including endogenous and exogenous protease activators, cofactor binding, substrate specificities, and protein inhibitors. We conclude by outlining pharmaceutical opportunities for unmet needs in hemophilia and thrombosis.

I. Introduction

A. Role of Coagulation Factors IX and X in Hemostasis

The blood coagulation cascade is an interdependent system of serine proteases, their cofactors, and inhibitors. This complex protease machinery rapidly stops life-threatening blood loss caused by vasculature lesions while avoiding uncontrolled clot formation. To this end, the involved proteins are produced on stock as inactive precursors, which are readily converted to the active form by proteolytic cleavage; however, they reveal their full catalytic potency only through further stimulating events like complex formation with cofactors.[1]

The blood coagulation cascade is formed by proteases and protein cofactors, which are highly homologous and share a similar, often identical, domain organization (Fig. 1). The multiplicity of homologous serine proteases, like factors II, VII, IX, and X, is however not a measure of redundancy but reflects the necessity for subtle regulation, substrate selectivity, and multiple control, as defects in any of these homologous proteases and cofactors manifest themselves in clinical pathologies like bleeding or clotting disorders.

The coagulation factor IX (fIX) is the most extreme example in many respects. While the trypsin-like serine protease fIX undergoes the classical proteolytic activation, "active" fIXa has very low amidolytic and proteolytic activity.[2,3] The *Open Sesame* of blood coagulation is multiple complex formation.

B. Aims of This Chapter

In this chapter, we review our knowledge and mechanistic understanding of key regulation steps in blood coagulation. We are not aiming at a complete overview on the blood coagulation cascade; instead, we focus on coagulation factors IX and X, which are positioned at the junction of the extrinsic and

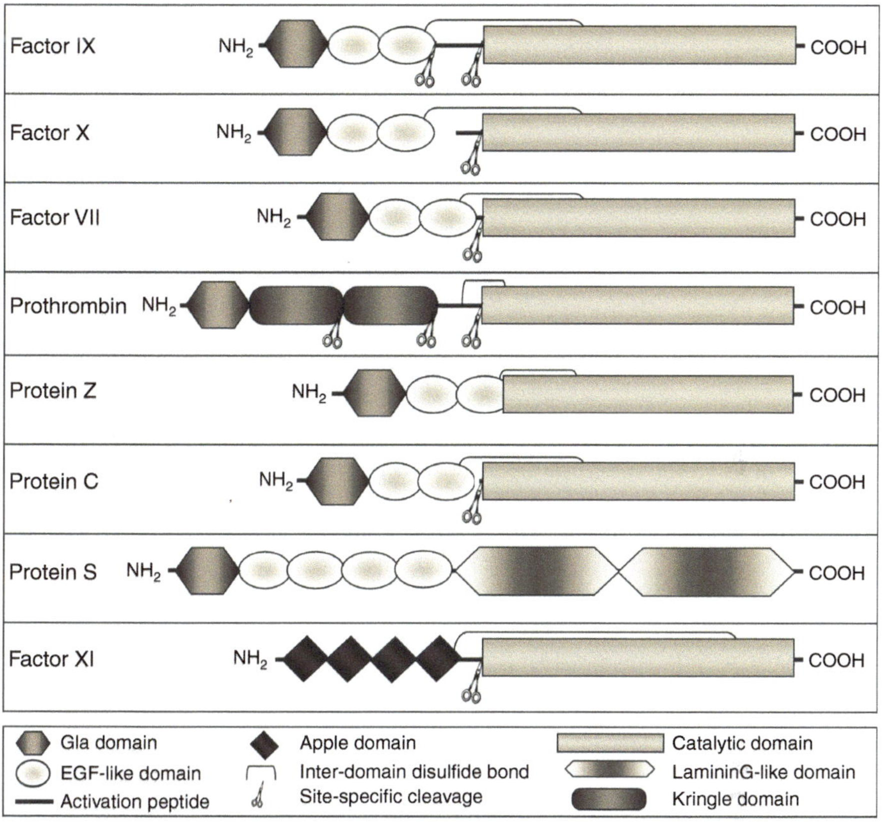

FIG. 1. Domain arrangement and activation pattern of coagulation factors. Schematic representation of the building blocks that compose the homologous family of coagulation factors. Factor IX-like coagulation factors share four modules: a Gla-domain anchoring the molecule to the membrane, two EGF-like domains involved in intermolecule interactions and serving as a spacer region to increase the distance to the membrane, and the serine-protease domain exerting its distinct (anti-)coagulative function. Several coagulation proteases deviate from this domain arrangement, as indicated. Gla, γ-carboxyglutamic acid-rich domain; EGF, epidermal growth factor-like domain.

intrinsic coagulation cascade. By this approach, we wish (aim #1) to elaborate the functional, structural, and mechanistic principles that apply to these factors and also to other components of the coagulation cascade.

Blood coagulation has been a field of intense work for the past four decades. We further wish (aim #2) to illustrate how vibrant the field is today, having lost none of its fascination but opening up grand perspectives. Current researchers are in the lucky situation to witness the beginning of the pharmaceutical harvest on blood coagulation research. Finally (aim #3), we wish to point out what we consider important future directions in the field, which will inevitably have to address the structure and mechanism of protein complexes, the machines that so delicately control the balance between bleeding and clotting. With great power comes great responsibility.

II. Proteolytic Activation of Trypsin-Like Serine Proteases

A. Activation Requires Correct N-Terminus Formation

To understand activity regulation in fIX and, in particular, its extraordinary low activity, one has to remember the fundamental mechanism underlying proteolytic activation in serine proteases. This mechanism has been uncovered by classic work on the conversion of (chymo)trypsinogen to (chymo)trypsin.[4–7] In the zymogen form of the protease, the propeptide is intact, represented in bold in Fig. 2A, and large parts of the protease, including the active site, is disordered. Upon proteolytic activation, the newly formed N-terminus, Ile16, forms a salt bridge with Asp194 and thus induces a disorder–order transition (Fig. 2B). Only upon this conformational transition is the active site completed, including the formation of the oxyanion hole by the amides of Gly193 and Ser195, the proper orientation and stabilization of the catalytic tetrade (Ser195-His57-Asp102-Ser214), and the short 215–216 strand that forms an antiparallel sheet with the substrate P3 backbone.

Similarly, fIX requires proteolytic cleavage to form the critical free N-terminus for activation. Details, however, differ with fIX and trypsin. Contrasting trypsinogen, in factor IX, an activation peptide needs to be released for full activation, requiring two proteolytic cleavages after $Arg145_{abs}$ and $Arg180_{abs}$ (cf. Fig. 1). Hereby, the subscript "abs" refers to the absolute sequence numbering, as compared to the chymotrypsinogen numbering that is used within the protease domain to highlight the structural homology with the (chymo-)trypsin family. The latter cleavage generates the new N-terminus $Val181_{abs}$ corresponding to Val16 in chymotrypsinogen numbering.

Depending on the course of blood coagulation, the physiological activators of the zymogen fIX are fVIIa and fXIa[11,12] within the extrinsic and intrinsic coagulation cascade, respectively.

The extrinsic coagulation is triggered by a vessel injury, inducing complex formation of factor VIIa with tissue factor (TF). The activation of fIX to fIXaβ by the TF–fVIIa complex proceeds through the obligate nonenzymatic fIXα intermediate, where the $Arg145_{abs}$–$Ala146_{abs}$ bond is initially cleaved.[13] Additional processing of the $Arg180_{abs}$-Val16 peptide bond converts fIXα to fIXaβ, releasing the activation peptide and generating the correct N-terminus; the latter induces the zymogen–enzyme conformational transition, similar to that described for trypsinogen–trypsin.

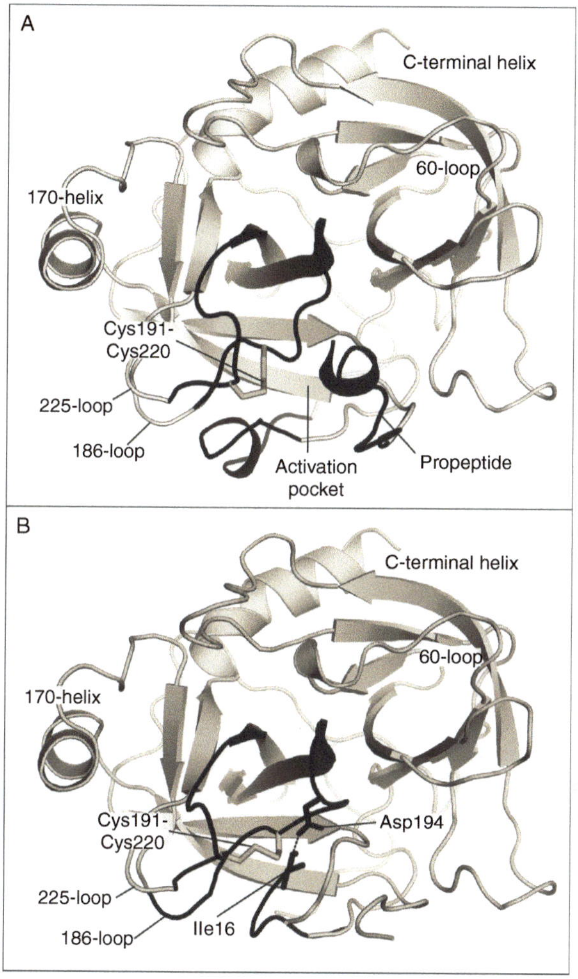

FIG. 2. Proteolytic activation of chymotrypsinogen to chymotrypsin. (A) The crystal structure of chymotrypsinogen reveals areas of increased disorder, indicated in dark color. This disorder is seen especially in the area around the activation pocket, and the close propeptide displays high mobility, even though the Cys191-Cys220 disulfide bond restricts the available conformations. The figure was prepared using PDB entry code, 2cga.[8] (B) The crystal structure of α-chymotrypsin (PDB entry code, 1yph) reveals the disorder–order transition associated with zymogen to protease conversion. Upon proteolytic cleavage of the activation peptide, the newly formed N-terminus Ile16 inserts into the activation pocket and forms a salt bridge with Asp194, indicated by a dashed line. This interaction aligns the catalytic triad, the oxyanion hole, and other segments flanking the entrance to the active site, indicated in black. All molecular figures were prepared using the molecular graphics program PyMOL.[9] Superposition of these and other structures was performed with the TopMatch-web server.[10]

Within the intrinsic coagulation cascade, fXIa activates the zymogen factor IX to fIXaβ. Factor XIa is a disulfide-linked homodimer consisting of four N-terminal apple domains, followed by a trypsin-like serine protease domain, the crystal structure of which has been determined.[14] FXIa converts the single-chain fIX preferentially to the intermediate-form fIXα by the above-described processing at the Arg145$_{abs}$-Ala146$_{abs}$ peptide bond.[15]

The physical interaction of the activator fXIa with its substrate fIX can be modeled by considering the substrate recognition sites on fIX, namely, the flanking regions of the activation peptide, Arg145$_{abs}$-Ala146$_{abs}$ and Arg180$_{abs}$-Val16 which bind to the active site of fXIa. A complementary restraint is given by the fXIa segment Asn145-Ala176, which could be mapped to bind fIX with an affinity of $K_d \sim 30$ nM.[16] Noteworthy and exceptional within blood coagulation proteases, fXIa exists as a disulfide-linked dimer; dimerization enables it to simultaneously bind to its substrate fIX and to the activated platelets.[12,17] Given this dual functionality, fXIa lacking an independent cofactor can be reconciled within the framework of cofactor-assisted catalysis that takes place on the surface of activated platelets. One subunit of the fXIa dimer serves as the cofactor to the other subunit that fulfills the protease task.

Depending on the substrate (fIX)–enzyme (fXIa) ratio, the processing of fIX to the double-cleaved fIXaβ may proceed processively, that is, without release of the single-cleaved intermediate fIXα or fIXaα.[18] Structural rearrangements in the apple 4 domain, which harbors the intersubunit disulfide bond, provide first clues of how fXIa accomplishes the processive, two-step fIX activation.[19] However, when the fXIa enzyme–fIX ratio is low (≤ 1000), single-cleaved fIXα or fIXaα intermediates are released from the enzyme.

An intriguing structural aspect with the activation of both factor IX and factor X is the apparent increase in size upon the first activation cleavage. Both fIXα and fIXaα undergo an apparent molecular mass shift on nonreducing SDS-PAGE corresponding to the effect of approximately one epidermal growth factor (EGF) domain, that is, 2–3 kDa (Fig. 3). Analogous size increases are observed when working with single-chain factor X upon cleavage at Arg142$_{abs}$-Ser143$_{abs}$ or Arg194$_{abs}$-Ile16 with furin or Russell's viper venom (RVV-X), respectively.

Apparently, the activation peptide is kept in a spring-like, compacted conformation in the zymogen state, which can expand upon the first cleavage (cf. Fig. 3C–F). The conformational strain on the activation peptide is provided by the structural framework of the protease where the flanking Ser136$_{abs}$ and Gly25 are in approximately 4 Å contact.

Along this line, there exists the possibility that incorrectly processed fIX variants, denoted as fIXa*, are produced. In the laboratory, this is indeed the case[20] (Fig. 4). Improper processing happens spontaneously either by trace contaminant proteases or by fIX autolysis. Significantly, improper processing of

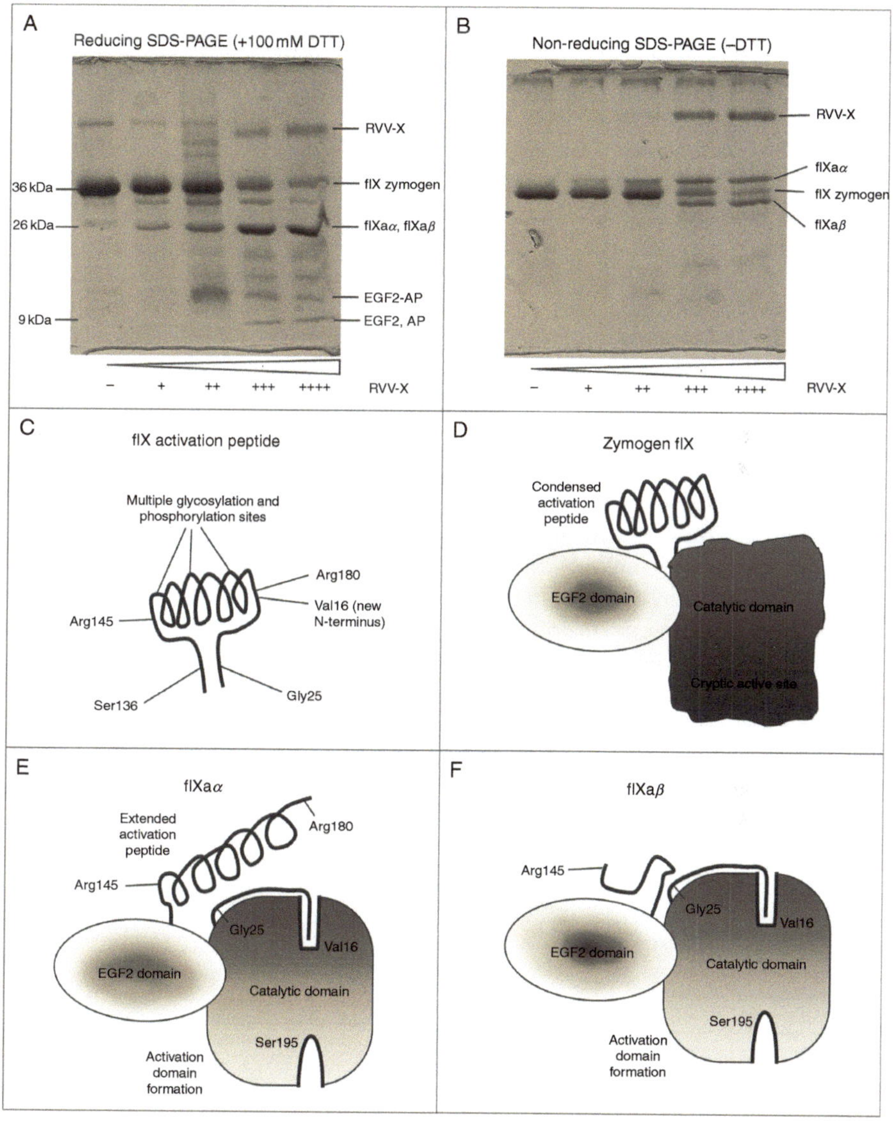

FIG. 3. Activation process of zymogen fIX and the role of the activation peptide. (A) Conversion of zymogen fIX to fIXa with varying RVV-X concentrations on a reducing SDS-PAGE gel. Activation is catalyzed by adding the RVV-X protease. The single-chain zymogen fIX (36 kDa) is processed at $Arg180_{abs}$ to form fIXaα; fIXaα consists of the heavy chain (26 kDa; catalytic domain) as well as the light chain (~ 10 kDa EGF2 domain plus activation peptide). A second cleavage at position $Arg145_{abs}$ separates the EGF2 domain (6.5 kDa) and the activation peptide (4 kDa).(B) Identical samples as in (A) are loaded on a nonreducing SDS-PAGE gel, whereby the fIXaα

the activation peptide appears to prevent correct cleavage at the Arg180$_{abs}$-Val16 bond, probably because the nonprimed substrate residues that would be needed for proper substrate recognition are no longer accessible. Thus, the improperly processed fIXa* is locked in an enzymatically inactive state with no return. The existence of fIXa* is consistent with earlier reports on the processing of factor IX by granulocyte-derived elastase, chymotrypsin, or thrombin in the absence of Ca^{2+} that resulted in an inactive factor IX isoform with an apparent weight closely resembling that of two-chain fIXaβ, as judged by SDS-PAGE analysis.[21,22] These fIXa* variants could not be turned to an enzymatically active fIXaβ form by postprocessing with fXIa, consistent with findings in our laboratory.

To further characterize the identity of fIXa*, we performed mass spectrometry. This analysis revealed that in situations with limiting activator enzyme (e.g., Russels viper venom activating protease, RVV-X), the processed factor IX proteins differ partly in their N-terminus, most of them carrying one additional residue resulting in Arg180$_{abs}$ as N-terminus, and some carrying up to five to six additional residues before the correct N-terminus, Ser174$_{abs}$. Clearly, these variants lack catalytic activity and also resist activation by postprocessing with RVV-X. Consequently, noncanonical cleavage of the activation peptide irreversibly arrests fIX in an enzymatically inactive state, fIXa*. Whether such pseudo-activation events play a physiological role awaits further investigations. In any case, this observation offers an interesting opportunity for biotechnological optimization, either when activating fIX *in vitro* or when designing recombinant fIX mutants as therapeutics, as described later (Section VI).

Biotechnologically, RVV-X is commonly used as an activator of fIX and fX[23,24] because it specifically cleaves the Arg180$_{abs}$-Val16 and Arg197$_{abs}$-Ile16 peptide bond in fIX and fX, respectively. Following the initial fIXaα cleavage, further processing of fIXaα depends partly on fIXaα autolytic activity, as an accumulation of the fIXaα species is observed for the fIX-S195A dead mutant, but not wild-type fIX.

---

intermediate can be discriminated from fIXaβ. Importantly, fIXaα migrates at significantly increased molecular mass compared to zymogen fIX. (C–F) The apparently increased molecular weight of fIXaα may be explained by alternative conformations of the activation peptide (AP) in the single-chain and two-chain form. A schematic representation of the AP topology is given in (C). In the fIXa crystal structure, Ser136$_{abs}$ and Gly25 are bridged via a water molecule, constraining the termini of the AP to form a compact, domain-like structure. (D) In zymogen fIX, the "AP-domain" comprising more than 50 amino acids is present in a compact, spring-like conformation. (E) Upon fIXaα generation, the spring is released and protrudes from the protein, explaining the increased apparent molecular weight of fIXaα. Additionally, by insertion of Val16 into the activation pocket, the catalytic domain becomes ordered and the active site is formed. (F) A second cleavage at Arg145$_{abs}$ releases the AP.

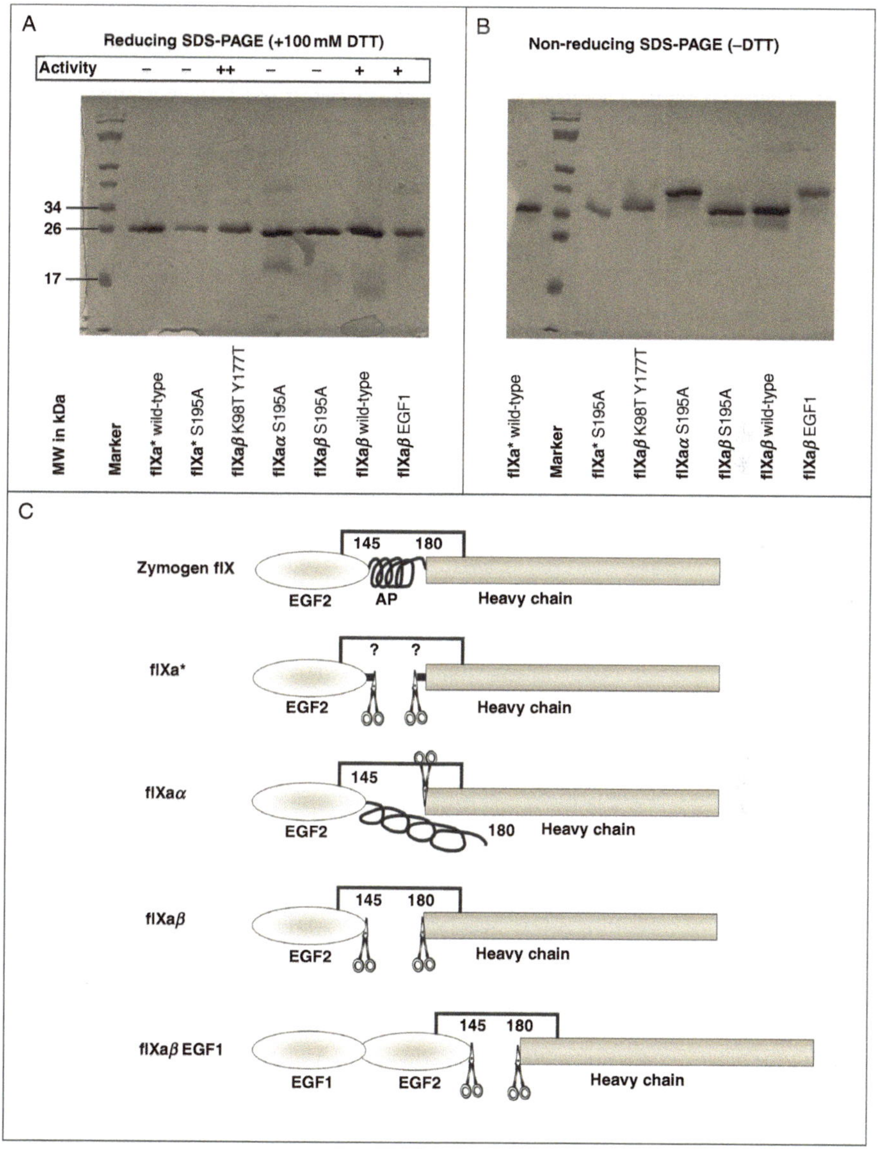

FIG. 4. Comparison of various (in)activated species of fIX. (A) Proteolytic processing of fIX results in species with strongly differing functional properties. These purified proteolytic fIX-derivatives were loaded on a reducing SDS-PAGE gel and migrate at the same molecular weight. Differences in amidolytic activity are indicated in the separate bar. In particular, fIXa* wild-type and fIXa* S195A are amidolytically inactive, despite their similarity to fIXaβ wild-type. FIXa* is processed close to, but not at, Arg180$_{abs}$ precluding the formation of a correct N-terminus.

Finally, the two isoforms of the cysteine protease gingipain R from the gram-negative anaerobic bacterium, *Porphyromonas gingivalis*, have been reported to efficiently activate factor IX to factor IXaβ via the intermediate factor IXα.[25] *P. gingivalis* has been implicated as a major pathogen in the development of chronic periodontitis, whereby gingipains are believed to act as major virulence factors. Recombinant constructs consisting of the catalytic domain of gingipain R may prove a valuable tool for blood coagulation research.

In contrast with fIX, fX becomes co-secretory processed at the furin recognition site, $R_{139abs}RKR_{142abs}$, whereby the tripeptide RKR gets excised by furin cleavage and postprocessing.[26,27] The zymogen thus circulates in a two-chain form in plasma, analogous to fIXα. Only this two-chain form is then efficiently converted to active fXa by the Xase complex. In the natural variant fX-Kurayoshi $Arg139_{abs}$ is replaced for Ser, leading to unusual hemophilic symptoms.[28] While platelet-supported coagulation processes were impaired, coagulation of the extrinsic pathway appeared intact. This finding implies that the intrinsic, but not (or at least less so) the extrinsic, Xase requires a two-chain factor X for activation.[28]

B. Release of the Activation Peptide

While the correct N-terminus formation (Val16 and Ile16 in fIX and fX, respectively) has a clear impact for activation, the relevance of the release of the complete activation peptide is not immediately obvious. *In vitro* activity of fIXaα toward small synthetic substrates is indeed as high as that of fIXaβ.[29] Why then is the two-step cleavage and subsequent release of a 35-(fIX) or a 542-(fX) amino acid long activation peptide necessary? There are several clues to this necessity. For one, the release of the activation peptide requires two cuts and thereby regulates both the specificity and efficiency of the fIX activation.[30] Additionally, the activation peptide release represents an additional control for proteolytic activity. FIXaα has full amidolytic activity, but lacks Xase activity, because only fIXaβ is able to engage in fVIIIa complex formation.[31] Furthermore, the activation peptide in factors IX and X is highly negatively charged

(B) The same samples, when loaded on a nonreducing SDS-PAGE gel, reveal some differences in their molecular mass. While fIXa°, as well as fIXaβ, variants migrate apparently identically, fIXaα, including the activation peptide (see Figs. 3E and 4C), shows increased molecular mass similar to the fIXaβ EGF1 construct. Importantly, the conversion of fIXaα to fIXaβ is slow only in the fIX S195A dead mutant. Remarkably, fIXaα S195A (MW ~ 36 kDa) and fIXaβ S195A (MW ~ 31.5 kDa) can be completely separated by gel filtration chromatography, which would not be expected on the basis of their molecular weights. (C) A schematic domain representation of the various constructs shown in (A) and (B). While the activation peptide in zymogen fIX exists in a compacted conformation, it can expand upon cleavage at $Arg180_{abs}$ in fIXaα.

and contributes to the high solubility of the zymogen, since release of the activation peptide significantly changes the biochemical properties, such as the isoelectric point. Clearly, these biochemical changes translate into changed functional properties. In this line, the release of the activation peptide is an important signal for the assembly of the activator complexes, that is, Xase or prothrombinase, on membrane surfaces.[32–34]

The negatively charged activation peptides of fIX and fX are strongly attracted by polylysine or polyarginine, very much resembling tissue plasminogen activator (t-PA), which gets strongly bound by polylysine segments.[35] In analogy, we might therefore speculate that zymogen factors IX and X could be colocalized with fibrin.[36]

III. Low-Molecular Weight Cofactors

Many proteases are known to be influenced by small compounds, both natural and synthetic in nature. Prominent examples in the context of coagulation proteases are certain alcohols and heparin, among others.

A. Ethylene Glycol

Similar to glycerol or sucrose, ethylene glycol (EG) is known to stabilize a broad spectrum of proteins. This effect is thought to be caused by the combination of favoring a more compact fold and the amphiphilic character of these alcohols that can interact with hydrophobic protein surface patches.[37] Surprisingly, however, EG was found to stimulate fIXa activity 20-fold. This effect was EG-specific, as other related alcohols did not boost fIXa activity or to a much lesser extent.[38] Moreover, we found EG to influence the activation process of fIX, such that production of futile fIXa* is minimized. To understand the structural basis for these EG-specific stimulations, we determined the crystal structures of fIXa in complex with EG at 2.2 Å resolution, which is a prerequisite to unambiguously distinguish an EG molecule from a cluster of four water molecules.

We found a single EG molecule positioned below the S4 site between Glu217 and the 170-helix (Fig. 5).[20] This area contributes to binding of the cofactor VIIIa,[39] explaining the observed interference of EG and cofactor binding.[40]

While there are no direct interactions of EG with the bound substrate analog (D-Phe-Pro-Arg-chloromethylketone), the comparison of the EG-bound structure with free structures provides a straightforward explanation of the mechanism of activity enhancement. Free structures exhibit a significant amount of disorder in the S3 site.[41,42] This disorder reflects an ensemble of fIXa conformations, most of them being catalytically unproductive. Binding of EG selects and thus accumulates the active conformer, leading to an up to

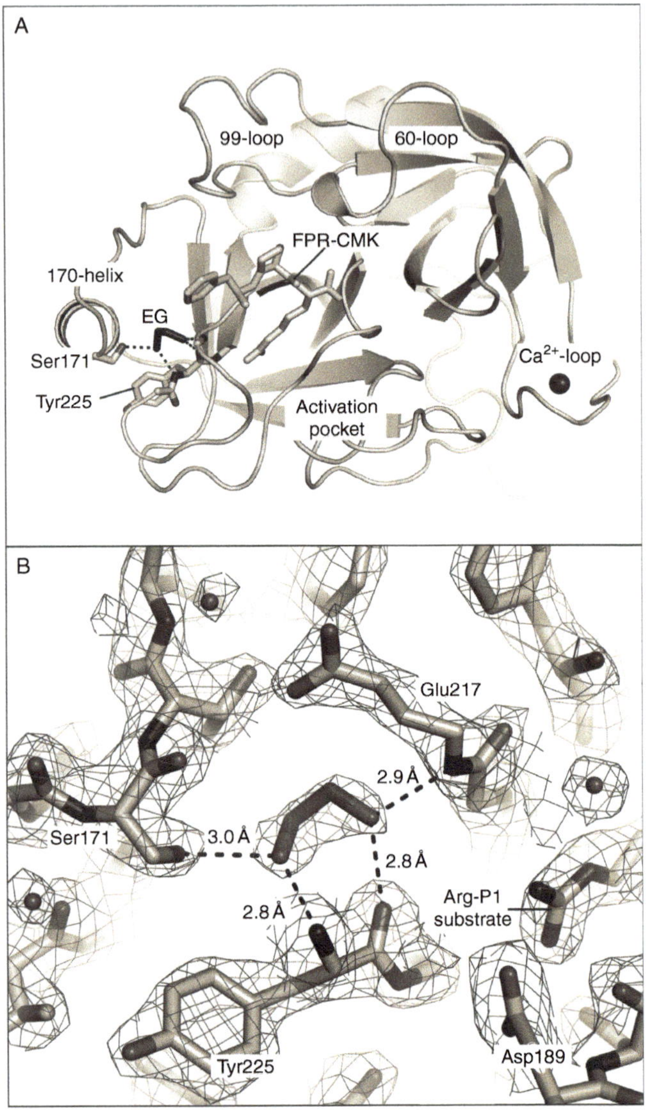

FIG. 5. Ethylene glycol reveals a distinct binding site within fIXa. (A) The ethylene glycol (EG) binding site in the context of the fIXa catalytic domain. EG binds close to the active site and engages the 170-helix, with the Oγ of Ser171 as coordinating residue. The covalent chloromethyl ketone inhibitor, FPR-CMK, Ser171, and the remaining two coordinating residues, Glu217 and Tyr225, are shown as sticks. Hydrogen bonds to EG are depicted as dashed lines. The figure was prepared using PDB entry code, 2wpk.[20] (B) A detailed view of the EG binding site, including a $2F_{obs} - F_{calc}$ density contoured at 1.0σ, reveals the close proximity to the nonprimed substrate binding site. The P1-Arg and Asp189 are indicated.

20-fold activity enhancement. Similarly, binding of covalent substrate-like inhibitors favors and accumulates the active species. An elegant, direct experiment to verify this interpretation would be by differential (± EG) NMR analysis.

While EG has no obvious physiological role *per se*, it serves us as a valuable tool to understand the physiological activation mechanisms in intrinsic Xase formation.

B. Heparin

Heparins belong to the family of glycosaminoglycans consisting of variably sulfated monosaccharide repeating units. The main repetitive element is composed of alternating 2-*O*-sulfated iduronic acid and 6-*O*-sulfated, *N*-sulfated glucosamine (Fig. 6). Starting with a minimal polymer length of five monosaccharides, heparins exhibit an anticoagulant effect. In fact, heparins are best known for their role as antithrombotics, both in acute therapy, such as ST-elevation myocardial infarction,[43] and in thrombosis prophylaxis.[44] The effectiveness of heparin's antithrombotic action results from a synergistic superposition of several modes of actions, described in more detail in Section V. Despite its medical use, the physiological role of heparin in the body remains controversial.[45,46]

It may therefore appear as a surprise that some heparins also act as activators to factor IXa amidolytic activity.[47,48] Neuenschwander and colleagues found an almost 20-fold increase in fIXa reactivity toward bovine

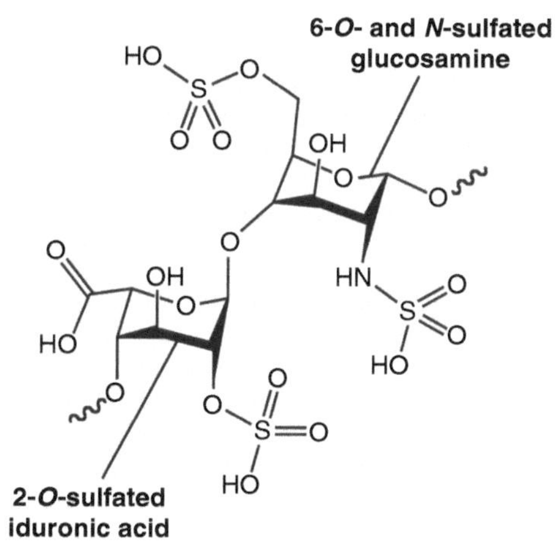

FIG. 6. *Molecular structure of heparin repetitive subunits*. Heparin is composed of multiple repeats of a basic unit consisting of two sulfated sugar moieties: 2-*O*-sulfated iduronic acid and 6-*O*- and *N*-sulfated glucosamine. Heparins can reach considerable length, with more than 50 module-repeats.

pancreatic trypsin inhibitor (BPTI), the archetype of a Kunitz protease inhibitor. Similarly, complex formation of factor IX with enoxaparin, a low-molecular weight heparin, induced an increase in amidolytic activity toward synthetic peptides.

Given structural and mechanistic data, it appears likely that these agonistic effects relate to the occupation of exosite II. Key residues contributing to heparin binding have been previously identified as His92, Lys230, Arg233, and Lys241 in exosite II as well as Arg165 and Arg170 residing on the 170-helix.[49,50] Together with structural and functional studies on superactive fIXa mutants, this interaction thus suggests that heparin exerts its stimulating effect by reordering the 99-loop and renders the nonprimed substrate recognition sites more accessible. Mediated by the hydrogen bond interaction of Y94 and E60, the S1′ and S2′ sites may be stabilized, as seen for the superactive fIXa double mutant (K98T-Y177T).[20,29] This notion is consistent with fluorescence studies on *Dansyl-Glu-Gly-Arg-chloromethylketone*-labeled fIXa, which indicate a change in the fIXa active site upon heparin binding.[47] The putative binding site of heparin on fIXa is visualized in Fig. 7A, which has been confirmed by the crystal structure of ternary fIXa–antithrombin (AT)–pentasaccharide complex (Fig. 7B).[51]

C. Metal Ions: Calcium and Sodium

Mono- and bivalent cations play prominent roles in the regulation of proteases. While this fact is obviously true for zinc metalloproteases, where the catalytic activity depends on the catalytic metal (usually Zn^{2+}), metals have an activity-modulating and/or structural effect on other protease classes. Consequently, zinc dissociation constants are typically in the subnanomolar range and may be as low as $K_D \sim 10^{-15}$ for dedicated metal sensor proteins.[52] With Zn^{2+} concentrations of approximately 100 μM in blood plasma, active sites of zinc proteases will be fully loaded, implying the need for different activity regulation mechanisms. By contrast, modulating metal sites are often tuned to match with the environmental metal content. This modulation is especially true for Na^+ and Ca^{2+} with plasma concentrations of approximately 135 and 2 mM, respectively.[53] The regulatory effect of calcium on the catalytic domain of trypsin-like serine proteases was known for more than half a century.[54] Additional important calcium binding sites are found in the noncatalytic domains of coagulation factors. Calcium (and magnesium) form ionic bridges between the highly negatively charged Gla-domain of coagulation factors and the surface of activated platelets. The calcium gradient toward the platelet membrane thus constitutes an important attractor and inducer of the blood coagulation process.[1] Calcium-binding sites in the catalytic and noncatalytic domains have been characterized by X-ray crystal structures in factors IXa and Xa, among others.[20,41,55]

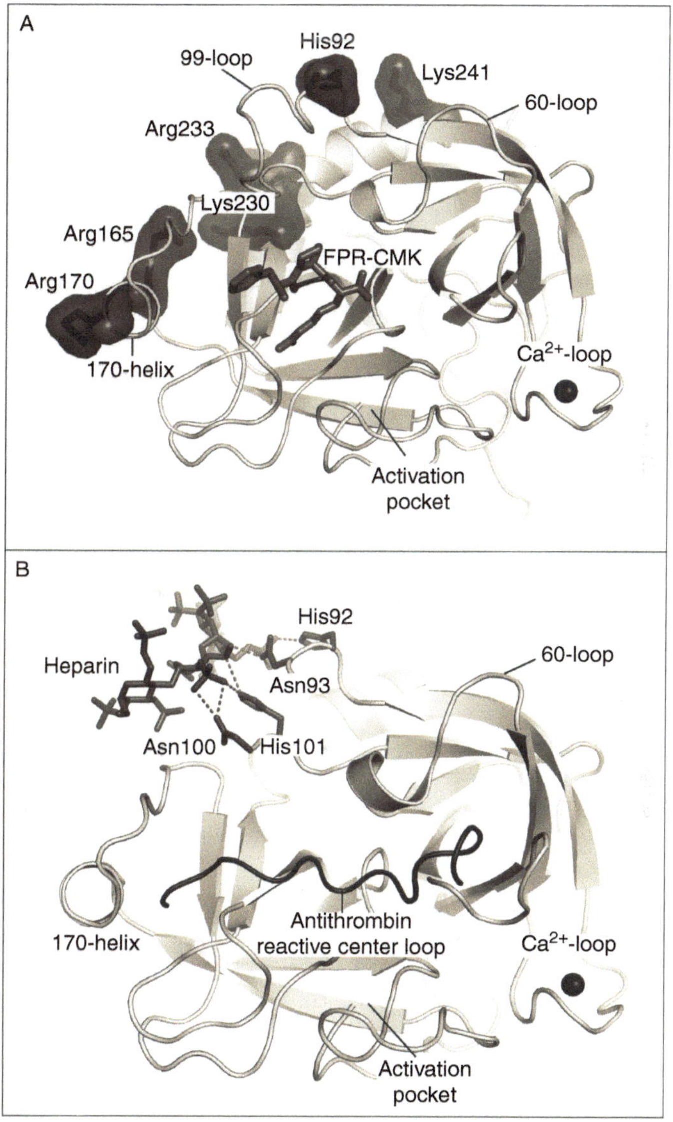

FIG. 7. *Mapping the heparin binding site on the fIXa catalytic domain.* (A) The heparin-binding exosite in fIXa consists of several positively charged residues, indicated by surface representation. The experimental evidence is based on mutagenesis studies. The active site-directed inhibitor (FPR-CMK) is shown in stick representation to aid orientation. (B) Crystal structure of a heparin pentasaccharide as determined in the antithrombin–fIXa complex. The small pentasaccharide binds to the 99-loop, confirming the mutagenesis data, as indicated in (A). The figure was prepared using the PDB entry, 3kcg.[51]

By contrast, the critical influence of sodium on blood coagulation enzymes was long overlooked and only discovered by pioneering work of the Di Cera laboratory. Initially identified in thrombin,[56–58] sodium was later shown to play important roles in factor Xa[59] and possibly also in factor VIIa,[60] although the hydrogen bonding network to the proposed sodium remained inconclusive.[61] In fact, the sodium-binding site in factor VIIa has to be considered controversial as it could not be confirmed in all high-resolution structures of factor VIIa.[62] In factor IXa, there is strong structural evidence that the sodium-binding site is not formed in the absence of the cofactor (Fig. 8).[20,51] The comparison of the homologous crystal structures in factors Xa and IXa explains that an EGG motif in fIXa (Glu186-Gly187-Gly188) induces flexibility in the backbone conformation, resulting in a reorientation of the Gly187 carbonyl oxygen. The latter critically contributes to sodium binding in fXa. This effect is further exaggerated by a hydrogen bond formed between His185 and Tyr225, reinforcing a carbonyl oxygen position that is incompatible with sodium coordination.

D. Long Distance-Communication Line Connects Ca^{2+} Exosite with the Active Site

It has been known for 70 years that Ca^{2+} allosterically influences the proteolytic activity of serine proteases.[54] Several elegant experiments have been performed to highlight the long-distance crosstalking that is present in most trypsin-like proteases. Intriguingly, thrombin also exhibits this allosteric effect, despite its lack of a Ca^{2+}-binding site within the catalytic domain.[64] However, the mechanism of signal transduction, including the physical mapping of the transducing amino acids, remained a challenge for a long time, because the allosteric signal is not transduced by conformational rearrangement. The lack of such a conformational fingerprint explains why even structure analyses to high resolution failed to detect the signal transduction pathway. Instead, the allosteric signal is transported by the change in dynamics of selected residues, as revealed by difference Fourier techniques.[20] The participating amino acids line up to form a continuous communication line between the Ca^{2+} site and the active site, as shown in Fig. 9. The electron density shown in Fig. 9 reflects the intensity in scattering power, which is inversely related to the flexibility and dynamics of these residues. Remarkably, the signal propagates nonisotropically along a wired-like path and utilizes backbone atoms to a large extent. Important signal check points are Val16-Gly193-Asp194 at the activation pocket and the Cys191-Cys220 disulfide bond that connects the oxyanion hole with the nonprimed substrate recognition sites.

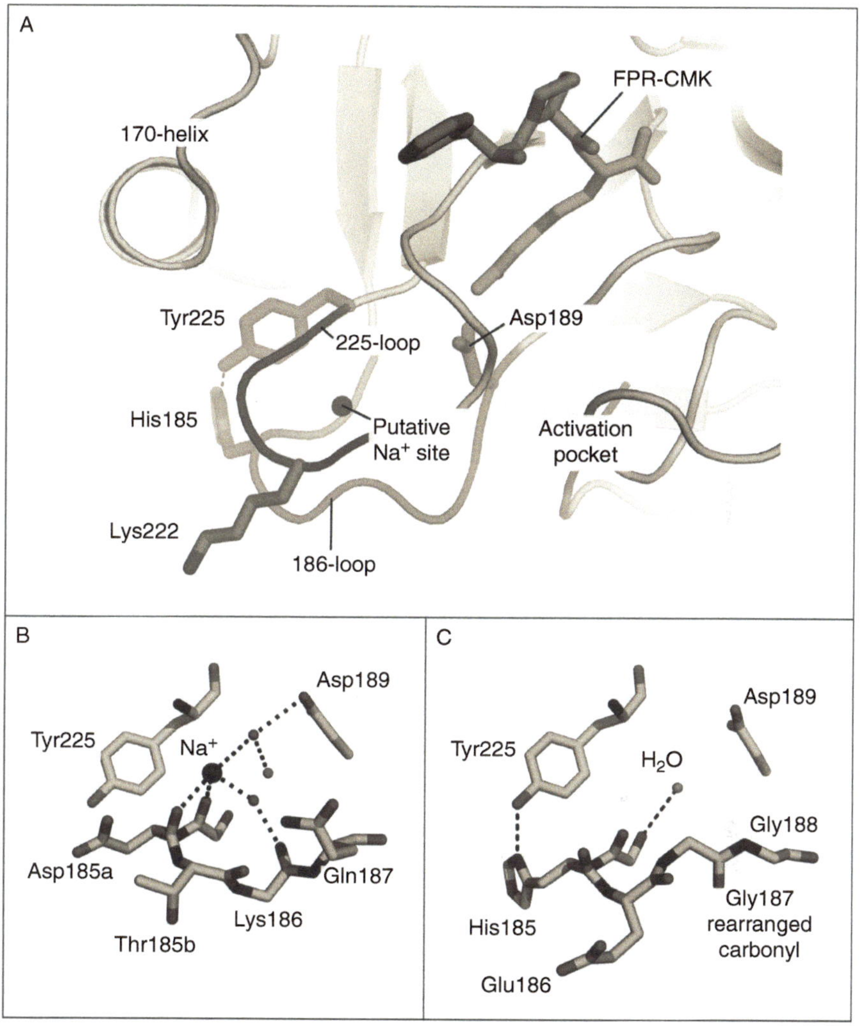

FIG. 8. Comparison of the putative sodium-binding sites in fIXa and fXa. (A) Overview of the 186- and the 225-loops, as seen in the fIXa, 2wph. The two loops are located near the nonprimed substrate binding sites and include Asp189, conferring the primary specificity toward the P1 residue. The active site-directed inhibitor is represented in sticks to aid the orientation. Residues His185 and Tyr225 form a hydrogen bond between the two loops. The coordinates of the putative Na^+ ion are derived from the superposed fXa structure with pdb entry code, 2bok.[63] (B and C) Comparison of the 186- and 225-loops in fXa versus fIXa. In (B), the sodium-binding site is identified in the fXa structure, 2bok, solved to 1.6 Å resolution. Sodium is bound by a total of four carbonyls and the coordination is completed by two water molecules. (C) An analogous sodium-binding environment is suppressed in the fIXa crystal structure determined at 1.6 Å resolution in the presence of > 2 M NaCl (pdb entry, 2wpj).[20] The backbone carbonyl of Gly187 is rotated > 120° as compared to the fXa structure. The flexibility induced by the Glu186-Gly187-Gly188 (EGG)-motif may prevent sodium binding in fIXa. Additionally, the His185-Tyr225 hydrogen bond increases the distance of the 186- and 225-loops by > 1 Å, which is incompatible with sodium binding by the backbone carbonyls.

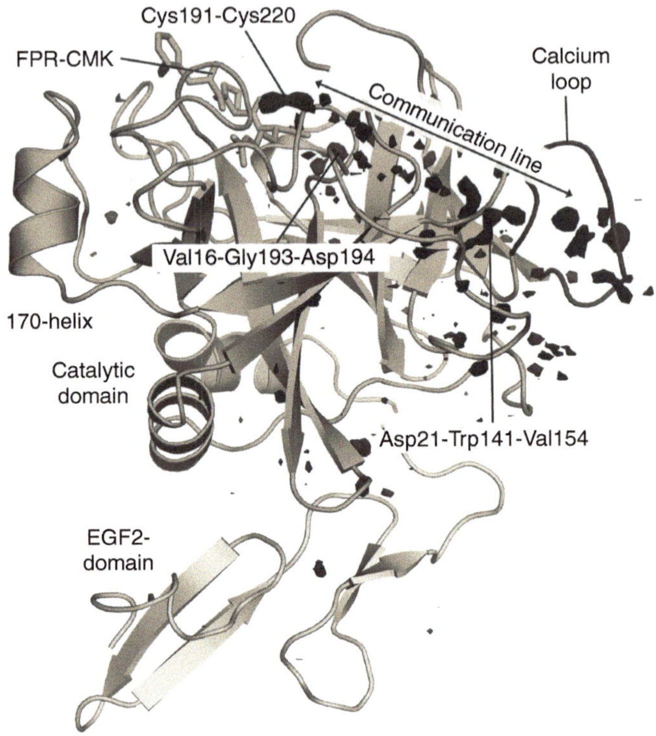

FIG. 9. *A communication line for exosite-active site mutual influence.* A difference Fourier electron density map ($|F_{\mathrm{obsCa}^{2+}} - F_{\mathrm{obsEDTA}}|$) reveals the changes in dynamics upon Ca^{2+} binding. The electron density provides an unbiased map of how the Ca^{2+} binding allosterically affects the active site. The isomorphous difference Fourier electron density map is contoured at 3.5σ, identifying changes in both the conformations and the scattering power of individual atoms that may occupy the same coordinates in both datasets. Hence, this method can be used to visualize changes in the order–disorder state of amino acids caused by Ca^{2+} binding. The figure was prepared from pdb entry, 2wpm.[20]

IV. Protein Cofactors

Within the family of blood coagulation proteases, it is the rule rather than the exception for a protease to require a protein cofactor in order to acquire full catalytic activity. This requirement is true for fVIIa, fIXa, fXa, and protein C with their protein cofactors, TF, factor VIIIa, factor Va, and protein S, respectively. Additionally, the specificity of the proteolytic activity is often tuned by cofactors, as exemplified by thrombin with thrombomodulin. For a more

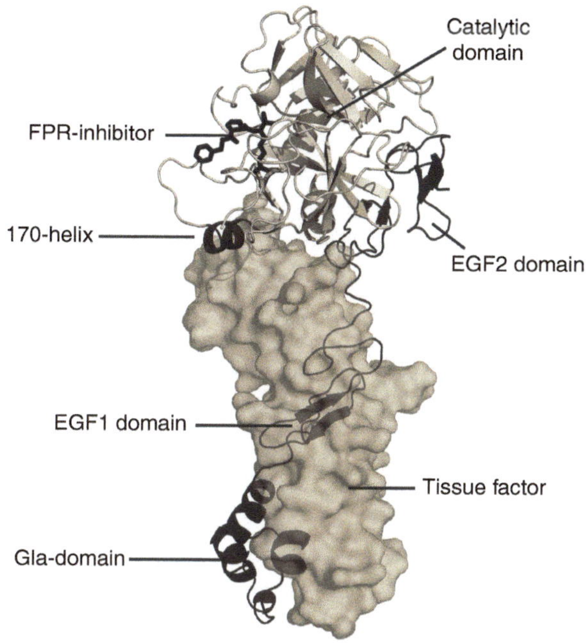

FIG. 10. The tissue factor (TF)–fVIIa complex. The structure of the TF–fVIIa complex is characterized by extended interactive surfaces throughout the Gla-, EGF-, and protease domains. The figure was prepared using pdb entry, 2fir.[60] Important domains and motifs are labeled.

complete review, please see Refs. 65–68. The factor VIIa–TF complex is the best studied example, with several crystal structures of isolated components, complex, and zymogen structures available, represented in Fig. 10.[62,69–71]

A. Factor VII: Tissue Factor

TF (CD142) is a transmembrane receptor consisting of a 219-amino acid long extracellular domain, followed by a transmembrane helix and a short cytosolic tail. Its ectodomain comprises two fibronectin type III modules, which mediate the complex formation with factor VIIa as well as its downstream substrates, factors X and IX.

Like other protein cofactors, TF serves two major physiological purposes: first, TF enhances fVIIa-catalyzed substrate turnover by several 1000-folds[72] and second, TF localizes the substrate turnover to the site of the vessel injury where TF gets exposed to the bloodstream.[65,73]

The mechanism of TF action in fVIIa activity enhancement is multifactorial; one important aspect consists of a conformational rearrangement of the fVIIa active site, such that a catalytically productive conformation is accumulated. Consistently, the affinity of TF to fVIIa is highest when its active site is covalently blocked by a peptidic chloromethyl ketone (CMK) inhibitor.[74] However, TF-induced conformational rearrangements are not restricted to the active site of fVIIa, but include the "intermediate" 170-helix,[75] which is indeed highly flexible in the free, uncomplexed fVIIa structure, but undergoes a disorder–order transition upon TF-binding.[62,70,71,76] This transition propagates to neighboring sites, including the putative Na^+-binding 186-loop and the 225-segment. It is also reflected by a stabilization of the salt bridge between the new N-terminus Ile16 and Asp194, the hallmark of serine protease activation.

In addition to this conformational rearrangement, TF exerts a template-type effect in providing additional binding sites to the macromolecular substrates, fIX and fX, thereby properly preorientating the substrate relative to the fVIIa active site.[77]

As a third factor, one has to emphasize the role of the correct substrate within the TF–fVIIa system. The fVIIa activity increase by TF toward the physiological substrate, fX, is more than 100-fold higher than toward peptidic substrates.[3,72,78] This substrate specificity is reflected when referring to the TF–fVIIa complex as the extrinsic Xase. In fact, the effects of substrate-assisted catalysis are even more pronounced in the intrinsic Xase system (fIXa–fVIIIa), where activity enhancement is limited almost exclusively toward the physiological substrate, fX, as will be discussed below.

Finally, as a fourth component in the extrinsic Xase system, the effect of the membrane has to be considered; it exerts an approximately 1000-fold stimulating effect toward the physiological substrate, but not toward peptidic substrates.[72,78] Much of the membrane's effect can be understood by a reduction of the enzyme–substrate diffusion from three to two dimensions, although membrane-anchored receptor proteins are likely to contribute in the physiological complex formation.

Thus, the TF–fVIIa system serves as the best model of a protein cofactor-stimulated activity modulation system available to date. It is frequently exploited to draw conclusions for the analogous intrinsic Xase and prothrombinase complexes, where high-resolution structural data are still missing.

B. Factor VIII

Coagulation factor IXa shares a homologous domain architecture with fVIIa. Together with factor VIIIa, the protease fIXa forms the intrinsic Xase complex that is analogous, but not homologous, to the extrinsic Xase complex formed by TF–fVIIa. While both the extrinsic and intrinsic Xase are membrane-supported, their prime location differs, as the intrinsic Xase assembles on the membrane of activated platelets, rather than endothelial cells. However, activated platelets are recruited to the site of vessel injury via specific molecular interactions that are mediated by platelet-anchored receptor proteins, as described in a following section.

Intrinsic Xase further distinguishes itself by its strict and multiple requirements for activation, which are necessary for every single component involved, namely, the protease fIX, the protein cofactor VIII, and the platelets providing the membrane support. This extra security level indicates that the consolidation phase of blood coagulation requires more accurate control than the initiation phase, which gets shut down rapidly by the anticoagulant TF pathway inhibitor (TFPI).[79]

While fVIIa or fXIa proteolytically activate the protease fIX to fIXa, as described in Section I, thrombin is the prime activator of both the cofactor VIII and platelets.[80–83] The procofactor VIII circulates in the bloodstream as a heterodimeric glycoprotein. The heavy chain consists of the homologous A1 (amino acids 1–329) and A2 (380–711) domains and the regulatory B domain (741–1648), and the light chain is composed of the A3 domain (1694–2021), followed by the C1 (2021–2169) and C2 (2174–2326) domains. The heterodimer is noncovalently linked by metal ions and forms a tight complex with von Willebrand factor (vWF) in plasma. Crystal structures of the procofactor VIII were reported recently by two groups, clarifying its intramolecular domain arrangement and providing clues to the interaction with other components of the Xase complex (Fig. 11).[84,86] Given the limiting resolution of approximately 4 Å of these structures, several details, such as metal interactions, remain ambiguous. However, the isolated C2 domain has been determined at high resolution,[87,88] allowing individual information to be combined and complemented.

The proteolytic activation of fVIII by thrombin occurs via the initial cleavage at Arg740-Ser741, followed by Arg372-Ser373 and Arg1689-Ser1690. The first cleavage releases the B domain and thereby frees fVIII from vWF.[89] The other two activating cleavages produce a free N-terminus at the A2 domain and trim the already existing N-terminus at the A3 domain.[81] The mechanism of activation mediated by these neo N-termini is unclear, even in the light of the pro-fVIII crystal structures where the corresponding loop regions are disordered.[84,86] Two scenarios appear reasonable, either alone or in combination: first, the cleavage unmasks an epitope that may be important for Xase assembly; second, the cleavage and formation of a new N-terminus induces a conformational rearrangement that allosterically contributes to Xase assembly.[90,91]

All the components of the intrinsic Xase work in concert to reach an approximate million-fold acceleration in fX activation.[34] Despite and because of the complexity of this molecular machinery, it is helpful to sketch a simplistic picture that helps to clarify the major contributions of the individual fVIIIa

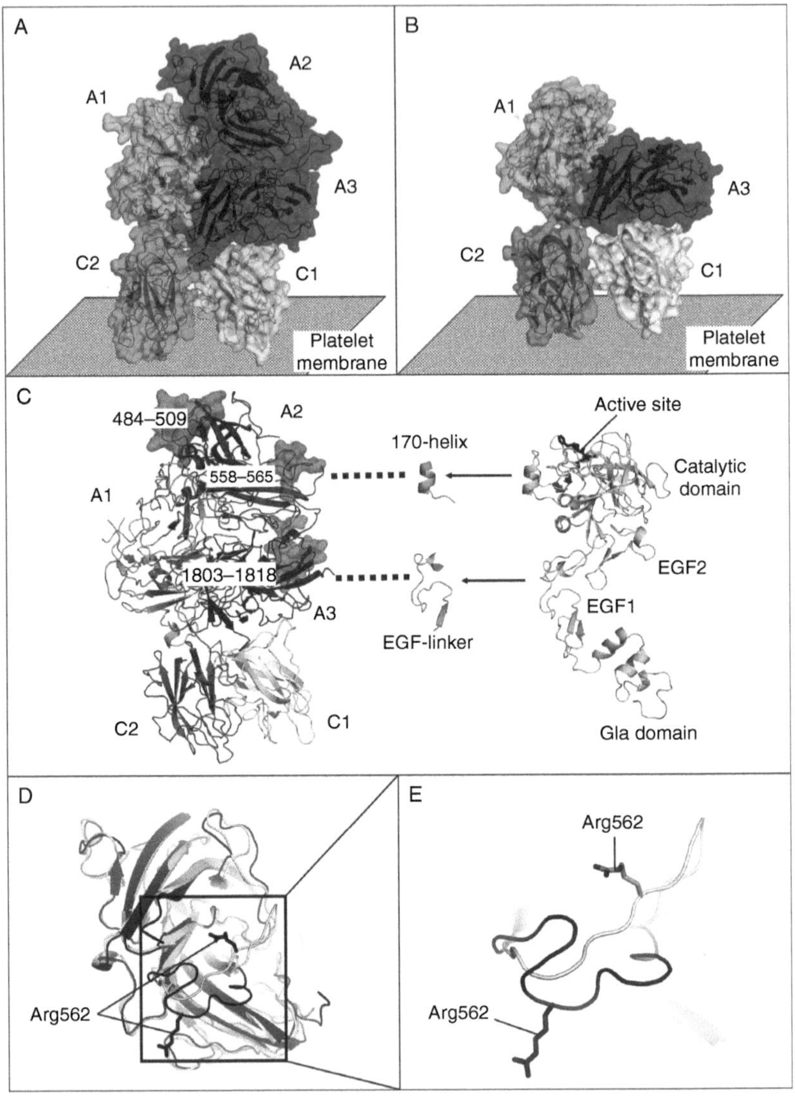

FIG. 11. FV and fVIII: unique proteins for unique functions. (A) Crystal structure of procofactor fVIII. Individual domains are highlighted by semitransparent surfaces and labeled. The C1 and C2 domains are known to mediate contacts with the phospholipid membrane, which is indicated schematically. The figure was prepared using pdb entry, 2r7e.[84] (B) The domain organization of bovine fVa-i, prepared using pdb entry, 1sdd.[85] The crystal structure lacks the A2 domain due to aPC inactivation. Domains are labeled and the position of the membrane is indicated schematically. (C) Modeling the fVIIIa–fIXa interactive sites based on the zymogen structure of fVIII (pdb entry, 3cdz[86]) and porcine fIXa (pdb entry, 1pfx[41]). The three main interactive sites residing on the A2 and A3 domains are indicated as surfaces. While the 558- to 565-area in the A2

domains to this concert (Fig. 11A and C). The A1 domain of fVIIIa was shown to act as a major interactive domain toward the substrate fX.[92] The A2 domain seems to induce most of the conformational rearrangements to the protease, fIXa, as also evidenced by its ability to stimulate fIXa's proteolytic activity by about 100-fold.[93] A3 serves as a high-affinity binding domain directed to the light chain of fIXa.[94–96] The C1 and C2 domains anchor fVIIIa to the platelet membrane and mediate vWF binding.[88,97–100]

Such a simplifying mapping of interaction partners provides a starting ground to rationalize the activity increase in the Xase complex. FVIIIa exerts a template effect such that all involved components fall into place, thereby turning stochastic-diffusive collisions into productive interactions. This picture explains not only the improved, that is, decreased, K_d/K_M values of the Xase components, but also an improved k_{cat} constant for the fX activation. Of course, this picture falls short of explaining conformational rearrangements that influence catalytic substrate turnover.

To this end, a refined and more detailed view on the component interaction is required. The combination of mutagenesis, enzyme kinetic, and structural studies indeed provided a more in-depth view. Specific interaction sites of the A2 domain on fIXa could be identified. Importantly, the connecting segment, $Leu84_{abs}$-$Thr87_{abs}$, between the EGF-like domains contributes to the binding and stimulation by the A2 domain (Fig. 11C).[101] Another A2 interaction site is the 170-helix; the latter is actually a conserved cofactor recognition motif in fIXa, fXa, and fVIIa.[102] Consequently, the fVIIIa A2 domain embraces the fIXa catalytic domain from opposite sides with a $K_d \sim 300$ nM, which helps to explain the observed fluorescence anisotropy fingerprint on an active site-labeled fIXa upon A2-domain binding.[93]

In a further attempt to sharpen our understanding of Xase's mode of action and to clarify the nature of the detected rearrangement near the active site, we employed a spectrum of methods to unravel the critical role of the 99-loop of fIXa. These studies revealed how the 99-loop in fIXa gets reordered and thus provides unhindered access to the nonprimed, as well as to primed, substrate recognition sites.[20,29,103,104] A more detailed description of the rearrangements in these Xase-like mutants is given below.

---

domain of fVIIIa interacts with the fIXa 170-helix, certain residues of the 1803–1818 area in the A3 domain constitute a binding site for the linker region between fIXa EGF1 and EGF2 domain. Additionally, fVIIIa residues 484–509 are known to bind to fIXa. (D) Superposition of the A2 domains from both available fVIII crystal structures, pdb entries, 3cdz and 2r7e. Importantly, the Arg562-loop mediating the key contact to the fIXa 170-helix occupies two distinct conformations in the two structures. (E) Close-up view of the Arg562-loops presented in (D).

Also, for the C2 domain of fVIIIa, a more differentiated view could be obtained. Besides its membrane anchoring role, C2 binds the fIXa Gla-domain with a dissociation constant of $K_d \sim 100$ nM.[105] However, a C-terminal truncated fVIII variant lacking the C2 domain retained significant *in vitro* Xase activity.[106]

While it is tempting to capitalize on the compiled structural and functional data and to extrapolate them to construct docking models of the full Xase complex, it is important to give a word of caution: current theoretical approaches cannot possibly reconcile all of the available data.[107] These limitations are exemplified when docking rigid domain structures of fIXa and fVIIIa, although both molecules are likely to undergo significant rearrangements upon complex formation. As illustrated in Fig. 11D and E, the A2 domain exhibits significant deviations within the two crystal structures of pro-fVIII and is likely to undergo further conformational changes upon activation and complex formation. A more complete understanding of how this marvelous molecular machine works awaits further experiments and, in particular, high-resolution crystal structures of the involved component complexes.

With the dramatic stimulation of fX activation comes the necessity to efficiently downregulate Xase activity. Nature developed several pathways for this purpose, two of them being specifically directed to inactivate fVIIIa, namely, proteolysis and subunit dissociation. In the presence of its cofactor, protein S, activated protein C (aPC) cleaves at the P1 residue Arg336 in the A1 domain, and subsequently, at Arg562 in the A2 domain, thus irreversibly inactivating fVIIIa.[108] In the proform, the complex of fVIII with vWF protects fVIII from premature inactivation by aPC.[109] To a lesser extent, other coagulation factors, such as fXa, also contribute to proteolytic inactivation of fVIIIa.[110] Additionally, fVIIIa exists as a noncovalent ternary complex composed of separate A1, A2, and A3–C1–C2 chains. Especially in the absence of factor IXa, the dissociation of the A2 domain within fVIIIa occurs quite rapidly, accounting for the steady decrease of cofactor activity over time.[91,111]

C. Substrate's Role in Xase Function

The enzymatic activity of the Xase complex is strictly specific toward fX as a substrate. In fact, even posttranslational modification of fX can severely affect its turnover. In the hemophilic fX mutation, Kurayoshi, $Arg139_{abs}$ is replaced by Ser, disrupting the furin recognition site, $R_{139}RKR$. Consequently, conversion of this fX variant to the two-chain form upon secretion is strongly prohibited.[28] The single-chain fX-Kurayoshi is poorly activated by the intrinsic Xase, but not the extrinsic Xase.[112]

Similarly, a furin-cleaved fX variant where most of the activation peptide has been removed by enzymatic postprocessing (fX-des143-191) reduces the k_{cat} of fX activation by the intrinsic Xase 100-fold, with the K_M being unaffected. By contrast and analogous to fX-Kurayoshi, the activation of fX-des143-191 by RVV-X is unaffected.[30] Both experimental findings underline that the substrate specificity of Xase is governed not only by the sequence of the substrate, but also by its three-dimensional conformation.

Even more so, the substrate fX takes an active part in its own turnover. There is evidence that the specificity toward fX recognition translates into a binding mode of fX unique to the intrinsic Xase. This specificity is illustrated by AT binding to fIXa.[51] Site-directed mutagenesis identified that Lys148 of fIXa strongly affects fX turnover, while it barely influences AT binding.[104] This observation suggests that the AT binding mode to fIXa differs from fX binding.

D. An Atomic Model of Xase-Like Activity Enhancement

Based on earlier findings on fIXa double and triple mutants, K98T-Y177T and Y94F-K98T-Y177T, with approximately 1000-fold amidolytic activity enhancement that clearly indicated synergistic effects, we determined the crystal structures of these Xase-like fIXa mutants.[29] The structures of these fIXa variants show striking rearrangements in the backbone chain trace from residues Tyr94 to Tyr99 within the 99-loop,[20] which was known to restrict substrate binding.[104] In stark contrast to the wild-type fIXa structures, the 99-loop in the Xase-like mutants is stabilized in a compacted conformation via new hydrogen bonds from Thr98-Oγ to the amide-nitrogen of Asn100 as well as Asn100-Oδ to Thr177-Oγ (Fig. 12A and B). Importantly, the observed 99-loop conformation is compatible with efficient and productive substrate binding, and thus improves both K_M (substrate affinity) and k_{cat} (productivity of substrate binding) of the catalyzed reaction.

Among the coagulation factors, tyrosine at position 94 is specific to fIXa. In the wild-type structures, the hydroxyl group of Tyr94 forms a hydrogen bond with the carbonyl oxygen of Lys98 in wild-type fIXa, thus locking the active site in an unproductive conformation. In contrast, reordering of the 99-loop in the Xase-like mutants is paralleled by a movement in the neighboring 60-loop. These concerted movements are stabilized by a newly formed hydrogen bond of the Tyr94 hydroxyl with the carboxylate group of Glu60 (Fig. 12C).

A straightforward translation of the underlying mechanism in the physiological Xase complex involves two synergistically operating steps, namely, (1) the destabilization of the unproductive 99-loop conformation by cofactor fVIIIa, followed by (2) the mutual stabilization of the 99-loop and the 60-loop in a productive Xase-like conformation (Fig. 13). The crystal structures of the Xase-like K98T-Y177T double mutant suggest that this coordinated 99- and 60-loop reordering can be triggered by the simultaneous action of the protein cofactor, VIIIa, and the scissile peptide bond-bridging of substrate fX.

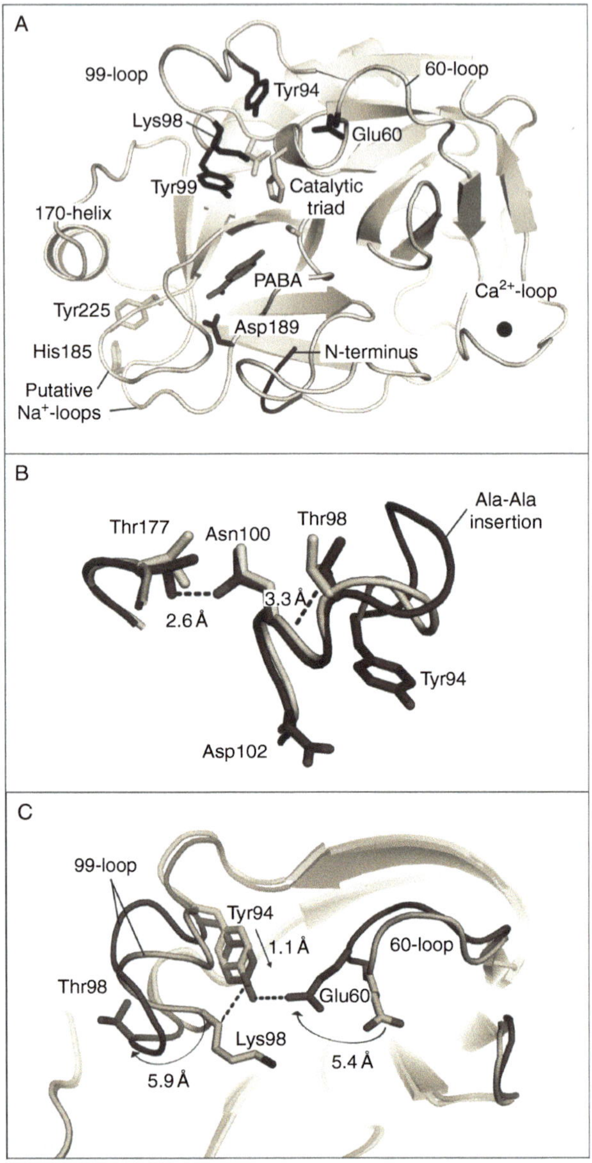

FIG. 12. Comparison of 99-loop conformations in wild-type fIXa, superactive fIXa mutants, and fXa. (A) Overview of the catalytic domain of the wild-type fIXa with the active site inhibitor, *para*-amino benzamidine (pdb entry, 1rfn).[42] (B) Superposition of the 99-loops of the fIXa K98T-Y177T double mutant (black; pdb entry, 2wpi) and the homologous fXa (gray; pdb entry, 2bok). Both structures show highly similar loop conformations, including a stabilizing hydrogen bonding pattern indicated by dashed lines. (C) Superposition of wild-type fIXa (gray; pdb entry, 1rfn) and the fIXa K98T-Y177T double mutant (black; pdb entry, 2wpi), highlighting the conformational changes in the 99- and 60-loops. In the highly active double mutant, Tyr94 forms a hydrogen bond with Glu60, contributing to order both loop conformations.

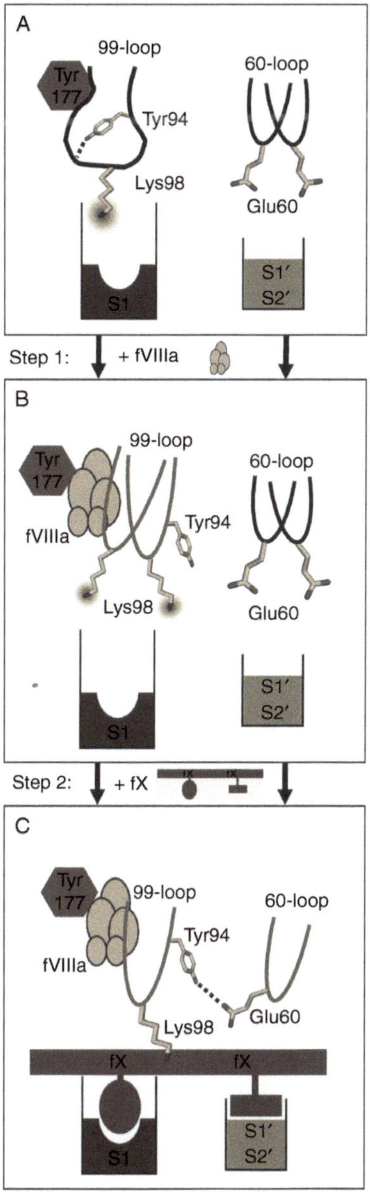

FIG. 13. Schematic model for a sequential Xase complex formation. (A) The 99-loop sterically and electrostatically hinders substrate access to the substrate binding pockets in the absence of fVIIIa and fX. A hydrogen bond between Tyr94 and the carbonyl of Lys98, together with structural restraints exerted by Tyr177, lock the 99-loop in an unproductive conformation. The highly flexible 60-loop further impairs primed substrate recognition sites. (B) In Step 1, fVIIIa binds to fIXa and liberates the 99-loop from its conformational lock. However, fVIIIa binding does not suffice to stabilize productive substrate recognition sites in fIXa. (C) In Step 2, addition of the physiological substrate, fX, completes Xase complex formation. By the concerted action of fVIIIa and fX, substrate recognition sites of fIXa are stabilized in a conformation enabling efficient fX turnover. One important contribution of the substrate fX is to provide simultaneous engagement of non-primed and primed site interactions, indicated by S1 and S1′ S2′, respectively.

While the mutant crystal structures have their limitations in modeling the Xase active site, they have the advantage of providing specific, testable predictions of key properties. This modeling can be exemplified by an array of three polar groups directed toward the S4 site, rendering it polar rather than hydrophobic.[20] The reorganization of the 99-loop thus explains the selectivity of the Xase complex for polar and basic P4-substrates, with natural substrate cleavage sites being N̲LTR in fX and K̲LTR in fIX, respectively (P4 residue underlined).

E. Factor V

Similar to Xase, much effort has been directed toward the elucidation of the ternary prothrombinase complex consisting of cofactor Va–factor Xa together with the substrate prothrombin (factor II). Cofactor fVa and protease fXa are highly homologous to the fVIIIa and protease fIXa in the intrinsic Xase complex. In particular, fV shares with fVIII the domain architecture, namely, A1–A2–B domain, followed by A3–C1–C2 domains. Like in fVIII, the A-domains are homologous to the copper-binding protein, ceruloplasmin, while the C-domains are related to the discoidin protein family.[113] Furthermore, processing during activation of fV to fVa by thrombin occurs similar to that described for fVIII activation. The same is true for the inactivation of fV by aPC, with the variation that the ternary fVa complex remains stable, and the A2 dissociation, which occurs in fVIIIa spontaneously, is induced in fVa by protein C cleavage.[114,115] It is therefore reasonable to expect a similar overall architecture of these complex protein machines. This expectation is confirmed by a large body of experimental data, most importantly by the crystal structures of the C2 domain of fV as well as the protein C-inactivated factor Va, lacking the A2 domain[85,116] (Fig. 11B). Thus, this similarity is used to justify the transfer of insights obtained for the intrinsic Xase system to the prothrombinase and vice versa, despite the subtle, yet important, differences between both systems.

From the perspective of the enzyme, several lines of evidence revealed an initial map of fXa-interactive sites within the prothrombinase complex. This evidence includes peptide mapping studies, which identified the fXa segment, Val231 to Thr244, as one cofactor-interactive site. Additionally, the fXa segment, Ser241-Lys252, was shown to contribute to the binding of its substrate, that is, prothrombin.[117]

From the alternative cofactor fVa perspective, important interactive surface patches could be mapped. Of particular interest, site-directed mutagenesis studies identified a charged, elongated cluster on the A2 domain of fVa formed by Arg501, Ar510, Ala511, Asp513, Asp577, and Asp578.[118] While it is in this respect particularly unfortunate that the A2 domain is missing from the crystal structure of aPC-inactivated fVa,[85] we can take advantage of the fVIII crystal structures to place the homologous A2 domain in the fVa-i correspondingly (Fig. 11A and B). Additionally, this comparison allows us to draw conclusions about the reorientation of A2 occurring during cofactor activation.

Despite the similarities between intrinsic Xase and prothrombinase, there are a number of differences between these highly homologous proteins. One of the most important differences is the strict cofactor and substrate dependence of fIXa. In fact, the only known function of fIX is its role as an activator of fX in the intrinsic Xase. By contrast, fXa is a highly active protease independent of its cofactor, fVa, toward a rather broad substrate panel, as long as certain sequence requirements are fulfilled. Consistently, fXa has physiological substrates beyond the activation of prothrombin to thrombin. Important physiological processes include the activation of fV,[119] the activation of fVII,[120] the inactivation of coagulation factor VIIIa,[121] the inactivation of cofactor protein S,[122] and multiple roles in protease-activated receptor (PAR) activation.[123]

V. Membranes: Nonhumoral Cofactors of Coagulation Factors

Binding of fIX and fIXa to cell membranes is mostly mediated by the N-terminal Gla domain, rich in γ-carboxylated glutamate residues. The mechanism of binding is partly electrostatic by divalent cation bridges, mostly Ca^{2+}, and partly hydrophobic by the Ω-loop in the Gla-domain, specifically by Leu6 and Phe9.[124] Two sites are of particular interest in blood coagulation. The first is the site of vessel injury where the extrinsic blood coagulation gets triggered. At this site, mostly endothelial cells are found, but collagen and other matrix proteins are also seen. The second site is at the surface of activated platelets, where the consolidation phase via the intrinsic coagulation cascade takes place.

A. The Injured Vasculature: Endothelial Cells

Blood coagulation is initiated at sites of vessel injury, where endothelial cells are exposed to the bloodstream. Binding sites on endothelial cells are dissimilar from those on activated platelets. Intrinsic Xase activity on endothelial cells is sustained, whereas the Xase activity is transient on activated platelets and also on synthetic, negatively charged phospholipid membranes. The mechanism underlying the observed difference in coagulation kinetics can be related to the 67-kDa light chain (A3–C1–C2 domains) of cofactor VIII, which remains bound to platelets after cofactor VIII dissociation and thereby competes with the binding of intact Xase to platelets.[125] By contrast, fVIII fragments were not seen to accumulate on endothelial cell surfaces, indicating a different Xase-receptor repertoire on these cell surfaces.

Several protein receptors on endothelial cells have been intensely studied with respect to their influence on blood coagulation in general and on activation of fIX and fX in particular. TF, the cofactor of coagulation factor VII, is probably best understood, including its atomic structure in complex with factor VIIa.[70] Another prominent protein receptor is the endothelial cell protein C receptor, EPCR, which was reported to interact not only with protein C but also with fVII.[126] Other protein receptors will certainly play important roles in different settings, as has been demonstrated for protease-activated receptor 1 (PAR1) activation by thrombin. As an example, in mouse embryos, PAR1 signaling on endothelial cells proved of more importance for vessel formation and hemostasis than on platelets.[127]

B. Activated Platelets and Phospholipid Membranes

The consolidation phase of blood coagulation cannot take place on endothelial cells, but only on platelets, simply because one key component of the intrinsic pathway, factor XIa, does not bind to endothelial cells.[128,129] To trigger the intrinsic pathway of blood coagulation, platelets are activated by small amounts of thrombin that have been generated by the extrinsic coagulation pathway. Platelet activation involves the specific cleavage of PARs, which are structurally related to the seven-transmembrane-helix receptors.[82] Among this four-membered family, PAR1 is believed to be the primary mediator of thrombin-stimulated procoagulant platelet activation,[83] but PAR4 cleavage by thrombin also contributes to platelet activation.[130,131] Therefore, blood coagulation is localized to the hemostatic thrombus and is partly shielded from the effect of plasma inhibitors.[132] Activation of platelets leads to the presentation of a catalytic phospholipid surface with an overall negative charge, mostly by anionic phosphatidylserine, as well as to the shedding of phospholipid microvesicles from the platelet membrane.[133]

Currently, the relevance of platelet receptor proteins is still insufficiently understood.[134] However, with clear evidence for several protein epitope-specific binding sites on activated platelets, it is likely that platelet-anchored receptors are key players in the coagulation process. As an important example, platelets were shown to possess a binding site specific to factor IXa that interacts with the Cys88-Cys99 loop of the EGF1/EGF2-connecting segment and with the Cys95-Cys109 loop of EGF2.[135,136] These EGF2-related sites showed a dissociation constant of $K_d \sim 1$ nM, and this affinity was considerably weakened in a hybrid fIXa molecule with the EGF2 domain exchanged with that of fX, resulting in a $K_d \sim 65$ nM. By contrast, rate enhancement of fIXa activity by cofactor VIIIa does not differentiate native and EGF2-swapped hybrid fIXa, suggesting that EGF2 is not a cofactor-interactive site.[137] A second platelet binding site that is shared by fIX and fIXa was shown to be located in the Ω-loop, Gly4-Gln11, in the Gla domain of fIX.[137] This interaction likely reflects a direct contact with the membrane.

Overall, our knowledge, and in particular, our detailed mechanistic understanding of platelet-anchored receptor proteins and their relevance for hemostasis was still limited. Heterodimeric glycoproteins of the integrin family were reported to bind to a plethora of ligands, thereby inducing a bidirectional signal transduction.[130] Contrasting with other glycoprotein receptors (GP), GPIb requires no stimulation to bind its target protein, vWF.[138] The prominent fibrinogen receptor, GPIIb/IIIa, is the most abundant platelet integrin and was recently shown to interact with fXIII.[139] The platelet membrane protein, glycoprotein VI (GPVI), binds to collagen proteins and thus directs platelets to the site of the vessel injury.[140] Finally, the glycoprotein, Ib-V-IX, represents a multifunctional receptor expressed on the surface of activated platelets. GPIb-V-IX is able to interact with a number of ligands, including vWF, thrombin, fXI, and fXII.[141,142]

Other protein receptors are likely to influence blood coagulation, such as platelet apolipoprotein E receptor 2 (ApoER2), which was shown to physically interact with fXIa.[143] In summary, we can conclude that protein receptors significantly affect the assembly of the cofactor-dependent coagulation complex, which is reflected by the observation that synthetic phospholipid vesicles cannot mimic important functional properties of the thrombin-activated platelet surface.[144]

VI. Protein Inhibitors of Coagulation Factors

Inhibition of a catalytic process is generally as critical as the catalysis itself. This regulation is even particularly relevant for blood coagulation, where proteolytic activity is usually very low or not detectable, but on demand, is boosted up to 1 million-fold. The case of high coagulation activity is the exception, not the rule. Therefore, nature developed a spectrum of pathways for rapid and efficient inactivation of the coagulation protease activity. One important line is directed toward cofactor inactivation, either by disintegration (mostly fVIIIa) or by proteolytic inactivation via protein C and other proteases (fVa and fVIIIa). These scenarios were described in Section III.

Protein inhibitors form the classical response in protease inactivation, and not surprisingly, they are also heavily used in blood coagulation. In fact, nature developed not only endogenous inhibitors optimized to keep hemostasis upright, but also exogenous inhibitors that are used in particular by blood-sucking animals.

A. Endogenous Inhibitors

Several protein inhibitor families specifically target the action of trypsin-like serine proteases. The interaction is mostly, but not always or exclusively, directed toward the active site, thus acting as competitive inhibitors to substrate turnover. Here, we summarize the most important protein inhibitor families in blood coagulation, namely, serpins and Kunitz-type inhibitors.

1. Serpins

The acronym, serpin, reflects the prime function of the members of this family, namely, to act as SERine Protease Inhibitors, although some serpin family members are known to inhibit cysteine proteases in addition to serine proteases.[145] Structurally, serpins constitute a highly conserved protein family with three β-sheets, termed A-, B-, and C-sheet, flanked by eight to nine α-helices (for a review, see Ref. 146).

Serpin inhibitory action occurs in two steps: in the first step or substrate recognition phase, the so-called *reactive center loop* (RCL) serves as a bait substrate for the target protease, leading to the formation of the Michaelis complex with the protease (Fig. 14). The sequence of the RCL is one important determinant of the protease specificity and distinguishes its primary specificity, such as tryptic versus chymotryptic.[147] Analogous to a substrate hydrolysis reaction, the tetrahedral transition state-intermediate is formed and the primed product dissociates.

In contrast to substrate hydrolysis, the hemiacetal is not hydrolyzed. Mechanistically, the serpin acts as a suicide substrate because it distorts the protease active site upon binding and thus presumably prevents the proper activation of the catalytic water, precluding hydrolysis of the nonprimed half-product.[148] At this point, the second phase of serpin inhibition comes in to play. The peptide bond cleavage results in an unstable two-chain conformation of the serpin, leading to a dramatic transition that is driven by the insertion of the cleaved nonprimed half-loop of the RCL into β-sheet A, leading to an approximately 70 Å movement of the tethered protease relative to the inhibitor.[148,149] This rearrangement further disrupts the classical alignment of the catalytic triad, which will prevent the activation of the catalytic water, resulting in stabilization of the tetrahedral enzyme half-product. Consequently, the protease–serpin complex is irreversibly trapped and will be recycled only by degradation.

2. Antithrombin, the Coagulation Inhibitor

The serpin AT *targets* most coagulation proteases, including thrombin, fIXa, fXa, and fXIa. This multitarget inhibition qualifies AT as a reliable diagnostic marker of the coagulation activation.[150,151] The specificity of AT for blood coagulation proteases is not routed in the serpin exclusively but

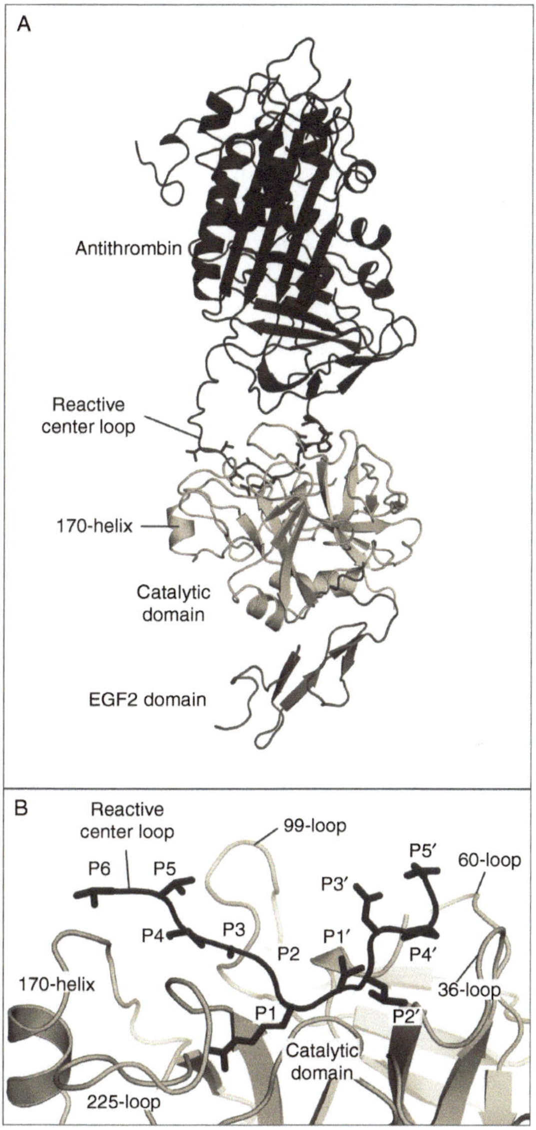

FIG. 14. Serpin-type serine protease inhibitors in the coagulation cascade. (A) Structure of the recognition complex of the serpin, antithrombin, and the protease, fIXa, pdb entry, 3kcg.[51] An fIXa S195A dead mutant forms the Michaelis complex with the intact antithrombin. The antithrombin reactive center loop (RCL) protrudes away to bind deep into the fIXa active site. (B) Close-up view of the interactions between the antithrombin RCL and fIXa, with the orientation identical as in (A).

conveyed by the cofactor dependence of this anticoagulant serpin: heparin directs AT toward the blood coagulation proteases, which exhibit dual affinity for both heparin and AT. Additionally, AT has remarkably adapted to a conserved motif on the autolysis loop in factors IXa and Xa, each harboring an arginine at position 150. Arg150 binds to AT's sheet C in a heparin-dependent manner and thereby stabilizes and strengthens the proper alignment of its RCL to match with the protease active site.[51,152–154]

This combination of specificity determinants of the AT-heparin system, together with heparin's several 1000-fold stimulating effect on AT's efficacy, is of particular physiological and therapeutic relevance, due to its ability to clean the bloodstream of active coagulation proteases away from the vessel injury.[155] This property explains why heparin is most commonly used for thrombosis prophylaxis in medicine.[156,157]

Three mechanisms combine to explain the stimulating effect of heparin on AT. For one, heparin binding to AT exposes parts of its RCL, making it more accessible to the target protease and thus accelerating Michaelis complex formation.[158,159] Synergistic with this AT-priming effect is a bridging effect of heparin which preorients and clamps AT relative to the target protease by employing both serpin- and protease-exosites.[51,154,160] A third contribution to heparin's stimulating effect is derived from its priming of the protease active site. As shown for fIXa, heparin binding induces a high-affinity conformation in the protease active site, thereby reducing the K_I for the serpin.[47] This latter effect reflects the particular plasticity of coagulation factor IXa, as was described earlier in Section II.

Today, there are crystal structures of three heparin-activated AT-protease complexes available: thrombin,[160] fIXa, and fXa,[51,154] each with the active site serine mutated to alanine, S195A. These complexes teach important lessons on the comparative specificity in these proteases, among others. It is further tempting to use these structures to derive information about the substrate recognition of these proteases, possibly in their activating complexes Xase or prothrombinase.

Thrombins' interaction with AT is widely dominated by its 60-insertion loop, which provides numerous unique contacts. The resulting binding mode of the RCL is noncanonical and induces it to fold back with its nonprimed residues to form a specific contact of P5-Val with the Trp60D of the thrombin insertion loop.[160]

In contrast, the AT-RCL binds in an extended, approximately canonical, conformation to the active sites of fIXa and fXa. The RCL residues, P4 to P2′, overlap in both complexes, but differ markedly in the primed recognition site. In the fIXa complex, the RCL is forced into a tight turn imposed by the fIXa 60- and 35-loops, a feature referred to as a wall by Huntington and colleagues.[51] However, whether this wall can explain fIXa's substrate recognition is unclear. First of all, AT is an inhibitor, not a substrate, which implies that the recognition in AT must differ in subtle, but critical ways from recognition in a substrate.

Secondly, by using crystal structures of superactive fIXa variants, we could detect a crosstalk between fIXa's 90- and 60-loop, both of which are very mobile in the wild-type ground state. This 99- to 60-loop stabilization unmasks sites between the 60- and 35-loops that differ from the protease ground state conformation.[20] These unmasked sites may be exploited by primed substrate interactions.

3. Protein Z-Dependent Protease Inhibitor

Another prominent serpin with preference for fXa and fXIa is the protein Z-dependent protease inhibitor (ZPI).[161,162] Analogous to AT, this serpin engages a protein cofactor, protein Z (PZ), which is circulating as a complex with the serpin ZPI. The cofactor PZ itself shows an fIX-like domain architecture (cf. Fig. 1), but lacks critical elements for catalysis, such as the activation peptide and residues Ser195 and His57 from the catalytic triad.[163]

The γ-carboxy-glutamate rich (Gla-) domain of PZ localizes the whole PZ–ZPI complex at the surface of platelet membranes in proximity to fXa or fXIa and thus positions it ideally for their inhibition. Upon complex formation of ZPI with fXa, PZ is released and recycled from the tight binding with ZPI.[164,165]

4. Kunitz-Type Inhibitors

BPTI is by far the best studied representative of this more than 60-membered family of disulfide-linked inhibitors. BPTI is 58 amino acids in length and forms an enormously robust folding module, which resists chemical unfolding beyond 6 M guanidinium with a thermal melting point of 95 °C.[166,167] These unusual biophysical properties relate to the stabilization by three disulfides with the characteristic bonding pattern of Cys1-Cys6, Cys2-Cys4, and Cys3-Cys5 (Fig. 15A). Cys2 is positioned within the reactive center loop (RCL) at the P2 position. The disulfide bridging of the P2 residue induces a conformational strain in the RCL, thereby establishing an extended RCL conformation with the amino acid side chains flanking the scissile peptide bond that is exposed to the solvent (Fig. 15B and C).[169,170]

These biophysical features explain BPTI's strong reactivity toward trypsin-like proteases[171] and, at the same time, why cleavage of the distorted, improperly activated scissile peptide bond is extremely slow, approximately 10^{-9} s^{-1}.[172] In contrast with serpins (see above), the active site of the protease remains intact; mechanistically, Kunitz-type inhibitors act as extremely slow substrates, with one formal turnover per 30 years.

The Kunitz class of inhibitors may be further divided into canonical and retro-binding inhibitors. Canonical binders form an antiparallel β-sheet at the S3-site (Gly216) with their RCL, whereas retro-binding inhibitors, mostly derived from hemophagous animals, form a parallel β-sheet interaction between the S3 site and the RCL.[170]

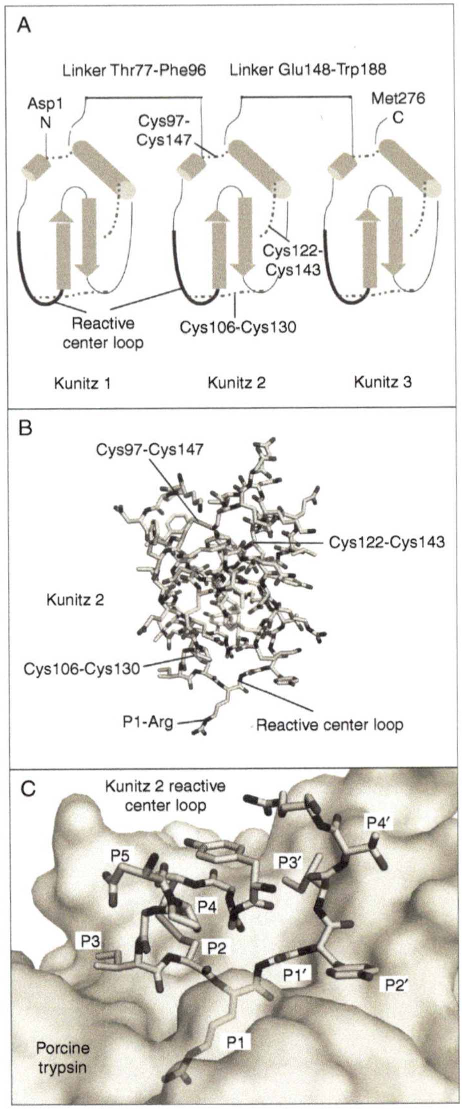

FIG. 15. Kunitz-type inhibitors in the coagulation cascade. (A) Schematic topology representation of the three Kunitz domains in tissue factor pathway inhibitor (TFPI). Disulfide bonds are indicated by dotted lines and confer extraordinary stability to these domains. The reactive center loop (RCL) is indicated in bold. TFPI consists of three tandem Kunitz domains, where the RCL of Kunitz domain 2 initially binds to fXa. TF–fVIIa can then bind the binary TFPI–fXa complex, resulting in a quaternary complex. (B) Structure of the TFPI Kunitz domain 2, which was crystallized in complex with porcine trypsin, pdb entry, 1tfx.[168] The overall orientation of the Kunitz domain is as in (A). The RCL is at the bottom of the figure with Arg in P1 position. (C) The TFPI Kunitz domain 2 in a stick model bound to the surface of bovine trypsin. The figure is a zoom-out of (B) with identical orientation.

5. TF Pathway Inhibitor

The only known mission of TFPI is to downregulate the extrinsic Xase, that is, the activation of factor X by the TF–factor VIIa complex. TFPI thus contributes to separating the intrinsic from the extrinsic coagulation signal.[173] TFPI is assembled from three tandem Kunitz domains, as shown in Fig. 15A. Kunitz domain 3 mediates its association with lipoproteins but has no known protease target.[174] It mediates multiple localization sites, including a GPI-anchor, causing more than 90% of TFPI to circulate in complex with lipoproteins.[175,176] This association refers to the synonym lipoprotein-associated coagulation inhibitor (LACI).

Inhibition of the extrinsic Xase is mediated by Kunitz domains 1 and 2, which primarily bind factors VIIa and Xa, respectively. The sequence of the binding events is critical. TF–fVIIa can bind to TFPI only after fXa has bound to the RCL of TFPI Kunitz domain 2.[177] This remarkable dependence implies that fXa, when bound to Kunitz domain 2, provides a necessary additional binding site to TF–fVIIa. Alternatively, fXa binding may allosterically enable TFPI to inhibit TF–fVIIa. The latter hypothesis is supported by the observation that fXa binding to Kunitz domain 2 is influenced by the Kunitz domain 1 and, analogously, fVIIa binding to Kunitz domain 1 is influenced by Kunitz domain 2.[174] The selectivity of TFPI's action to the extrinsic pathway depends on the location of TFPI (mediated by Kunitz domain 3) and the molar excess of fXa over TFPI. It is therefore critical to rapidly produce enough fXa to trigger the self-sustaining intrinsic pathway; otherwise, TFPI will suppress any spontaneous coagulation events.

The anticoagulant cofactor, protein S, which is known to stimulate aPC, was recently shown to strengthen the TFPI–fXa inhibition by about 10-fold, resulting in a K_I value of 0.5 nM. This increase is significant, as it matches the plasma concentration of TFPI, such that a physiologically significant regulation can be expected.[176] While the exact mechanism of the cofactor action is still unclear, Kunitz domain 3 was shown to bind protein S.[178]Protein S was reported to exhibit a direct anticoagulant effect by inhibiting the extrinsic ternary Xase complex (TF–fVIIa–fX(a)) in a Zn^{2+}-dependent manner.[179]

6. Amyloid β Precursor Protein aka Protease Nexin 2 (PN2)

Despite its clear relation to Alzheimer's disease, the physiological function of amyloid β precursor protein (AβPP) is incompletely understood. There is a significant body of evidence that relates AβPP to coagulation; major AβPP reservoirs are present in α-granules of platelets, which get released upon thrombin activation.[180–183]

The multimodular AβPP contains a 50-amino acid spanning, Kunitz-type domain, similar to the homolog amyloid precursor-like protein 2 (APLP-2). Both Kunitz domains were shown to exert protease inhibitory activity directed against fXa, fIXa, fVIIa, and fXIa, among others.[184,185] The crystal structure information of the AβPP Kunitz inhibitor domain in complex with fXIa confirms a canonical binding mode into the fXIa active site.[186] Binding of fIXa is very tight and heparin-dependent, as reflected by kinetic parameters, $K_I = 3.9 \times 10^{-11}$ and 7.9×10^{-11} M, in the presence and absence of heparin, respectively.[187] The AβPP plasma concentrations of approximately 30 nM are approximately 100-fold lower as compared to AT, matching the approximately 100-fold higher affinity of the AβPP Kunitz inhibitor as compared to AT. Thus, the physiological relevance of AβPP for blood coagulation remains controversial,[180] but may simply result from the AβPP location in, and release from, platelets upon activation.

Interestingly, and different from fIXa, the binding mode of AβPP to fXa appears not to be active site-directed. Despite inhibition constants of approximately 10 and 1000 nM toward fXa alone and in the prothrombinase complex, respectively, the activity reduction remains incomplete, even with excess AβPP. Also, in contrast to fIXa, heparin appears to weaken AβPP inhibition of fXa.[188]

B. Exogenous Inhibitors

Endogenous inhibitors have to finely balance blood coagulation and therefore invoke a network of pro- and anticoagulant molecules. In contrast, exogenous inhibitors typically require a single target to efficiently shutdown the host's blood coagulation system. Typical sources of exogenous coagulation protease inhibitors are the saliva of hemophagous (blood sucking) animals or the venom of snakes. Consistently, many blood-sucking animals target thrombin, the final protease in the blood coagulation cascade. The most prominent representative is hirudin from the leech, *Hirudo medicinalis*. Its high specificity is based on its simultaneous interaction with thrombin's active site and fibrinogen exosite.[189,190]

Whereas the majority of exogenous protease inhibitors target proteases more downstream in the coagulation cascade, nature also developed ways and means to efficiently shut down blood coagulation by targeting other components of the coagulation cascade, including fIX(a) and fXa. Correspondingly, several highly potent coagulation factor inhibitors have been isolated from different sources, an overview of which is given below.

The insect salivary protein, prolixin S, also known as nitrophorin 2, from *Rhodnius prolixus*, interferes with Xase formation by targeting the Gla domain of fIX and fIXa with nanomolar affinity.[191,192] Nitrophorin is a multifunctional protein carrying a ferric heme group in its β-barrel structure.[193] The Ca^{2+}-dependent interaction of nitrophorin with fIX presumably exerts its anticoagulant effect by a combination of mechanisms, including interfering with the proper localization of fIX(a) to activated platelets, in addition to antagonizing the productive interaction with the components of Xase.

The principle of localization inhibition is exploited by a family of related lectin-like proteins, the fIX-binding protein,[194] the fIX/X binding protein,[195] and the fX binding protein,[196] all of which were isolated from snake venom. This class of protein inhibitors binds to the Gla-domain of the respective coagulation factors in a Ca^{2+}-dependent manner.

The 140-amino acid protein, ixolaris, from the hard tick, *Ixodes scapularis*, is composed of two tandem Kunitz-like domains, thus partly mimicking the architecture and also the function of TFPI.[197] Details of the subnanomolar interaction differ from those in the TFPI complex. In contrast to TFPI, ixolaris binds both zymogen fX and active fXa via its heparin-binding exosite, leaving the active site unoccupied. This interaction interferes with fX binding to cofactor VIIIa and thus inhibits the activation of zymogen fX by the intrinsic Xase.[198] The binary ixolaris–fX(a) complex is able to bind the TF–fVIIa complex. Interestingly, this interaction is dependent on the Gla-domain of fX. The best known exogenous fXa inhibitor is the tick-derived anticoagulant peptide (TAP) from *Ornithodoros moubata*. TAP exhibits a distorted Kunitz-like domain structure. Contrasting other Kunitz inhibitors, TAP is strictly selective toward fXa with a K_I of 0.6 nM, but no reactivity against other proteases.[199] Interestingly, the crystal structure of the fXa–TAP complex reveals the N-terminal three TAP residues bound to the fXa active site in a retro-binding mode, with the N-terminal Tyr binding to the S1 site. This binding mode was expected, based on the ornithodorin–thrombin complex structure.[200,201] As seen in the ornithodorin–thrombin complex, the C-terminus of TAP exploits an exosite binding site. The fXa exosite is located on the autolysis loop, which is largely disordered in the uncomplexed fXa structure.[55] Notably, there is a 30-fold stronger inhibition by TAP toward the prothrombinase complex ($K_I \sim 31$ pM) when compared to free fXa ($K_I \sim 1$ nM). This effect is paralleled by an approximate 30-fold strengthening of the fVa complex toward TAP-inhibited fXa as compared to free fXa, indicating a crosslinking function within this ternary complex.[202]

NAP5 from the hematophagous hookworm, *Ancylostoma caninum*, is another Kunitz-type fXa inhibitor with a $K_I \sim 43$ pM.[203] The crystal structure of the NAP5–fXa complex shows an overall canonical, that is, substrate-like, binding to the active site with an arginine at the P1 position. An unexpected feature was the "hijacking" interaction of the P2 serine of NAP5 toward His57 of fXa, which triggers its rotation away from the catalytic Ser195, rendering the protease inactive.[204] Possibly a crystal artifact, the crystal structure also reveals an exosite interaction of the NAP5 C-terminus with a symmetry-related fXa molecule in the crystal lattice.

VII. Pharmaceutical Opportunities and Challenges

Bleeding or clotting disorders can be caused by multiple mechanisms, either through genetic predisposition or acquired by venom, autoimmune reactions, or certain cancers. Due to their prevalence, bleeding/clotting disorders constitute a major need of society. Factors IX and X are positioned in the control center of blood coagulation and therefore represent an important approach in developing therapy strategies for the broad field of coagulation-related diseases. Often, intact natural proteins can be used directly as therapeutics. In the case of coagulation factors and their inhibitors, these proteins are either plasma-derived or recombinantly overexpressed.

Plasma-derived coagulation factors are attractive for a number of reasons, particularly for complex posttranslational modifications, such as γ-carboxylation, diverse glycosylation, or sulfation, as is required for coagulation factors that are difficult to produce recombinantly. Due to their correct posttranslational modification pattern, plasma-derived proteins may exhibit the more physiological activity and bear a reduced risk of an immune response directed against the therapeutic protein.[205]

Plasma-derived therapeutic proteins carry one critical disadvantage, however: the risk of infecting a patient with a severe or even deadly disease. This risk came to the full public awareness with HIV-contamination disaster of fVIII samples in the 1980s. Since then, a panel of quality assurance measures was installed, including careful donor selection, virus inactivation, and diagnostics. These measures helped to reduce the risk of viral infections to less than 1 per million.[206]

Of the known disease-classes, protein-based diseases, such as prion-related bovine spongiform encephalopathy (BSE) and variant Creutzfeldt Jakob disease (vCJD), pose a remaining risk factor that is not fully controlled.[207] Standard inactivation procedures as applied to viral infections are not applicable in these cases; therefore, one has to rely on diagnostic methods such as immune detection, possibly combined with automatic separation methods, although this challenge is as yet unsettled.[208]

A. Antihemophilic Drug Discovery

In most cases, excess bleeding disorders are due to a variant of hemophilia A and B, which are caused by defects in coagulation factors VIII and IX, respectively. fVIII and fIX are encoded on the X chromosome, explaining why males are affected by deficiencies of these genes with a probability of 1 per approximately 5000. Current standard therapies to hemophilia A (fVIII deficiency) and B (fIX deficiency) are the intravenous administration of plasma-derived or recombinant fVIII and fIX, respectively.

With the increased awareness of contamination risks and improved technology in recombinant protein expression, more than 10 different recombinant rfVIII and rfIX protein variants have successfully reached the market. Several technical hurdles had to be overcome, including the need for mammalian expression systems suited to produce the correct posttranslational modifications, coexpression of furin for efficient pre-pro-protein processing, vitamin K dependent γ-carboxylation, and separation of incompletely modified protein variants.[209–211] Detectable differences of recombinant fIX as compared to the plasma-derived fIX are due to incomplete tyrosine sulfation and serine phosphorylation, correlating with a reduced *in vivo* recovery after intravenous rfIX administration.[212]

Despite the success of current antihemophilic drugs, there is plenty of opportunity to improve on the current prophylaxis and treatment options. Examples for such improvements are (i) prolonged half-life of the administered protein and (ii) alternatives to the intravenous administration form. Prolonged half-life of the administered drug is obviously related to the current way the drug is administered. Intravenous injection is painful, particularly for small children (mostly boys). Standard strategies to improve on the bioavailability of biologicals involve PEGylation or dimerization. These modifications increase the apparent molecular size and reduce the accessibility for proteolytic degradation and their diffusive elimination via the kidney filter.

Dimerization can be induced in different ways, including the use of low-molecular weight compounds that may be applied independently from the therapeutic protein. Such applications aim at developing drugs that can be administered nonintravenously, for example, orally or by inhalation. Depending on the patients' coagulation factor deficiencies, it may be possible to develop low-molecular weight compounds that help to accumulate the concentration of the patients' impaired coagulation factors by prolonging their half-life, specifically by inducing homo- or heterodimerization. Alternatively, low-molecular weight compounds may act as stimulating agonists to the functionally impaired coagulation factor of the patient. Finally, low-molecular weight compounds might mimic the effect of a coagulation cofactor.[213,214]

In the light of the current structural information on blood coagulation factors, improvements can also be achieved by rational engineering of modified coagulation factor variants. Such improvements might address the base-level activity of the coagulation factor, their stability with respect to dissociation, and the avoidance of proteolytic inactivation.

B. Antithrombotic Drugs

The tendency to thrombotic disorders does not usually depend on single-point mutations but on a complex background; therefore, it is difficult to identify clear prognostic factors for this disease.[215] It is therefore an intriguing and nontrivial observation that there exists a clear antirisk factor to thrombosis. Hemophiliacs show a significantly reduced incidence of venous thrombosis, with only 1 in about 30,000 hemophiliacs being affected as compared to 1 in about 1500 "healthy" persons from a control group.[216] These data constitute an excellent validation of the intrinsic Xase as a thrombosis target, though for technical reasons, this opportunity has not been exploited so far. Technical challenges specific to Xase-directed drug discovery programs are the lack of a Xase crystal structure and the difficulty in the development of a physiologically relevant assay suitable for high-throughput testing of compound libraries.

Successful development of thrombin and fXa inhibitors recently generated optimism in the feasibility of novel, active site-directed drug discovery programs.[217–220] With these important accomplishments, we can hope for the discovery of a new generation of drugs that address the urgent need posed by the manifold clotting diseases.

Acknowledgments

We are thankful for financial support by the Austrian Science Foundation FWF and the Land Salzburg.

References

1. Zögg T, Brandstetter H. Activation mechanisms of coagulation factor IX. *Biol Chem* 2009;**390**:391–400.
2. Byrne R, Link RP, Castellino FJ. A kinetic evaluation of activated bovine blood coagulation factor IX toward synthetic substrates. *J Biol Chem* 1980;**255**:5336–41.
3. McRae BJ, Kurachi K, Heimark RL, Fujikawa K, Davie EW, Powers JC. Mapping the active sites of bovine thrombin, factor IXa, factor Xa, factor XIa, factor XIIa, plasma kallikrein, and trypsin with amino acid and peptide thioesters: development of new sensitive substrates. *Biochemistry* 1981;**20**:7196–206.
4. Matthews BW, Sigler PB, Henderson R, Blow DM. Three-dimensional structure of tosyl-alpha-chymotrypsin. *Nature* 1967;**214**:652–6.
5. Bode W, Huber R. Induction of the bovine trypsinogen-trypsin transition by peptides sequentially similar to the N-terminus of trypsin. *FEBS Lett* 1976;**68**:231–6.
6. Bode W, Schwager P, Huber R. The transition of bovine trypsinogen to a trypsin-like state upon strong ligand binding. The refined crystal structures of the bovine trypsinogen-pancreatic trypsin inhibitor complex and of its ternary complex with Ile-Val at 1.9 Å resolution. *J Mol Biol* 1978;**118**:99–112.
7. Fehlhammer H, Bode W, Huber R. Crystal structure of bovine trypsinogen at 1-8 Å resolution. II. Crystallographic refinement, refined crystal structure and comparison with bovine trypsin. *J Mol Biol* 1977;**111**:415–38.
8. Wang D, Bode W, Huber R. Bovine chymotrypsinogen A X-ray crystal structure analysis and refinement of a new crystal form at 1.8 Å resolution. *J Mol Biol* 1985;**185**:595–624.

9. DeLano WL. The case for open-source software in drug discovery. *Drug Discov Today* 2005;**10**:213–7.
10. Sippl MJ, Wiederstein M. A note on difficult structure alignment problems. *Bioinformatics* 2008;**24**:426–7.
11. Bajaj SP, Rapaport SI, Russell WA. Redetermination of the rate-limiting step in the activation of factor IX by factor XIa and by factor VIIa/tissue factor. Explanation for different electrophoretic radioactivity profiles obtained on activation of 3H- and 125I-labeled factor IX. *Biochemistry* 1983;**22**:4047–53.
12. Gailani D. Activation of factor IX by factor XIa. *Trends Cardiovasc Med* 2000;**10**:198–204.
13. Lawson JH, Mann KG. Cooperative activation of human factor IX by the human extrinsic pathway of blood coagulation. *J Biol Chem* 1991;**266**:11317–27.
14. Papagrigoriou E, McEwan PA, Walsh PN, Emsley J. Crystal structure of the factor XI zymogen reveals a pathway for transactivation. *Nat Struct Mol Biol* 2006;**13**:557–8.
15. Fujikawa K, Legaz ME, Kato H, Davie EW. The mechanism of activation of bovine factor IX (Christmas factor) by bovine factor XIa (activated plasma thromboplastin antecedent). *Biochemistry* 1974;**13**:4508–16.
16. Baglia FA, Jameson BA, Walsh PN. Identification and chemical synthesis of a substrate-binding site for factor IX on coagulation factor XIa. *J Biol Chem* 1991;**266**:24190–7.
17. Gailani D, Ho D, Sun MF, Cheng Q, Walsh PN. Model for a factor IX activation complex on blood platelets: dimeric conformation of factor XIa is essential. *Blood* 2001;**97**:3117–22.
18. Wolberg AS, Morris DP, Stafford DW. Factor IX activation by factor XIa proceeds without release of a free intermediate. *Biochemistry* 1997;**36**:4074–9.
19. Samuel D, Cheng H, Riley PW, Canutescu AA, Nagaswami C, Weisel JW, et al. Solution structure of the A4 domain of factor XI sheds light on the mechanism of zymogen activation. *Proc Natl Acad Sci USA* 2007;**104**:15693–8.
20. Zögg T, Brandstetter H. Structural basis of the cofactor- and substrate-assisted activation of human coagulation factor IXa. *Structure* 2009;**17**:1669–78.
21. Takaki A, Enfield DL, Thompson AR. Cleavage and inactivation of factor IX by granulocyte elastase. *J Clin Invest* 1983;**72**:1706–15.
22. Enfield DL, Thompson AR. Cleavage and activation of human factor IX by serine proteases. *Blood* 1984;**64**:821–31.
23. Takeya H, Nishida S, Miyata T, Kawada S, Saisaka Y, Morita T, et al. Coagulation factor X activating enzyme from Russell's viper venom (RVV-X). A novel metalloproteinase with disintegrin (platelet aggregation inhibitor)-like and C-type lectin-like domains. *J Biol Chem* 1992;**267**:14109–17.
24. Lindquist PA, Fujikawa K, Davie EW. Activation of bovine factor IX (Christmas factor) by factor XIa (activated plasma thromboplastin antecedent) and a protease from Russell's viper venom. *J Biol Chem* 1978;**253**:1902–9.
25. Imamura T, Tanase S, Hamamoto T, Potempa J, Travis J. Activation of blood coagulation factor IX by gingipains R, arginine-specific cysteine proteinases from *Porphyromonas gingivalis*. *Biochem J* 2001;**353**:325–31.
26. Wallin R, Stanton C, Ross RP. Intracellular proteolytic processing of the two-chain vitamin K-dependent coagulation factor X. *Thromb Res* 1994;**73**:395–403.
27. Himmelspach M, Pfleiderer M, Fischer BE, Plaimauer B, Antoine G, Falkner FG, et al. Recombinant human factor X: high yield expression and the role of furin in proteolytic maturation in vivo and in vitro. *Thromb Res* 2000;**97**:51–67.
28. Iijima K, Murakami M, Kimura O, Murakami F, Shimomura T, Ikawa S. A dysfunctional factor X (factor X Kurayoshi) with a substitution of Arg 139 for Ser at the carboxyl-terminus of the light chain. *Thromb Res* 2001;**101**:311–6.

29. Sichler K, Kopetzki E, Huber R, Bode W, Hopfner KP, Brandstetter H. Physiological fIXa activation involves a cooperative conformational rearrangement of the 99-loop. *J Biol Chem* 2003;**278**:4121–6.
30. Duffy EJ, Lollar P. Intrinsic pathway activation of factor X and its activation peptide-deficient derivative, factor Xdes-143-191. *J Biol Chem* 1992;**267**:7821–7.
31. Lenting PJ, ter Maat H, Clijsters PP, Donath MJ, van Mourik JA, Mertens K. Cleavage at arginine 145 in human blood coagulation factor IX converts the zymogen into a factor VIII binding enzyme. *J Biol Chem* 1995;**270**:14884–90.
32. Berrettini M, Schleef RR, Heeb MJ, Hopmeier P, Griffin JH. Assembly and expression of an intrinsic factor IX activator complex on the surface of cultured human endothelial cells. *J Biol Chem* 1992;**267**:19833–9.
33. Nemerson Y, Furie B. Zymogens and cofactors of blood coagulation. *CRC Crit Rev Biochem* 1980;**9**:45–85.
34. van Dieijen G, van Rijn JL, Govers-Riemslag JW, Hemker HC, Rosing J. Assembly of the intrinsic factor X activating complex—Interactions between factor IXa, factor VIIIa and phospholipid. *Thromb Haemost* 1985;**53**:396–400.
35. Gyzander E, Eriksson E, Teger-Nilsson AC. A sensitive assay for tissue plasminogen activator activity in plasma, using adsorption on lysine-sepharose. *Thromb Res* 1984;**35**:547–58.
36. Allen RA. An enhancing effect of poly-lysine on the activation of plasminogen. *Thromb Haemost* 1982;**47**:41–5.
37. Vagenende V, Yap MG, Trout BL. Mechanisms of protein stabilization and prevention of protein aggregation by glycerol. *Biochemistry* 2009;**48**:11084–96.
38. Stürzebecher J, Kopetzki E, Bode W, Hopfner KP. Dramatic enhancement of the catalytic activity of coagulation factor IXa by alcohols. *FEBS Lett* 1997;**412**:295–300.
39. Kolkman JA, Lenting PJ, Mertens K. Regions 301-303 and 333-339 in the catalytic domain of blood coagulation factor IX are factor VIII-interactive sites involved in stimulation of enzyme activity. *Biochem J* 1999;**339**(Pt 2):217–21.
40. Sheehan JP, Phan TM. Phosphorothioate oligonucleotides inhibit the intrinsic tenase complex by an allosteric mechanism. *Biochemistry* 2001;**40**:4980–9.
41. Brandstetter H, Bauer M, Huber R, Lollar P, Bode W. X-ray structure of clotting factor IXa: active site and module structure related to Xase activity and hemophilia B. *Proc Natl Acad Sci USA* 1995;**92**:9796–800.
42. Hopfner KP, Lang A, Karcher A, Sichler K, Kopetzki E, Brandstetter H, et al. Coagulation factor IXa: the relaxed conformation of Tyr99 blocks substrate binding. *Structure* 1999;**7**:989–96.
43. Rubboli A. Efficacy and safety of low-molecular-weight heparins as an adjunct to thrombolysis in acute ST-elevation myocardial infarction. *Curr Cardiol Rev* 2008;**4**:63–71.
44. Camporese G, Bernardi E. Low-molecular-weight heparin for thromboprophylaxis. *Curr Opin Pulm Med* 2009;**15**:443–54.
45. Bosner MS, Gulick T, Riley DJ, Spilburg CA, Lange LG. Heparin-modulated binding of pancreatic lipase and uptake of hydrolyzed triglycerides in the intestine. *J Biol Chem* 1989;**264**:20261–4.
46. Spilburg CA, Cox DG, Wang X, Bernat BA, Bosner MS, Lange LG. Identification of a species specific regulatory site in human pancreatic cholesterol esterase. *Biochemistry* 1995;**34**:15532–8.
47. Neuenschwander PF. Exosite occupation by heparin enhances the reactivity of blood coagulation factor IXa. *Biochemistry* 2004;**43**:2978–86.
48. Misenheimer TM, Buyue Y, Sheehan JP. The heparin-binding exosite is critical to allosteric activation of factor IXa in the intrinsic tenase complex: the role of arginine 165 and factor X. *Biochemistry* 2007;**46**:7886–95.

49. Yang L, Manithody C, Rezaie AR. Localization of the heparin binding exosite of factor IXa. *J Biol Chem* 2002;**277**:50756–60.
50. Yuan QP, Walke EN, Sheehan JP. The factor IXa heparin-binding exosite is a cofactor interactive site: mechanism for antithrombin-independent inhibition of intrinsic tenase by heparin. *Biochemistry* 2005;**44**:3615–25.
51. Johnson DJ, Langdown J, Huntington JA. Molecular basis of factor IXa recognition by heparin-activated antithrombin revealed by a 1.7-A structure of the ternary complex. *Proc Natl Acad Sci USA* 2010;**107**:645–50.
52. Hitomi Y, Outten CE, O'Halloran TV. Extreme zinc-binding thermodynamics of the metal sensor/regulator protein, ZntR. *J Am Chem Soc* 2001;**123**:8614–5.
53. Morena M, Bosc JY, Jaussent I, Dupuy AM, Terrier N, Leray-Moragues H, et al. The role of mineral metabolism and inflammation on dialysis vascular access failure. *J Vasc Access* 2006;**7**:77–82.
54. McDonald MR, Kunitz M. The effect of calcium and other ions on the autocatalytic formation of trypsin from trypsinogen. *J Gen Physiol* 1941;**25**:53–73.
55. Brandstetter H, Kuhne A, Bode W, Huber R, von der Saal W, Wirthensohn K, et al. X-ray structure of active site-inhibited clotting factor Xa. Implications for drug design and substrate recognition. *J Biol Chem* 1996;**271**:29988–92.
56. Dang QD, Di Cera E. Residue 225 determines the Na(+)-induced allosteric regulation of catalytic activity in serine proteases. *Proc Natl Acad Sci USA* 1996;**93**:10653–6.
57. Di Cera E, Guinto ER, Vindigni A, Dang QD, Ayala YM, Wuyi M, et al. The Na+ binding site of thrombin. *J Biol Chem* 1995;**270**:22089–92.
58. Wells CM, Di Cera E. Thrombin is a Na(+)-activated enzyme. *Biochemistry* 1992;**31**:11721–30.
59. Monnaie D, Arosio D, Griffon N, Rose T, Rezaie AR, Di Cera E. Identification of a binding site for quaternary amines in factor Xa. *Biochemistry* 2000;**39**:5349–54.
60. Bajaj SP, Schmidt AE, Agah S, Bajaj MS, Padmanabhan K. High resolution structures of p-aminobenzamidine- and benzamidine-VIIa/soluble tissue factor: unpredicted conformation of the 192-193 peptide bond and mapping of Ca^{2+}, Mg^{2+}, Na^{+}, and Zn^{2+} sites in factor VIIa. *J Biol Chem* 2006;**281**:24873–88.
61. Harding MM. Metal-ligand geometry relevant to proteins and in proteins: sodium and potassium. *Acta Crystallogr D Biol Crystallogr* 2002;**58**:872–4.
62. Sichler K, Banner DW, D'Arcy A, Hopfner KP, Huber R, Bode W, et al. Crystal structures of uninhibited factor VIIa link its cofactor and substrate-assisted activation to specific interactions. *J Mol Biol* 2002;**322**:591–603.
63. Scharer K, Morgenthaler M, Paulini R, Obst-Sander U, Banner DW, Schlatter D, et al. Quantification of cation-pi interactions in protein-ligand complexes: crystal-structure analysis of factor Xa bound to a quaternary ammonium ion ligand. *Angew Chem Int Ed Engl* 2005;**44**:4400–4.
64. Gandhi PS, Chen Z, Mathews FS, Di Cera E. Structural identification of the pathway of long-range communication in an allosteric enzyme. *Proc Natl Acad Sci USA* 2008;**105**:1832–7.
65. Butenas S, Orfeo T, Mann KG. Tissue factor activity and function in blood coagulation. *Thromb Res* 2008;**122**(Suppl. 1):S42–46.
66. Bock PE, Panizzi P, Verhamme IM. Exosites in the substrate specificity of blood coagulation reactions. *J Thromb Haemost* 2007;**5**(Suppl 1):81–94.
67. Spronk HM, Govers-Riemslag JW, ten Cate H. The blood coagulation system as a molecular machine. *Bioessays* 2003;**25**:1220–8.
68. Davidson CJ, Tuddenham EG, McVey JH. 450 million years of hemostasis. *J Thromb Haemost* 2003;**1**:1487–94.

69. Harlos K, Martin DM, O'Brien DP, Jones EY, Stuart DI, Polikarpov I, et al. Crystal structure of the extracellular region of human tissue factor. *Nature* 1994;**370**:662–6.
70. Banner DW, D'Arcy A, Chene C, Winkler FK, Guha A, Konigsberg WH, et al. The crystal structure of the complex of blood coagulation factor VIIa with soluble tissue factor. *Nature* 1996;**380**:41–6.
71. Eigenbrot C, Kirchhofer D, Dennis MS, Santell L, Lazarus RA, Stamos J, et al. The factor VII zymogen structure reveals reregistration of beta strands during activation. *Structure* 2001;**9**:627–36.
72. Neuenschwander PF, Branam DE, Morrissey JH. Importance of substrate composition, pH and other variables on tissue factor enhancement of factor VIIa activity. *Thromb Haemost* 1993;**70**:970–7.
73. Mandal SK, Pendurthi UR, Rao LV. Cellular localization and trafficking of tissue factor. *Blood* 2006;**107**:4746–53.
74. Higashi S, Matsumoto N, Iwanaga S. Molecular mechanism of tissue factor-mediated acceleration of factor VIIa activity. *J Biol Chem* 1996;**271**:26569–74.
75. Bode W, Mayr I, Baumann U, Huber R, Stone SR, Hofsteenge J. The refined 1.9 Å crystal structure of human alpha-thrombin: interaction with D-Phe-Pro-Arg chloromethylketone and significance of the Tyr-Pro-Pro-Trp insertion segment. *EMBO J* 1989;**8**:3467–75.
76. Pike AC, Brzozowski AM, Roberts SM, Olsen OH, Persson E. Structure of human factor VIIa and its implications for the triggering of blood coagulation. *Proc Natl Acad Sci USA* 1999;**96**:8925–30.
77. Zhong D, Bajaj MS, Schmidt AE, Bajaj SP. The N-terminal epidermal growth factor-like domain in factor IX and factor X represents an important recognition motif for binding to tissue factor. *J Biol Chem* 2002;**277**:3622–31.
78. Neuenschwander PF, Morrissey JH. Roles of the membrane-interactive regions of factor VIIa and tissue factor. The factor VIIa Gla domain is dispensable for binding to tissue factor but important for activation of factor X. *J Biol Chem* 1994;**269**:8007–13.
79. Mann KG, Orfeo T, Butenas S, Undas A, Brummel-Ziedins K. Blood coagulation dynamics in haemostasis. *Hamostaseologie* 2009;**29**:7–16.
80. Lollar P, Knutson GJ, Fass DN. Activation of porcine factor VIII:C by thrombin and factor Xa. *Biochemistry* 1985;**24**:8056–64.
81. Lollar P, Parker CG. Subunit structure of thrombin-activated porcine factor VIII. *Biochemistry* 1989;**28**:666–74.
82. London FS, Marcinkiewicz M, Walsh PN. PAR-1-stimulated factor IXa binding to a small platelet subpopulation requires a pronounced and sustained increase of cytoplasmic calcium. *Biochemistry* 2006;**45**:7289–98.
83. Andersen H, Greenberg DL, Fujikawa K, Xu W, Chung DW, Davie EW. Protease-activated receptor 1 is the primary mediator of thrombin-stimulated platelet procoagulant activity. *Proc Natl Acad Sci USA* 1999;**96**:11189–93.
84. Shen BW, Spiegel PC, Chang CH, Huh JW, Lee JS, Kim J, et al. The tertiary structure and domain organization of coagulation factor VIII. *Blood* 2008;**111**:1240–7.
85. Adams TE, Hockin MF, Mann KG, Everse SJ. The crystal structure of activated protein C-inactivated bovine factor Va: implications for cofactor function. *Proc Natl Acad Sci USA* 2004;**101**:8918–23.
86. Ngo JC, Huang M, Roth DA, Furie BC, Furie B. Crystal structure of human factor VIII: implications for the formation of the factor IXa-factor VIIIa complex. *Structure* 2008;**16**:597–606.
87. Pratt KP, Shen BW, Takeshima K, Davie EW, Fujikawa K, Stoddard BL. Structure of the C2 domain of human factor VIII at 1.5 Å resolution. *Nature* 1999;**402**:439–42.

88. Liu Z, Lin L, Yuan C, Nicolaes GA, Chen L, Meehan EJ, et al. Trp2313-His2315 of factor VIII C2 domain is involved in membrane binding: structure of a complex between the C2 domain and an inhibitor of membrane binding. *J Biol Chem* 2010;**285**:8824–9.
89. Newell JL, Fay PJ. Proteolysis at Arg740 facilitates subsequent bond cleavages during thrombin-catalyzed activation of factor VIII. *J Biol Chem* 2007;**282**:25367–75.
90. Fay PJ, Mastri M, Koszelak ME, Wakabayashi H. Cleavage of factor VIII heavy chain is required for the functional interaction of a2 subunit with factor IXA. *J Biol Chem* 2001;**276**:12434–9.
91. Fay PJ, Jenkins PV. Mutating factor VIII: lessons from structure to function. *Blood Rev* 2005;**19**:15–27.
92. Lapan KA, Fay PJ. Localization of a factor X interactive site in the A1 subunit of factor VIIIa. *J Biol Chem* 1997;**272**:2082–8.
93. Fay PJ, Koshibu K. The A2 subunit of factor VIIIa modulates the active site of factor IXa. *J Biol Chem* 1998;**273**:19049–54.
94. O'Brien LM, Medved LV, Fay PJ. Localization of factor IXa and factor VIIIa interactive sites. *J Biol Chem* 1995;**270**:27087–92.
95. Lenting PJ, van de Loo JW, Donath MJ, van Mourik JA, Mertens K. The sequence Glu1811-Lys1818 of human blood coagulation factor VIII comprises a binding site for activated factor IX. *J Biol Chem* 1996;**271**:1935–40.
96. Patsialas K, Koutsas C, Makris P, Liakopoulou-Kyriakides M. Peptide analogues of 1811-1818 loop of the A3 subunit of the light chain A3-C1-C2 of FVIII of blood coagulation: biological evaluation. *Amino Acids* 2010;**39**:481–8.
97. Meems H, Meijer AB, Cullinan DB, Mertens K, Gilbert GE. Factor VIII C1 domain residues Lys 2092 and Phe 2093 contribute to membrane binding and cofactor activity. *Blood* 2009;**114**:3938–46.
98. Hsu TC, Pratt KP, Thompson AR. The factor VIII C1 domain contributes to platelet binding. *Blood* 2008;**111**:200–8.
99. Jacquemin M, Benhida A, Peerlinck K, Desqueper B, Vander Elst L, Lavend'homme R, et al. A human antibody directed to the factor VIII C1 domain inhibits factor VIII cofactor activity and binding to von Willebrand factor. *Blood* 2000;**95**:156–63.
100. Gilbert GE, Kaufman RJ, Arena AA, Miao H, Pipe SW. Four hydrophobic amino acids of the factor VIII C2 domain are constituents of both the membrane-binding and von Willebrand factor-binding motifs. *J Biol Chem* 2002;**277**:6374–81.
101. Celie PH, Van Stempvoort G, Fribourg C, Schurgers LJ, Lenting PJ, Mertens K. The connecting segment between both epidermal growth factor-like domains in blood coagulation factor IX contributes to stimulation by factor VIIIa and its isolated A2 domain. *J Biol Chem* 2002;**277**:20214–20.
102. Bajaj SP. Region of factor IXa protease domain that interacts with factor VIIIa: analysis of select hemophilia B mutants. *Thromb Haemost* 1999;**82**:218–25.
103. Hopfner KP, Brandstetter H, Karcher A, Kopetzki E, Huber R, Engh RA, et al. Converting blood coagulation factor IXa into factor Xa: dramatic increase in amidolytic activity identifies important active site determinants. *EMBO J* 1997;**16**:6626–35.
104. Kolkman JA, Mertens K. Insertion loop 256-268 in coagulation factor IX restricts enzymatic activity in the absence but not in the presence of factor VIII. *Biochemistry* 2000;**39**:7398–405.
105. Soeda T, Nogami K, Nishiya K, Takeyama M, Ogiwara K, Sakata Y, et al. The factor VIIIa C2 domain (residues 2228-2240) interacts with the factor IXa Gla domain in the factor xase complex. *J Biol Chem* 2009;**284**:3379–88.
106. Wakabayashi H, Griffiths AE, Fay PJ. Factor VIII lacking the C2 domain retains cofactor activity in vitro. *J Biol Chem* 2010;**285**:25176–84.

107. Autin L, Miteva MA, Lee WH, Mertens K, Radtke KP, Villoutreix BO. Molecular models of the procoagulant factor VIIIa-factor IXa complex. *J Thromb Haemost* 2005;**3**:2044–56.
108. O'Brien LM, Mastri M, Fay PJ. Regulation of factor VIIIa by human activated protein C and protein S: inactivation of cofactor in the intrinsic factor Xase. *Blood* 2000;**95**:1714–20.
109. Fay PJ, Coumans JV, Walker FJ. von Willebrand factor mediates protection of factor VIII from activated protein C-catalyzed inactivation. *J Biol Chem* 1991;**266**:2172–7.
110. Plantier JL, Rolli V, Ducasse C, Dargaud Y, Enjolras N, Boukerche H, et al. Activated factor X cleaves factor VIII at arginine 562, limiting its cofactor efficiency. *J Thromb Haemost* 2010;**8**:286–93.
111. Fay PJ, Beattie TL, Regan LM, O'Brien LM, Kaufman RJ. Model for the factor VIIIa-dependent decay of the intrinsic factor Xase. Role of subunit dissociation and factor IXa-catalyzed proteolysis. *J Biol Chem* 1996;**271**:6027–32.
112. Girolami A, Scarparo P, Scandellari R, Allemand E. Congenital factor X deficiencies with a defect only or predominantly in the extrinsic or in the intrinsic system: a critical evaluation. *Am J Hematol* 2008;**83**:668–71.
113. Fuentes-Prior P, Fujikawa K, Pratt KP. New insights into binding interfaces of coagulation factors V and VIII and their homologues lessons from high resolution crystal structures. *Curr Protein Pept Sci* 2002;**3**:313–39.
114. Hockin MF, Jones KC, Everse SJ, Mann KG. A model for the stoichiometric regulation of blood coagulation. *J Biol Chem* 2002;**277**:18322–33.
115. Mann KG, Hockin MF, Begin KJ, Kalafatis M. Activated protein C cleavage of factor Va leads to dissociation of the A2 domain. *J Biol Chem* 1997;**272**:20678–83.
116. Macedo-Ribeiro S, Bode W, Huber R, Quinn-Allen MA, Kim SW, Ortel TL, et al. Crystal structures of the membrane-binding C2 domain of human coagulation factor V. *Nature* 1999;**402**:434–9.
117. Yegneswaran S, Mesters RM, Griffin JH. Identification of distinct sequences in human blood coagulation factor Xa and prothrombin essential for substrate and cofactor recognition in the prothrombinase complex. *J Biol Chem* 2003;**278**:33312–8.
118. Steen M, Tran S, Autin L, Villoutreix BO, Tholander AL, Dahlback B. Mapping of the factor Xa binding site on factor Va by site-directed mutagenesis. *J Biol Chem* 2008;**283**:20805–12.
119. Thorelli E, Kaufman RJ, Dahlback B. Cleavage requirements for activation of factor V by factor Xa. *Eur J Biochem* 1997;**247**:12–20.
120. Rao LV, Rapaport SI. The effect of platelets upon factor Xa-catalyzed activation of factor VII in vitro. *Blood* 1988;**72**:396–401.
121. Nogami K, Wakabayashi H, Fay PJ. Mechanisms of factor Xa-catalyzed cleavage of the factor VIIIa A1 subunit resulting in cofactor inactivation. *J Biol Chem* 2003;**278**:16502–9.
122. Heeb MJ, Griffin JH. Activated protein C-dependent and -independent anticoagulant activities of protein S have different structural requirements. *Blood Cells Mol Dis* 2002;**29**:190–9.
123. Becker RC. Factor Xa-protease activated receptor interface: a paradigm for combination pharmacotherapy. *J Thromb Thrombolysis* 2004;**18**:153–5.
124. Freedman SJ, Blostein MD, Baleja JD, Jacobs M, Furie BC, Furie B. Identification of the phospholipid binding site in the vitamin K-dependent blood coagulation protein factor IX. *J Biol Chem* 1996;**271**:16227–36.
125. Brinkman HJ, Koster P, Mertens K, van Mourik JA. Dissimilar interaction of factor VIII with endothelial cells and lipid vesicles during factor X activation. *Biochem J* 1997;**323** (Pt 3):735–40.
126. Pendurthi UR, Rao LV. Factor VIIa interaction with tissue factor and endothelial cell protein C receptor on cell surfaces. *Semin Hematol* 2008;**45**:S21–24.
127. Griffin CT, Srinivasan Y, Zheng YW, Huang W, Coughlin SR. A role for thrombin receptor signaling in endothelial cells during embryonic development. *Science* 2001;**293**:1666–70.

128. Baird TR, Walsh PN. The interaction of factor XIa with activated platelets but not endothelial cells promotes the activation of factor IX in the consolidation phase of blood coagulation. *J Biol Chem* 2002;**277**:38462–7.
129. Baird TR, Walsh PN. Activated platelets but not endothelial cells participate in the initiation of the consolidation phase of blood coagulation. *J Biol Chem* 2002;**277**:28498–503.
130. Jennings LK. Role of platelets in atherothrombosis. *Am J Cardiol* 2009;**103**:4A–10A.
131. Busso N, Chobaz-Peclat V, Hamilton J, Spee P, Wagtmann N, So A. Essential role of platelet activation via protease activated receptor 4 in tissue factor-initiated inflammation. *Arthritis Res Ther* 2008;**10**:R42.
132. Walsh PN. Platelet coagulation-protein interactions. *Semin Thromb Hemost* 2004;**30**:461–71.
133. Sletnes KE. Role of phospholipids in hemostasis. *Tidsskr Nor Laegeforen* 1993;**113**:2238–41.
134. Yang X, Walsh PN. An ordered sequential mechanism for factor IX and Factor IXa binding to platelet receptors in the assembly of the factor X-activating complex. *Biochem J* 2005;**390**:157–67.
135. Yang X, Chang YJ, Lin SW, Walsh PN. Identification of residues Asn89, Ile90, and Val107 of the factor IXa second epidermal growth factor domain that are essential for the assembly of the factor X-activating complex on activated platelets. *J Biol Chem* 2004;**279**:46400–5.
136. Ahmad SS, London FS, Walsh PN. The assembly of the factor X-activating complex on activated human platelets. *J Thromb Haemost* 2003;**1**:48–59.
137. Wong MY, Gurr JA, Walsh PN. The second epidermal growth factor-like domain of human factor IXa mediates factor IXa binding to platelets and assembly of the factor X activating complex. *Biochemistry* 1999;**38**:8948–60.
138. Reininger AJ. Coagulation activity of platelets. *Hamostaseologie* 2007;**27**:247–50.
139. Nagy Jr. B, Simon Z, Bagoly Z, Muszbek L, Kappelmayer J. Binding of plasma factor XIII to thrombin-receptor activated human platelets. *Thromb Haemost* 2009;**102**:83–9.
140. Schulz C, Leuschen NV, Frohlich T, Lorenz M, Pfeiler S, Gleissner CA, et al. Identification of novel downstream targets of platelet glycoprotein VI activation by differential proteome analysis: implications for thrombus formation. *Blood* 2010;**115**:4102–10.
141. Canobbio I, Balduini C, Torti M. Signalling through the platelet glycoprotein Ib-V-IX complex. *Cell Signal* 2004;**16**:1329–44.
142. Dormann D, Clemetson KJ, Kehrel BE. The GPIb thrombin-binding site is essential for thrombin-induced platelet procoagulant activity. *Blood* 2000;**96**:2469–78.
143. White-Adams TC, Berny MA, Tucker EI, Gertz JM, Gailani D, Urbanus RT, et al. Identification of coagulation factor XI as a ligand for platelet apolipoprotein E receptor 2 (ApoER2). *Arterioscler Thromb Vasc Biol* 2009;**29**:1602–7.
144. Larson PJ, Camire RM, Wong D, Fasano NC, Monroe DM, Tracy PB, et al. Structure/function analyses of recombinant variants of human factor Xa: factor Xa incorporation into prothrombinase on the thrombin-activated platelet surface is not mimicked by synthetic phospholipid vesicles. *Biochemistry* 1998;**37**:5029–38.
145. Schick C, Pemberton PA, Shi GP, Kamachi Y, Cataltepe S, Bartuski AJ, et al. Cross-class inhibition of the cysteine proteinases cathepsins K, L, and S by the serpin squamous cell carcinoma antigen 1: a kinetic analysis. *Biochemistry* 1998;**37**:5258–66.
146. Rau JC, Beaulieu LM, Huntington JA, Church FC. Serpins in thrombosis, hemostasis and fibrinolysis. *J Thromb Haemost* 2007;**5**(Suppl. 1):102–15.
147. O'Malley KM, Nair SA, Rubin H, Cooperman BS. The kinetic mechanism of serpin-proteinase complex formation. An intermediate between the michaelis complex and the inhibited complex. *J Biol Chem* 1997;**272**:5354–9.
148. Huntington JA, Read RJ, Carrell RW. Structure of a serpin-protease complex shows inhibition by deformation. *Nature* 2000;**407**:923–6.

149. Engh RA, Huber R, Bode W, Schulze AJ. Divining the serpin inhibition mechanism: a suicide substrate 'springe'? *Trends Biotechnol* 1995;**13**:503–10.
150. Egeberg O. Inherited antithrombin deficiency causing thrombophilia. *Thromb Diath Haemorrh* 1965;**13**:516–30.
151. Lindo VS, Kakkar VV, Melissari E. Cleaved antithrombin (ATc): a new marker for thrombin generation and activation of the coagulation system. *Br J Haematol* 1995;**89**:157–62.
152. Manithody C, Yang L, Rezaie AR. Role of basic residues of the autolysis loop in the catalytic function of factor Xa. *Biochemistry* 2002;**41**:6780–8.
153. Yang L, Manithody C, Olson ST, Rezaie AR. Contribution of basic residues of the autolysis loop to the substrate and inhibitor specificity of factor IXa. *J Biol Chem* 2003;**278**:25032–8.
154. Johnson DJ, Li W, Adams TE, Huntington JA. Antithrombin-S195A factor Xa-heparin structure reveals the allosteric mechanism of antithrombin activation. *EMBO J* 2006;**25**:2029–37.
155. Olson ST, Bjork I. Regulation of thrombin activity by antithrombin and heparin. *Semin Thromb Hemost* 1994;**20**:373–409.
156. Izaguirre G, Zhang W, Swanson R, Bedsted T, Olson ST. Localization of an antithrombin exosite that promotes rapid inhibition of factors Xa and IXa dependent on heparin activation of the serpin. *J Biol Chem* 2003;**278**:51433–40.
157. Rogers SJ, Pratt CW, Whinna HC, Church FC. Role of thrombin exosites in inhibition by heparin cofactor II. *J Biol Chem* 1992;**267**:3613–7.
158. Skinner R, Abrahams JP, Whisstock JC, Lesk AM, Carrell RW, Wardell MR. The 2.6 Å structure of antithrombin indicates a conformational change at the heparin binding site. *J Mol Biol* 1997;**266**:601–9.
159. Chuang YJ, Swanson R, Raja SM, Olson ST. Heparin enhances the specificity of antithrombin for thrombin and factor Xa independent of the reactive center loop sequence. Evidence for an exosite determinant of factor Xa specificity in heparin-activated antithrombin. *J Biol Chem* 2001;**276**:14961–71.
160. Dementiev A, Petitou M, Herbert JM, Gettins PG. The ternary complex of antithrombin-anhydrothrombin-heparin reveals the basis of inhibitor specificity. *Nat Struct Mol Biol* 2004;**11**:863–7.
161. Han X, Fiehler R, Broze Jr. GJ. Characterization of the protein Z-dependent protease inhibitor. *Blood* 2000;**96**:3049–55.
162. Han X, Huang ZF, Fiehler R, Broze Jr. GJ. The protein Z-dependent protease inhibitor is a serpin. *Biochemistry* 1999;**38**:11073–8.
163. Wei Z, Yan Y, Carrell RW, Zhou A. Crystal structure of protein Z-dependent inhibitor complex shows how protein Z functions as a cofactor in the membrane inhibition of factor X. *Blood* 2009;**114**:3662–7.
164. Huang X, Dementiev A, Olson ST, Gettins PG. Basis for the specificity and activation of the serpin ZPI as an inhibitor of membrane-associated factor Xa. *J Biol Chem* 2010;**285**:20399–409.
165. Huang X, Swanson R, Broze Jr. GJ, Olson ST. Kinetic characterization of the protein Z-dependent protease inhibitor reaction with blood coagulation factor Xa. *J Biol Chem* 2008;**283**:29770–83.
166. Moses E, Hinz HJ. Basic pancreatic trypsin inhibitor has unusual thermodynamic stability parameters. *J Mol Biol* 1983;**170**:765–76.
167. Schwarz H, Hinz HJ, Mehlich A, Tschesche H, Wenzel HR. Stability studies on derivatives of the bovine pancreatic trypsin inhibitor. *Biochemistry* 1987;**26**:3544–51.
168. Burgering MJ, Orbons LP, van der Doelen A, Mulders J, Theunissen HJ, Grootenhuis PD, et al. The second Kunitz domain of human tissue factor pathway inhibitor: cloning, structure determination and interaction with factor Xa. *J Mol Biol* 1997;**269**:395–407.

169. Bode W, Huber R. Proteinase-protein inhibitor interaction. *Biomed Biochim Acta* 1991;**50**:437–46.
170. Bode W, Huber R. Natural protein proteinase inhibitors and their interaction with proteinases. *Eur J Biochem* 1992;**204**:433–51.
171. Hanson WM, Beeser SA, Oas TG, Goldenberg DP. Identification of a residue critical for maintaining the functional conformation of BPTI. *J Mol Biol* 2003;**333**:425–41.
172. Vincent JP, Lazdunski M. Trypsin-pancreatic trypsin inhibitor association. Dynamics of the interaction and role of disulfide bridges. *Biochemistry* 1972;**11**:2967–77.
173. Broze Jr. GJ, Warren LA, Novotny WF, Higuchi DA, Girard JJ, Miletich JP. The lipoprotein-associated coagulation inhibitor that inhibits the factor VII-tissue factor complex also inhibits factor Xa: insight into its possible mechanism of action. *Blood* 1988;**71**:335–43.
174. Girard TJ, Warren LA, Novotny WF, Likert KM, Brown SG, Miletich JP, et al. Functional significance of the Kunitz-type inhibitory domains of lipoprotein-associated coagulation inhibitor. *Nature* 1989;**338**:518–20.
175. Broze Jr. GJ, Lange GW, Duffin KL, MacPhail L. Heterogeneity of plasma tissue factor pathway inhibitor. *Blood Coagul Fibrinolysis* 1994;**5**:551–9.
176. Hackeng TM, Maurissen LF, Castoldi E, Rosing J. Regulation of TFPI function by protein S. *J Thromb Haemost* 2009;**7**(Suppl. 1):165–8.
177. Baugh RJ, Broze Jr. GJ, Krishnaswamy S. Regulation of extrinsic pathway factor Xa formation by tissue factor pathway inhibitor. *J Biol Chem* 1998;**273**:4378–86.
178. Ndonwi M, Tuley EA, Broze GJ. The Kunitz-3 domain of TFPI{alpha} is required for protein S-dependent enhancement of factor Xa inhibition. *Blood* 2010; **116**:1344–51.
179. Fernandes N, Mosnier LO, Tonnu L, Heeb MJ. Zn^{2+}-containing protein S inhibits extrinsic factor X activating complex independently of tissue factor pathway inhibitor (TFPI). *J Thromb Haemost* 2010; **8**:1976–85.
180. Van Nostrand WE, Schmaier AH, Farrow JS, Cines DB, Cunningham DD. Protease nexin-2/amyloid beta-protein precursor in blood is a platelet-specific protein. *Biochem Biophys Res Commun* 1991;**175**:15–21.
181. Van Nostrand WE, Schmaier AH, Farrow JS, Cunningham DD. Protease nexin-II (amyloid beta-protein precursor): a platelet alpha-granule protein. *Science* 1990;**248**:745–8.
182. Van Nostrand WE, Wagner SL, Farrow JS, Cunningham DD. Immunopurification and protease inhibitory properties of protease nexin-2/amyloid beta-protein precursor. *J Biol Chem* 1990;**265**:9591–4.
183. Wolfe MS, Guenette SY. APP at a glance. *J Cell Sci* 2007;**120**:3157–61.
184. Xu F, Davis J, Miao J, Previti ML, Romanov G, Ziegler K, et al. Protease nexin-2/amyloid beta-protein precursor limits cerebral thrombosis. *Proc Natl Acad Sci USA* 2005;**102**:18135–40.
185. Xu F, Previti ML, Nieman MT, Davis J, Schmaier AH, Van Nostrand WE. AbetaPP/APLP2 family of Kunitz serine proteinase inhibitors regulate cerebral thrombosis. *J Neurosci* 2009;**29**:5666–70.
186. Navaneetham D, Jin L, Pandey P, Strickler JE, Babine RE, Abdel-Meguid SS, et al. Structural and mutational analyses of the molecular interactions between the catalytic domain of factor XIa and the Kunitz protease inhibitor domain of protease nexin 2. *J Biol Chem* 2005;**280**:36165–75.
187. Schmaier AH, Dahl LD, Rozemuller AJ, Roos RA, Wagner SL, Chung R, et al. Protease nexin-2/amyloid beta protein precursor. A tight-binding inhibitor of coagulation factor IXa. *J Clin Invest* 1993;**92**:2540–5.
188. Mahdi F, Van Nostrand WE, Schmaier AH. Protease nexin-2/amyloid beta-protein precursor inhibits factor Xa in the prothrombinase complex. *J Biol Chem* 1995;**270**:23468–74.

189. Grutter MG, Priestle JP, Rahuel J, Grossenbacher H, Bode W, Hofsteenge J, et al. Crystal structure of the thrombin-hirudin complex: a novel mode of serine protease inhibition. *EMBO J* 1990;**9**:2361–5.
190. Rydel TJ, Ravichandran KG, Tulinsky A, Bode W, Huber R, Roitsch C, et al. The structure of a complex of recombinant hirudin and human alpha-thrombin. *Science* 1990;**249**:277–80.
191. Isawa H, Yuda M, Yoneda K, Chinzei Y. The insect salivary protein, prolixin-S, inhibits factor IXa generation and Xase complex formation in the blood coagulation pathway. *J Biol Chem* 2000;**275**:6636–41.
192. Gudderra NP, Ribeiro JM, Andersen JF. Structural determinants of factor IX(a) binding in nitrophorin 2, a lipocalin inhibitor of the intrinsic coagulation pathway. *J Biol Chem* 2005;**280**:25022–8.
193. Andersen JF, Montfort WR. The crystal structure of nitrophorin 2. A trifunctional antihemostatic protein from the saliva of *Rhodnius prolixus*. *J Biol Chem* 2000;**275**:30496–503.
194. Mizuno H, Fujimoto Z, Koizumi M, Kano H, Atoda H, Morita T. Crystal structure of coagulation factor IX-binding protein from habu snake venom at 2.6 Å: implication of central loop swapping based on deletion in the linker region. *J Mol Biol* 1999;**289**:103–12.
195. Mizuno H, Fujimoto Z, Koizumi M, Kano H, Atoda H, Morita T. Structure of coagulation factors IX/X-binding protein, a heterodimer of C-type lectin domains. *Nat Struct Biol* 1997;**4**:438–41.
196. Mizuno H, Fujimoto Z, Atoda H, Morita T. Crystal structure of an anticoagulant protein in complex with the Gla domain of factor X. *Proc Natl Acad Sci USA* 2001;**98**:7230–4.
197. Francischetti IM, Valenzuela JG, Andersen JF, Mather TN, Ribeiro JM. Ixolaris, a novel recombinant tissue factor pathway inhibitor (TFPI) from the salivary gland of the tick, *Ixodes scapularis*: identification of factor X and factor Xa as scaffolds for the inhibition of factor VIIa/tissue factor complex. *Blood* 2002;**99**:3602–12.
198. Monteiro RQ, Rezaie AR, Bae JS, Calvo E, Andersen JF, Francischetti IM. Ixolaris binding to factor X reveals a precursor state of factor Xa heparin-binding exosite. *Protein Sci* 2008;**17**:146–53.
199. Waxman L, Smith DE, Arcuri KE, Vlasuk GP. Tick anticoagulant peptide (TAP) is a novel inhibitor of blood coagulation factor Xa. *Science* 1990;**248**:593–6.
200. van de Locht A, Stubbs MT, Bode W, Friedrich T, Bollschweiler C, Hoffken W, et al. The ornithodorin-thrombin crystal structure, a key to the TAP enigma? *EMBO J* 1996;**15**:6011–7.
201. Wei A, Alexander RS, Duke J, Ross H, Rosenfeld SA, Chang CH. Unexpected binding mode of tick anticoagulant peptide complexed to bovine factor Xa. *J Mol Biol* 1998;**283**:147–54.
202. Krishnaswamy S, Vlasuk GP, Bergum PW. Assembly of the prothrombinase complex enhances the inhibition of bovine factor Xa by tick anticoagulant peptide. *Biochemistry* 1994;**33**:7897–907.
203. Stassens P, Bergum PW, Gansemans Y, Jespers L, Laroche Y, Huang S, et al. Anticoagulant repertoire of the hookworm *Ancylostoma caninum*. *Proc Natl Acad Sci USA* 1996;**93**:2149–54.
204. Rios-Steiner JL, Murakami MT, Tulinsky A, Arni RK. Active and exo-site inhibition of human factor Xa: structure of des-Gla factor Xa inhibited by NAP5, a potent nematode anticoagulant protein from *Ancylostoma caninum*. *J Mol Biol* 2007;**371**:774–86.
205. Ehrenforth S, Kreuz W, Scharrer I, Linde R, Funk M, Gungor T, et al. Incidence of development of factor VIII and factor IX inhibitors in haemophiliacs. *Lancet* 1992;**339**:594–8.
206. Hollinger FB, Dolana G, Thomas W, Gyorkey F. Reduction in risk of hepatitis transmission by heat-treatment of a human factor VIII concentrate. *J Infect Dis* 1984;**150**:250–62.
207. Cervia JS, Sowemimo-Coker SO, Ortolano GA, Wilkins K, Schaffer J, Wortham ST. An overview of prion biology and the role of blood filtration in reducing the risk of transfusion-transmitted variant Creutzfeldt-Jakob disease. *Transfus Med Rev* 2006;**20**:190–206.
208. Houston F, McCutcheon S, Goldmann W, Chong A, Foster J, Siso S, et al. Prion diseases are efficiently transmitted by blood transfusion in sheep. *Blood* 2008;**112**:4739–45.

209. Galeffi P, Brownlee GG. The propeptide region of clotting factor IX is a signal for a vitamin K dependent carboxylase: evidence from protein engineering of amino acid -4. *Nucleic Acids Res* 1987;**15**:9505–13.
210. Suttie JW. Mechanism of action of vitamin K: synthesis of gamma-carboxyglutamic acid. *CRC Crit Rev Biochem* 1980;**8**:191–223.
211. Wasley LC, Rehemtulla A, Bristol JA, Kaufman RJ. PACE/furin can process the vitamin K-dependent pro-factor IX precursor within the secretory pathway. *J Biol Chem* 1993;**268**:8458–65.
212. White 2nd GC, Pickens EM, Liles DK, Roberts HR. Mammalian recombinant coagulation proteins: structure and function. *Transfus Sci* 1998;**19**:177–89.
213. Blostein MD, Rigby AC, Furie BC, Furie B, Gilbert GE. Amphipathic helices support function of blood coagulation factor IXa. *Biochemistry* 2000;**39**:12000–6.
214. Ganopolsky JG, Charbonneau S, Peng HT, Shek PN, Blostein MD. Characterization of an ideal amphipathic peptide as a procoagulant agent. *Biochem J* 2008;**412**:545–51.
215. Lopez JA, Chen J. Pathophysiology of venous thrombosis. *Thromb Res* 2009;**123**(Suppl. 4): S30–34.
216. Girolami A, Scandellari R, Zanon E, Sartori R, Girolami B. Non-catheter associated venous thrombosis in hemophilia A and B. A critical review of all reported cases. *J Thromb Thrombolysis* 2006;**21**:279–84.
217. Chen T, Lam S. Rivaroxaban: an oral direct factor Xa inhibitor for the prevention of thromboembolism. *Cardiol Rev* 2009;**17**:192–7.
218. Batorova A. Advances in antithrombotic treatment—Antithrombotics with anti-Xa effect. *Vnitr Lék* 2009;**55**:295–301.
219. Turpie AG, Bauer KA, Davidson BL, Fisher WD, Gent M, Huo MH, et al. A randomized evaluation of betrixaban, an oral factor Xa inhibitor, for prevention of thromboembolic events after total knee replacement (EXPERT). *Thromb Haemost* 2009;**101**:68–76.
220. Agnelli G, Eriksson BI, Cohen AT, Bergqvist D, Dahl OE, Lassen MR, et al. Safety assessment of new antithrombotic agents: lessons from the EXTEND study on ximelagatran. *Thromb Res* 2009;**123**:488–97.

Structure and Proteolytic Properties of ADAMTS13, A Metalloprotease Involved in the Pathogenesis of Thrombotic Microangiopathies

STEFANO LANCELLOTTI AND
RAIMONDO DE CRISTOFARO

Institute of Internal Medicine and Geriatrics, Physiopathology of Haemostasis Research Center, Catholic University School of Medicine, Rome, Italy

I. Introduction 106
A. Brief History of ADAMTS13 106
B. Von Willebrand Factor: The Only Recognized Substrate of ADAMTS13 107
II. Gene, Biosynthesis and Secretion of ADAMTS13 109
III. ADAMTS13 Structure and Function: General Aspects 109
IV. ADAMTS13: A Multidomain Enzyme 111
A. The Propeptide Domain (Residues 34–74) 111
B. The Metalloprotease Domain (Residues 75–289) 112
C. The Calcium Binding Sites 114
D. The Disintegrin-Like Domain (Residues 290–385) 115
E. The Thrombospondin Type 1-Like Repeat (TSR) 115
F. The Cysteine-Rich Domain (Residues 440–555) 116
G. The Spacer Domain (Residues 556–685) 116
H. The CUB Domains (Residues 1192–1408) 117
V. Regulation of ADAMTS13 Activity 118
VI. ADAMTS13 and Its Role in the Pathogenesis of Thrombotic Microangiopathy, A Pleiomorphic Clinical Setting 120
A. Relationship Between ADAMTS13 and Occurrence of TMAs 122
B. Congenital ADAMTS13 Deficiency 123
C. Inhibitors of ADAMTS13 130
VII. Are ADAMTS13 Changes Always Associated with Thrombotic Microangiopathies? 132
VIII. Issues Concerning Laboratory Methods Used to Measure the ADAMTS13 Activity in Clinical Samples 133
IX. Future Directions 134
References 134

ADAMTS13 is a 190-kDa zinc protease encoded by a gene located on chromosome 9q34. This protease specifically hydrolyzes von Willebrand factor (VWF) multimers, thus causing VWF size reduction. ADAMTS13 belongs to

Progress in Molecular Biology
and Translational Science, Vol. 99
DOI: 10.1016/S1877-1173(11)99003-6

the *A Disintegrin And Metalloprotease with ThromboSpondin* type 1 repeats (ADAMTS) family, involved in proteolytic processing of many matrix proteins. ADAMTS13 consists of numerous domains, including a metalloprotease domain, a disintegrin domain, several thrombospondin type 1 (TSP1) repeats, a cysteine-rich domain, a spacer domain, and two CUB (*C*omplement c1r/c1s, sea *U*rchin epidermal growth factor, and *B*one morphogenetic protein) domains. ADAMTS13 cleaves a single peptide bond (Tyr_{1605}–Met_{1606}) in the central A2 domain of the VWF molecule. This proteolytic cleavage is essential to reduce the size of ultralarge VWF polymers, which, when exposed to high shear stress in the microcirculation, are prone to form platelets clumps, which cause severe syndromes called thrombotic microangiopathies (TMAs). In this chapter, we (a) discuss the current knowledge of structure–function aspects of ADAMTS13 and its involvement in the pathogenesis of TMAs, (b) address the ongoing controversies, and (c) indicate the direction of future investigations.

I. Introduction

A. Brief History of ADAMTS13

The discovery of the metalloprotease referred to as ADAMTS13 (*A Disintegrin-like And Metalloprotease with ThromboSpondin type 1 motif 13*), as many other examples in biomedical research, found its way in the attempt to address the issue concerning the pathogenesis of severe forms of thrombotic microangiopathies (TMAs). The latter are a group of severe diseases characterized by deposition of blood platelet thrombi in the microcirculation, responsible for potentially fatal multiorgan failure. Moake *et al.*[1] reported in 1982 the first evidence that the pathogenesis of the main form of microangiopathy, that is, Thrombotic Thrombocytopenic Purpura (TTP), arises from a defect in proteolytic processing of von Willebrand factor (VWF), a multimeric glycoprotein with very high molecular weight that plays an essential role in platelet-dependent hemostasis. VWF mediates platelet adhesion to damaged blood vessels through interactions with glycoprotein Ibα on the platelet membrane and collagen in the subendothelial matrix and contributes to platelet aggregation through interactions with integrin $\alpha_{IIb}\beta_3$, the mechanisms of primary haemostasis.[2,3] Moake *et al.* observed the presence of potentially thrombogenic forms of VWF, characterized by ultralarge multimers of VWF (UL-VWF) in the plasma of four patients with chronic relapsing TTP. In 1996, two groups independently reported a metalloprotease that specifically cleaves VWF at the Tyr_{1605}–Met_{1606} bond in the A2 domain[4,5] (Fig. 1). The proteolytic activity required VWF in a denatured conformation, achieved by preincubation with either low-concentration guanidine-HCl[5] or urea,[4] or by exposure to high shear stress *in vitro*.[5] The proteolysis also required divalent cations such as

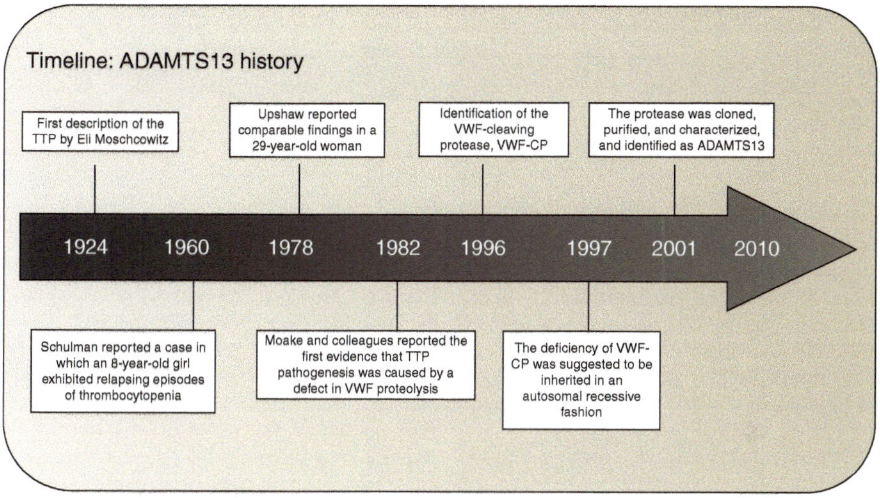

FIG. 1. Timeline of ADAMTS13 history concerning its discovery and involvement in thrombotic microangiopathies.

Ba^{2+}, Zn^{2+}, Ca^{2+}, or Co^{2+}.[5] A few years later, the protease was cloned, purified, and characterized, and several groups identified the VWF-cleaving protease as ADAMTS13, a novel member of the ADAMTS family of metalloproteases.[6–10] Before analyzing the structure and functional properties of ADAMTS13, a brief description of the enzyme's substrate, that is VWF, will be provided in Section I.B.

B. Von Willebrand Factor: The Only Recognized Substrate of ADAMTS13

Considerable evidence now implicates the haemostatic protein VWF as a key component in TTP pathogenesis.[11] VWF is an abundant plasma glycoprotein synthesized in all vascular endothelial cells and megakaryocytes as a precursor containing a signal peptide and large propeptide.[2,12] The prepro-VWF undergoes a maturation process in the rough endoplasmatic reticulum and in the Golgi complex. In the rough endoplasmatic reticulum, a monomer of proVWF (275 kDa) forms dimers via disulfide bonds at the carboxyl terminus, and it is known as "tail-to-tail" dimerization. After transport to the Golgi apparatus, proVWF dimers form homopolymers or multimers through an additional disulfide bond near the amino terminus of the mature subunit, a process termed "head-to-head" multimerization. After multimerization, the propeptide is cleaved by furin, a calcium-dependent serine protease. The multimeric VWF is stored in highly ordered tubules within storage organelles

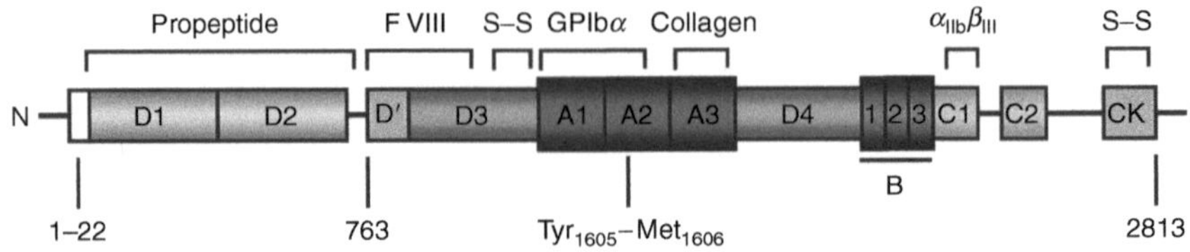

FIG. 2. Scheme of von Willebrand factor monomer molecule with its functional domains. The prepro-VWF polypeptide is indicated with amino acids numbered from the amino- (aa 1) to carboxy-terminal portions (aa 2813). Binding sites are indicated for factor VIII (D′ and D3 domains), platelet glycoprotein Ibα (GPIbα) (A1 domain), collagen (A1 and A3 domains), and integrin $\alpha_{IIb}\beta_{III}$ (RGDS sequence within the C1 domain). The cleavage site (Tyr_{1605}–Met_{1606}) for ADAMTS13 is located at the central A2 domain of von Willebrand factor. The locations of intersubunit disulfide bonds (S–S) are shown in the CK and D3 domains, which are important for the formation of VWF dimers and multimers, respectively.

referred to as Weibel–Palade bodies of endothelial cells and α-granules of platelets.[13] Endothelial cell VWF is secreted via both constitutive and regulated pathways. In response to a variety of stimuli, VWF is released from endothelial cells as UL-VWF, which can be up to ~20,000 kDa in size[14,15] and are the most adhesive and reactive forms of VWF. UL-VWF form string-like structures attached to the endothelial cell surface, perhaps through interaction with P-selectin.[16] Under fluid shear stress, the UL-VWF strings are cleaved by ADAMTS13 at the Tyr_{1605}–Met_{1606} bond in the A2 domain[5] to generate the range of VWF multimer sizes that normally circulate in the blood.

VWF serves as the primary adhesive link between platelets and subendothelium, and it also carries and stabilizes coagulation factor VIII (FVIII) in the circulation. These hemostatic functions depend upon the ability of VWF to bind circulating factor VIII, subendothelial collagens, platelet glycoprotein Ibα (GPIbα), and integrin $\alpha_{IIb}\beta_{III}$, but the regulation of platelet adhesion depends upon cleavage of VWF multimers by ADAMTS13[17] (Fig. 2). However, VWF in plasma adopts a compact conformation that does not bind to platelet GPIbα and is not cleaved by ADAMTS13. Fluid shear stress,[18] or binding to certain surfaces, changes the conformation of VWF so that it assumes an elongated form, disclosing the buried binding site for platelet GPIbα, localized in the A1 domain of the protein. Upon this physically induced conformational transition, VWF multimers bind tightly to platelet GPIbα and, at the same time, can be recognized by ADAMTS13. A similar modulating effect *in vitro* is achieved by including antibiotic ristocetin or by denaturing reagents such as urea and guanidine–HCl.[4,5] There are also experimental evidences showing that *in vivo* the stretched conformer of VWF, more prone to ADAMTS13 proteolysis, is stabilized through the interaction with P-selectin,[19] a glycoprotein secreted by endothelial cells. In addition, certain mutations in the VWF protein increase the proteolysis of VWF by ADAMTS13 and reduce the size of VWF multimers, leading to the compromised hemostatic function in patients with von Willebrand disease.[17,20] Conversely, inability to cleave the newly released UL-VWF multimers[1,21,22] owing to hereditary or acquired deficiency of plasma ADAMTS13 activity may induce spontaneous VWF-dependent platelet adhesion and aggregation,[23] leading to disseminated microvascular thrombosis as seen in patients with TTP.

II. Gene, Biosynthesis and Secretion of ADAMTS13

The human *ADAMTS13* gene is located on chromosome 9 at position 9q34. It spans 37 kb in length and contains 29 exons.[8,10] ADAMTS13 mRNA is approximately 5 kb and encodes a 1427 amino acid protein. Several alternatively spliced mRNA variants have been characterized; their significance remains unknown.[8,10] The predicted molecular weight of 145 kDa differs from the observed molecular mass of purified plasma ADAMTS13 ($\sim$ 190 kDa),[6,7] and this difference is likely due to its extensive glycosylation.[24]

ADAMTS13 is synthesized predominantly in the liver,[7–10] although variable expression has been observed in endothelial cells,[25,26] megakaryocytes, or platelets[27,28] and secreted into plasma as an already active enzyme.

Mutations in the *ADAMTS13* gene[25] may result in a reduced or an aberrant secretion of ADAMTS13 protein into the circulation. Various truncated forms of ADAMTS13 are detectable in plasma,[29] perhaps owing to alternative splicing of ADAMTS13 mRNA or proteolysis of ADAMTS13 by serine proteases such as thrombin[30] and leukocyte elastase.[31] Human placenta and skeletal muscle synthesize a 2.4-kb ADAMTS13 mRNA.[10] There are some evidences from *in vivo*[32,33] and *in vitro*[33,34] studies that ADAMTS13 mRNA and protein are produced in liver hepatic stellate cells, which lie between the hepatocytes and endothelial cells. However, the contribution of hepatic stellate cells to plasma levels of ADAMTS13 remains to be determined. Considering the large surface area of vascular endothelial beds, plasma ADAMTS13 might be derived mainly from endothelial cells even though each endothelial cell produces little amounts of ADAMTS13 compared to hepatic stellate cells.[25,26]

III. ADAMTS13 Structure and Function: General Aspects

ADAMTS13 is the 13th member of the ADAMTS family of zinc proteases, which is related to the large ADAM (*A* *D*isintegrin *A*nd *M*etalloprotease) family. The ADAMTS family of zinc metalloproteases contains 19 members that share the common structure of a hydrophobic signal sequence, a

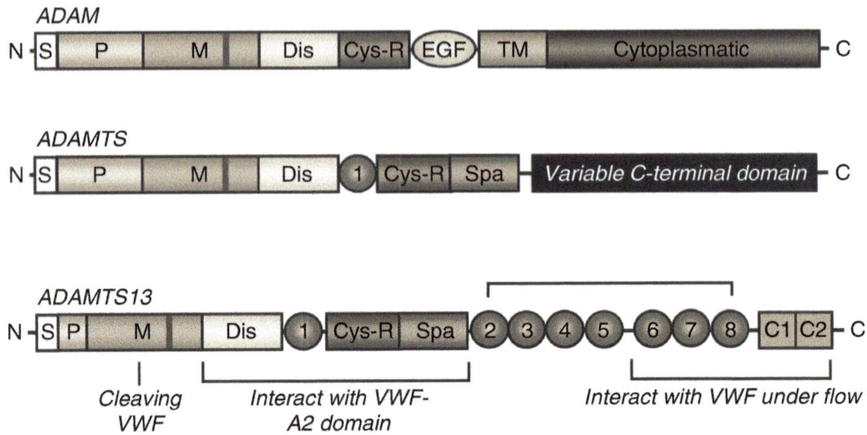

FIG. 3. Schematic diagram of ADAM, ADAMTS, and ADAMTS13 structure. The structural domains are indicated: signal peptide (S), propeptide (P), metalloprotease (M) (location of zinc-binding motif shown in red), disintegrin domain (Dis), first thrombospondin type 1 (TSP1) repeat (1), cysteine-rich domain (Cys-R), spacer domain (Spa), the second to eighth TSP1 repeats(2) through (8), and two CUB domains (C1 and C2). The metalloprotease domain is the catalytic center that cleaves von Willebrand factor (VWF). The proximal carboxyl-terminal domains from Dis to Spa interact with the A2 domain of VWF. More distal carboxyl-terminal domains (TSP1 2–8) interact with VWF under fluid shear stress. EGF indicates epidermal growth factor-like repeat and TM, transmembrane domain. (See Color Insert.)

propeptide, a metalloprotease domain, a thrombospondin type 1 (TSP1) repeat, a disintegrin-like domain, a cysteine-rich domain, and a spacer domain.[8,10] In contrast to ADAM proteases, ADAMTSs lack EGF-like repeats and a transmembrane domain and therefore are secreted rather than membrane-bound enzymes. In addition, all ADAMTS family members possess one or more thrombospondin type 1 (TSP1) motifs[35] and variable additional C-terminal domains. The carboxyl terminus of ADAMTS13 contains seven more TSP1 repeats and two CUB domains, which are named after motifs first identified in *C*omplement components C1r and C1s, sea urchin protein *U*egf, and *B*one morphogenetic protein-1[36] (Fig. 3).

Globally, the family of ADAMTS is composed of extracellular, multidomain enzymes whose main functions include (1) collagen processing as procollagen N-proteinase; (2) cleavage of the matrix proteoglycans aggrecan, versican, and brevican; (3) inhibition of angiogenesis; and (4) blood coagulation homoeostasis as the von Willebrand factor cleaving protease. Roles in organogenesis, inflammation, and fertility are also apparent. Some *ADAMTS* genes have been found to show altered expression in arthritis and various types of cancer. For instance,

ADAMTS2 cleaves the propeptide of collagen II, and mutations in this protein are responsible for the Ehlers–Danlos syndrome type VII C.[37] Mutations in *ADAMTS10* cause autosomal recessive Weill–Marchesani syndrome, a connective tissue disorder characterized by abnormalities of the lens of the eye, proportionate short stature, brachydactyly, and joint stiffness.[38] Notably, some mutations in the fibrillin gene can cause the autosomal dominant form of this disease,[39] suggesting that fibrillin may be a substrate of ADAMTS10. ADAMTS3 and ADAMTS14, which show homology to ADAMTS2, can also cleave procollagen, suggesting possible redundant functions or as yet unrecognized differences in substrate specificity.[37] ADAMTS1, ADAMTS4, and ADAMTS5/11 (also known as aggrecanases) cleave the cartilage proteoglycan aggrecan and may play a role in inflammatory joint disease.[40–42] Interestingly, an anti-inflammatory role has also been recently attributed to ADAMTS13.[43] In mice, disruption of the *ADAMTS1* gene results in growth retardation and several developmental defects, including adipose tissue malformation, histological changes in the uterus and ovaries, enlarged renal calices with fibrotic changes, and abnormal adrenal medullary architecture.[41] Since the isolation and cloning of the ADAMTS13 cDNA, several laboratories have expressed recombinant ADAMTS13 in cell culture. Recombinant ADAMTS13 cleaves VWF *in vitro*, providing a formal demonstration that ADAMTS13 is indeed the VWF-cleaving protease identified in earlier studies.[7,24] Furthermore, several groups have gone on to perform structure/function analyses of ADAMTS13 to determine the structural requirements for its activity.

IV. ADAMTS13: A Multidomain Enzyme

The detailed structure of the full-length ADAMTS13 molecule has not yet been solved. Only recently, the X-ray diffraction map of the recombinant ADAMTS13 fragment composed of the thrombospondin-1 (TSP-1) type-1 repeat domain (T), the cysteine-rich (C) region, and the spacer domain (S) has been reported.[44] A considerable wealth of data showed the relevance of these noncatalytic exosites of ADAMTS13 for the molecular recognition and proteolytic processing of VWF multimers. Thus, the knowledge of the TCS fragment is extremely important for understanding the role of these domains in substrate recognition by the entire family of ADAMTS enzymes.

A. The Propeptide Domain (Residues 34–74)

The propeptide of ADAMTS13 contains 41 amino acids, in contrast to the ~200 amino acids that comprise the propeptides of most other members of ADAM and ADAMTS family.[45,46] Like other proteases, ADAMTS13 propeptide presents a typical proprotein processing site (RQRR), which has been shown to be a furin cleavage site.[10] In addition, the propeptide of ADAMTS13 does not appear to contain a potential "cysteine-switch" motif. This motif functions in other metalloproteases to inhibit protease activity by coordinating with the active site Zn^{2+} in the protease domain to keep the molecule in an inactive zymogen form until cleavage of the propeptide (by a furin-like enzyme), concurrent with secretion from the cell.[47] At variance with what has been observed for other metalloproteases, deletion of the ADAMTS13 propeptide does not impair secretion or enzymatic activity, demonstrating that the propeptide is not required for folding or secretion and likely does not confer enzymatic latency.[46] Moreover, it has been shown that a mutation in the furin consensus recognition site leads to secretion of an active pro-ADAMTS13.[46] Detection of antipropeptide antibodies in some patients with TTP suggests that not all plasma ADAMTS13 have this sequence removed.[48] However, the ADAMTS13 propeptide sequence has been conserved across a wide evolutionary distance (from fish to mammals), suggesting that it may serve a yet unknown function.

B. The Metalloprotease Domain (Residues 75–289)

As anticipated above, no detailed knowledge of the atomic structure of the metalloprotease domain of ADAMTS13 is available. The metalloprotease domain of ADAMTSs consists of about 200 amino acids. The structural relationship of ADAMTSs to other zinc matrix metalloproteinases (MMPs) is shown in Fig. 4. ADAMTSs are reprolysin-like proteins, which, together with ADAMs, MMPs, astacins, and serralysins, constitute the metzincin superfamily. The catalytic domains of ADAMTS proteinases share a high degree of similarity and contain the zinc-binding sequence, in which the catalytic Zn^{2+} ion is coordinated by the three histidine residues, "$H_{224}EXXHXXGXXHD_{235}$," where "X" represents any amino acid residue and the conserved aspartic acid residue distinguishes the ADAMs and ADAMTSs from other metalloproteinases. The glutamate following the first zinc-binding histidine has a catalytic role,[49] polarizing a water molecule through hydrogen bonding, which is stabilized by coordination with the Zn^{2+} ion and is responsible for the nucleophilic attack on the carbonyl of the substrate scissile peptide bond.[50] The arrangement of this active site motif is facilitated by the conserved glycine 231, which permits a tight hairpin loop and enables the third histidine to occupy its correct position.[51,52] As in all MMPs and adamalysins, the zinc-binding sequence is followed a short distance from the C-terminal end (10–20 amino acids after the third histidine[53]), by a conserved methionine residue, an active-site arrangement that has been termed "metzincin-type." This methionine constitutes the "Met-turn," a tight turn arranged as a right-handed screw that seems to serve an important function in the structure of the active site.[51] It could form indeed a hydrophobic base beneath the catalytic Zn^{2+}. Different studies using

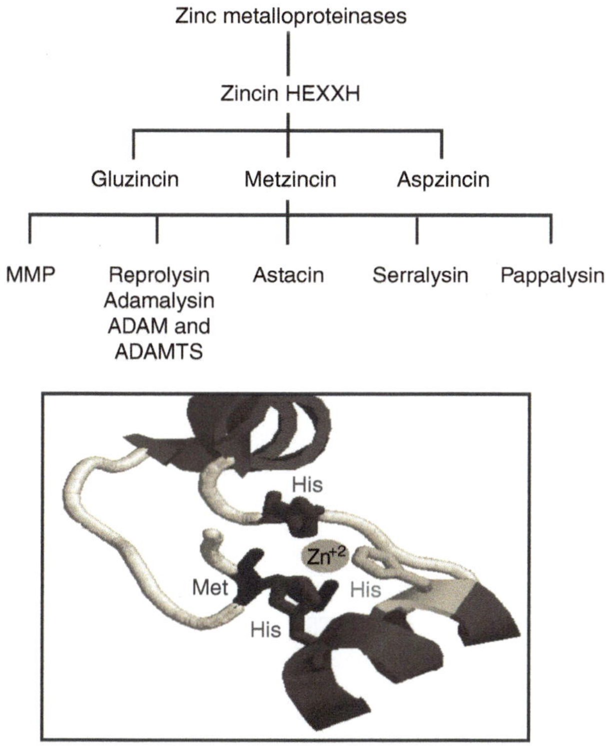

FIG. 4. The zinc metalloproteinases of the zincin type that have the minimal catalytic zinc-binding motif containing two histidine residues flanking the catalytic glutamate, HEXXH, comprise three superfamilies: the gluzincins, the aspzincins, and the metzincins. Within the metzincins, the major families are the matrixins or matrix metalloproteinases (MMPs), the reprolysins, also known as adamalysins, which include some ADAM (a disintegrin and metalloproteinase) and ADAMTs (ADAMs with thrombospondin repeats proteins), and the astacins. Metzincins have an HEXXHXXGXXHZ...M motif with three histidine residues binding the zinc ion and an invariant methionine turn in the active site that generates the name metzincins. X represents any amino acid residue and Z indicates a subfamily specific conserved residue, which is D for both ADAM and ADAMTS members. (Inset) Homology modeling of the metalloproteinase (M) domain of ADAMTS-13. The structure was generated using the program RasMol versus. 2.7.5. The structure of the polypeptide chain 80–290, corresponding to the M-domain of ADAMTS13, was modeled by homology on the crystallographic structure of ADAMTS4 solved at 2.80 Å (PDB entry code: 2RJP). Zinc ion (green) is shown together with the three catalytic His-residues. The "Met-turn" typical of the metzincin family is also indicated. (See Color Insert.)

C-terminal truncations of recombinant ADAMTS13 have shown that the metalloprotease domain alone was not able to cleave plasma VWF[54,55] or a short peptide VWF73, which contains 73 amino acid residues from the central

A2 domain of VWF.[56] Truncation of ADAMTS13 within or distal to TSP1 results in generation of enzymes that retain VWF-cleaving activity *in vitro*, while truncations proximal to TSP1 (within the protease, TSP1, cysteine-rich or spacer domains) result in an inactive protein. However, it was found that the metalloprotease domain alone could cleave the substrate after long incubation times (16–24 h), but at a different site than the predicted Tyr_{1605}–Met_{1606}.[56] These results indicate that the protease domain alone, even if functional, is not sufficient to recognize and specifically cleave the VWF cleavage site, suggesting that sequences within the region spanning the protease domain to the spacer domain of ADAMTS13 are sufficient for VWF-cleaving activity, at least *in vitro*.

C. The Calcium Binding Sites

Another divalent cation, which plays an essential role in ADAMTS13 activity, is calcium. The conserved residues Glu_{83}, Asp_{173}, Cys_{281}, and Asp_{284} are predicted to coordinate a structural Ca^{2+} ion.[10] From the structure-based alignment of ADAMTS13 with other ADAMTS enzymes of known three-dimensional structure, it is apparent that most of the amino acids involved in the calcium binding sites (CaBSs) of ADAMTS1, ADAMTS4, and ADAMTS5 are also conserved in ADAMTS13. Hence, two putative CaBSs, CaBS-I and CaBS-II, can be identified in ADAMTS13. Using the geometry of CaBSs in the ADAMTS5 structure (pdb file 3b8z), our research group recently found an ADAMTS13 model showing that CaBS-I is formed by highly conserved amino acids (Asp165, Asp173, Glu83) and by a connector loop consisting of Cys281 and Asp284. It is positioned at about 25 Å from the active site and, as observed in ADAMTS1, ADAMTS4, and ADAMTS5, can accommodate two Ca^{2+} ions (Pozzi *et al.*, submitted for publication). Alternatively, CaBS-II binds a single Ca^{2+} ion, is positioned opposite to CaBS-I, and is adjacent to the active site ($\sim$13 Å from the catalytic zinc). CaBS-II is formed by the carboxylate of Asp182, the carbonyl oxygens of Leu183, Arg190, and Val192 and the carboxylate of Glu212 (Pozzi *et al.*, submitted for publication). Of these, only Asp182, Leu183, and Glu212 are conserved in the ADAMTS family. Our model is consistent with recent biochemical and mutagenesis studies[57] showing that ADAMTS13 contains a low-affinity CaBS-I ($EC_{50} = 880$ μM), which is occupied by two Ca^{2+} ions and mainly exerts a structural role, and a high-affinity site (K_D(app) = 70 μM), CaBS-II, which plays a functional role. As observed with ADAMTS1,[58] CaBS-I might enforce the interaction between the "upper" and the "lower" domain and orient the connector loop that positions the disintegrin domain close to the active site at the other end of the metalloprotease domain. However, calcium binding to CaBS-II enhances the catalytic efficiency of ADAMTS13 by more than 10-fold, likely by ordering the region around the S1/S1′ sites.[46,57] Notably, the model also explains the results of recent mutagenesis data, indicating that Glu184Ala exchange only moderately reduces the affinity for calcium (i.e., from 72 to 242 μM) without appreciably affecting V_{max}. On the contrary, Asp187Ala or Glu212Ala replacements decrease Ca^{2+} binding affinity by 7- and 10-fold, respectively, and reduce V_{max} by approximately fourfold. Indeed, mutation of Glu212 in the Ca^{2+} coordination shell can be easily predicted to dramatically alter both the structure and metal-binding properties of CaBS-II. However, mutation of Glu184 is expected to affect Ca^{2+} binding only indirectly, by disrupting the salt bridge with Arg190.[57] Finally, the presence of binding sites for both Ca^{2+} and Zn^{2+} is consistent with data showing that enzymatic activity is inhibited by chelation of either cation.[54,55] There is evidence that ADAMTS13 activity is cooperatively enhanced by the two divalent cations, and it is likely that ADAMTS13 circulating in normal plasma is fully saturated by both ions.[46]

D. The Disintegrin-Like Domain (Residues 290–385)

The catalytic domain is followed by a region with 25–45% identity to the snake venom disintegrins, although it does not contain the cysteine arrangement of the latter.[13] This domain has therefore been termed disintegrin-like, though currently there is no published evidence that this ADAMTS domain interacts with integrins. The precise role that the ADAMTS13 disintegrin-like domain plays in its function remains uncertain.[10] Truncated ADAMTS13 variants suggested the importance of the disintegrin-like domain for both enzyme activity and specificity. Targeted mutagenesis of nonconserved regions (among ADAMTS family members) in this domain identified ADAMTS13 mutants with reduced proteolytic activity.[59] However, several mutations have been found in the disintegrin-like domain in some patients with inherited ADAMTS13 deficiency, but whether they actually affect ADAMTS13 activity or just impair secretion is not clear.[8,60]

E. The Thrombospondin Type 1-Like Repeat (TSR)

Unlike ADAM proteins, ADAMTS proteinases possess a well-conserved thrombospondin type 1-like repeat (TSR), homologous to the type I repeats of thrombospondins 1 and 2,[61] between the disintegrin-like and cysteine-rich domain. ADAMTS13 has eight TSRs: the first one (residues 386–439) is located as in other ADAMTS proteinases and the other seven (residues 686–1131) are located between the spacer domain and the two C-terminal CUB domains. By analogy to thrombospondins 1 and 2,[62] the central TSR of ADAMTS proteinases is believed to function as a sulfated glycosaminoglycan-binding domain. In fact, in ADAMTS13, TSRs may play a role in binding glycosaminoglycans and/or CD36, localizing ADAMTS13 to the surface of endothelial cells. Supporting this hypothesis, anti-CD36 antibodies have frequently been found in patients with thrombotic thrombocytopenic purpura.[63,64] TSP1 motifs are also known to mediate protein–protein interactions, especially among proteins in the extracellular matrix.[65]

F. The Cysteine-Rich Domain (Residues 440–555)

This domain is a well-conserved, cysteine-rich sequence containing 10 cysteine residues. The importance of this domain is highlighted by the identification of a polymorphism, P475S, in the Japanese population that altered activity but not secretion of ADAMTS13.[24] Although it is a peculiarity of the disintegrin-like domains, an RGD sequence is present within the ADAMTS13 cysteine-rich domain.[66] Additionally, mutation of the RGD sequence within the cysteine-rich region to RGE had no effect on secretion or VWF-cleaving activity *in vitro*.[54]

G. The Spacer Domain (Residues 556–685)

All ADAMTS proteinases possess a cysteine-free spacer region. This domain varies in length and contains several conserved hydrophobic residues in the N-terminal portion and an extremely variable C-terminal portion. As mentioned previously, the ADAMTS13 spacer domain has been shown to be essential for efficient ADAMTS13 VWF-cleaving activity.[54,55] ADAMTS13 fragments truncated after the metalloprotease domain, the disintegrin domain, the first TSR, or the Cys-rich domain were not able to cleave full-length VWF, whereas addition of the spacer region restored protease activity.[54,55] *In vitro* experiments have reported that a monoclonal antibody against the Cys-rich/spacer domain completely abolished ADAMTS13 binding and cleavage of both VWF and the peptide substrate VWF73.[29] It is likely that this domain plays a crucial role in VWF–ADAMTS13 interaction. Other studies reported that the spacer domain is able to bind a complementary site that includes residues Glu_{1660}–Arg_{1668} at the carboxyl terminus of the A2 domain of VWF.[67,68] This interaction could play an allosteric and regulatory role in VWF–ADAMTS13 catalytic interaction, representing a positive element for the initial recognition between the protease and VWF. Deleting the spacer domain from ADAMTS13 or deleting the carboxyl-terminal end of the A2 domain did indeed reduce the rate of cleavage by $\sim$20-fold.[67] However, once a sufficient amount of cleaved product is generated, occupation of this domain by the C-terminal product of VWF hydrolysis may trigger a negative feedback mechanism for the interaction itself.[68] ADAMTS13 truncated after the first TSR-bound VWF with a K_D of 206 nM, whereas ADAMTS13 truncated after the spacer domain-bound VWF with a K_D of 23 nM, comparable to the K_D value of 14 nM normally found for this interaction.[46]

H. The CUB Domains (Residues 1192–1408)

CUB [complement C1r/C1s, Uegf (EGF-related sea urchin protein) and BMP-1 (bone morphogenic protein-1)] domains are present at the C-terminus of ADAMTS13. This domain is also present in many proteins involved in development regulation (e.g., the dorsoventral patterning protein tolloid, bone morphogenetic protein-1, and spermadhesins), and there is evidence to suggest that these domains mediate protein–protein interactions with other CUB domain-containing proteins.[36] To date, the function of the ADAMTS13 CUB domains is controversial. The ADAMTS13 synthesized in endothelial cells appears to be delivered directly to the apical domain of polarized cells (i.e., toward the lumen of the vessels).[25] Deletion of the CUB domains, unique to ADAMTS13, abolishes the apical secretion of ADAMTS13 in cell culture,[25] suggesting that the CUB domains may interact with intracellular sorting receptors that direct apical targeting. Moreover, Zhou and colleagues have recently demonstrated that naturally occurring mutations in the CUB-1 domain result in normal rates of synthesis, but lower rates of secretion to the luminal surface of transfected Hela cells.[69] Three different studies using C-terminal truncations of ADAMTS13 have shown that these domains are not required for VWF cleaving activity under static[54,55] and flow conditions.[70] However, mutations in these distal domains of the *ADAMTS13* gene are reported in congenital TTP.[71,72] Autoantibodies directed against these mutations occur in acquired idiopathic TTP,[48] but it is unknown whether they compromise ADAMTS13 function or lead to ADAMTS13 depletion by the formation of antibody–antigen complexes.

In conclusion, the domains between the metalloprotease and the spacer domain are critical for substrate recognition and cleavage because the mutants lacking one or more of these domains do not cleave multimeric VWF.[56] The addition of the remaining TSP1 domains (TSP1 2–8) and two CUB domains does not further increase the rate of proteolytic cleavage of full length VWF or VWF73,[54,55,67] suggesting that these more distal carboxyl-terminal domains may be dispensable for substrate recognition, at least under static conditions. Substrate recognition under fluid shear stress appears to be more complex. Competitive binding studies using a parallel flow chamber system have found that recombinant CUB1 domain (and not CUB2) inhibited UL-VWF cleaving activity and VWF–ADAMTS13 interaction, suggesting a potentially important physiological role for this region under flow.[73] This is also suggested by Zhang *et al.*[74] Using a surface plasmon resonance on a BIAcore system, they have shown that the cooperative activity between the middle carboxyl-terminal TSP1 repeats and the distal carboxyl-terminal CUB domains may be crucial for the recognition and cleavage of VWF under flow. Together, these data support a hypothesis that the middle TSP1 repeats and distal CUB domains may be critical for productive recognition of VWF under flow shear stress.

V. Regulation of ADAMTS13 Activity

Modulation of the ADAMTS13–VWF interaction is critical for an efficient proteolysis and involves both VWF and ADAMTS13. The latter binds to VWF under static conditions and under both venous (2.5 dyn/cm^2) and arterial (30 dyn/cm^2) shear stress. This interaction, however, is unproductive for proteolysis unless shear stress is high enough to stretch VWF and expose the buried A2 domain for cleavage.[18,75] Under static conditions, ADAMTS13 cleaves VWF only under denaturing conditions[4,5] or in the presence of the antibiotic ristocetin,[76] whereas under conditions of high shear stress found in the microvasculature, VWF proteolysis is extremely rapid and occurs in the absence of any chemical effector.[5,18,77] Fluid shear stress alters the conformation of VWF so that the binding and catalysis of ADAMTS13 takes place at the VWF A2 domain.[78] High shear stress causes micro- and macroconformational changes in VWF.[79] Multimers with ultra large size (UL-VWF) are secreted in response to thrombogenic stimuli. The secreted UL-VWF is in part bound locally to endothelial cells and partly to collagen at sites of tissue injury through its A3 domain. These hydrodynamic forces cause conformational changes in VWF that expose a binding site in the A1 domain for the platelet glycoprotein Ib (GPIb) molecule,[46] facilitating the process of platelet adhesion to the subendothelium. It has to be noted that, once secreted by endothelial cells, UL-VWF is trimmed by ADAMTS13 with the production of smaller VWF fragments that have a lower prohemostatic potential, having a lower number of interacting sites to platelet GpIb.[46] Notably, VWF multimers extracted from platelet α-granules, which do not undergo any proteolytic activity by ADAMTS13 during their storage, have multimers with higher molecular weight than plasma-derived VWF.[80] This is additional evidence concerning the relevance of ADAMTS13 activity in VWF fragmentation, which limits its thrombogenic potential. In the absence of ADAMTS13 activity, either due to genetic mutations or formation of anti-ADAMTS13 autoantibodies, a life-threatening disease, referred to as thrombotic thrombocytopenic purpura (TTP), does occur causing an uncontrolled microvascular thrombosis (see below).[81] Contradictory to this clinical setting, some mutations in the A2 domain that likely destabilize it cause both excessive cleavage by ADAMTS13 and a shift in the size distribution to smaller VWF multimers with less hemostatic potential.[82] This effect causes a severe bleeding disorder defined as type 2A von Willebrand disease.[82] VWF is cleaved by ADAMTS13 within the A2 domain at its Tyr_{1605}–Met_{1606} bond.[4,6,77] The latter is buried at the middle of VWF A2 domain,[83,84] but is exposed to solvent and can be attacked by ADAMTS13 under high shear stress (>30 dyn/cm^2) when A2 is present in full-length VWF multimers, whereas it is proteolyzed with lower specificity as an isolated domain.[4,5,75] This phenomenon is likely due to the tensile forces that act on proteins in shear flow increase as a function of protein length.[84,85] From the seminal work of Shankaran,[85] we know that, considering a protein multimer composed of N monomers, the tensile force on a monomer increases as a function of the distance from the nearest end of the multimer and that at the mid-part of the multimer chain, the force is proportional to N^2. In particular, following the Shankaran theory, tensile force increases with the square length of the multimer chain, as both its size and the difference in velocity between the shear lamina of the fluid in which the two ends of the multimer are present, increase with length. The unique requirement of shear forces, which permit the cleavage by ADAMTS13 of the Tyr_{1605}–Met_{1606} peptide bond, finely regulates ADAMTS13 activity and impedes an uncontrolled VWF proteolysis from taking place. Moreover, the VWF-cleaving activity may be positively or negatively modulated by the other structural elements of VWF[86]: heparin sulfate, platelet GPIbα, sodium chloride,[76] and inflammatory cytokines.[87] Most studies have focused on domain A2, but the adjacent A1 and A3 domains also might participate in VWF recognition by ADAMTS13. Domain A1 binds platelet GPIbα and heparin and domain A3 binds collagen; the proximity of these binding sites to domain A2 suggests that their ligands could regulate ADAMTS13 activity independent of fluid shear stress. For example, one study suggested that the A1 domain in VWF inhibits the cleavage of the A2 domain, and that the interaction of platelet GPIbα or heparin with the A1 domain terminates such inhibition, thus making the A2 domain more susceptible to cleavage by ADAMTS13.[86]

On the contrary, binding of chloride ions to the A1 domain of VWF inhibits proteolysis by ADAMTS13, inducing conformational transitions in the A1–A2 VWF domains, which make the cleavable peptide bond in the A2 domain unavailable to proteolysis.[76] The cleavage of VWF in the absence of sodium chloride at pH 8.0 occurs with apparent k_{cat}/k_m of $3.4 \times 10^4\ M^{-1} s^{-1}$, but this value decreases ~10-fold in the presence of 0.15-M sodium chloride. The potency of the inhibitory effect varies with different anions and decreases in the order $ClO^-_4 > Cl^- > F^-$.[76]

Other factors may influence ADAMTS13 and VWF interactions, such as inflammatory cytokines[87] and hemolysis products.[48] The inflammatory cytokine interleukin-6 (IL-6; when present at levels considerably higher than typically observed *in vivo*), but not interleukin-8 or tumor necrosis factor α (TNFα), greatly abrogates the ability of ADAMTS13 to cleave UL-VWF under flowing conditions in the parallel plate chamber.[87] The physiological significance of both IL-6 and hemoglobin is uncertain, and they might be involved in ADAMTS13 regulation only under pathological conditions.

Thrombin, factor Xa, leukocyte elastase, and plasmin can cleave and inactivate ADAMTS13, and therefore might regulate ADAMTS13 activity at the site of thrombus formation. One study found that thrombin cleaved ADAMTS13 at different sites, thus inactivating it.[30] The authors observed that added full-length recombinant ADAMTS13 is proteolytically cleaved during blood coagulation in human plasma *in vitro*, and thrombin (0.9–90 nM) degrades

ADAMTS13 in a time- and concentration-dependent manner. This loss was inhibited by hirudin. This cleavage is also prevented when thrombin binds soluble thrombomodulin but not heparin, suggesting the involvement of thrombin exosite I but not exosite II in ADAMTS13 recognition.[30] Plasmin also cleaved ADAMTS13 into similar fragments, resulting in the loss of ADAMTS13 activity. Plasmin cleaves more rapidly than thrombin or factor Xa,[30] indicating that plasmin might enhance VWF-dependent platelet adhesion by degrading ADAMTS13 at sites of tissue repair. In a subsequent study,[88] the same authors could see no clear evidence for fragmentation of endogenous ADAMTS13. The major reason for this was due to the comparatively low affinity of thrombin for ADAMTS13 ($K_D \sim 95$ nM). The high K_D and K_M for this reaction likely protects ADAMTS13 from rapid proteolytic inactivation by thrombin in normal plasma, and it is consistent with the observed presence of unproteolyzed ADAMTS13 in serum. However, it must be considered that the local concentrations of ADAMTS13 at sites of vessel damage are likely to be elevated, as VWF is recruited to the injured region. Despite these findings, it is clear that ADAMTS13 is susceptible to proteolysis *in vivo*. This is particularly evident in patients with sepsis-induced disseminated intravascular coagulation (DIC), in which leukocyte elastase activity is increased, suggesting that cleavage of ADAMTS13 by leukocyte elastase might contribute to the relatively severe deficiency of ADAMTS13 activity.[31]

VI. ADAMTS13 and Its Role in the Pathogenesis of Thrombotic Microangiopathy, A Pleiomorphic Clinical Setting

Thrombotic macroangiopathies (TMAs) refer to the disorder of diffuse microvascular thrombosis involving the capillary and arteriolary bed of the brain, kidney, and other organs. The patients typically present with (1) severe thrombocytopenia ($< 50{,}000$ plts/μL), (2) nonimmune hemolysis with the presence of schistocytes on blood smears, and (3) variable neurologic abnormalities reaching even coma and/or acute renal failure.[89] Thrombocytopenia results from peripheral consumption of platelets in the microvasculature, whereas erythrocyte fragmentation and hemolysis stem from mechanical injury induced by the passage of erythrocytes through platelet thrombi under abnormally high shear stress in the microvasculature (Fig. 5). TMAs constitute a group of severe clinical settings that, without treatment, undertake a rapid worsening and death in most cases. Plasma exchange or infusion is the mainstay of treatment for most TMAs. As anticipated in the above paragraphs, the pathogenesis of complex syndromes such as TMAs is mostly explicable on the basis of the deficiency of ADAMTS13. However, it should be noted that TMAs are not monogenetic diseases. Therefore, the clinical manifestations of this group of disorders are highly variable and heavily affected by the coexistence of other genetic and environmental modifiers. This group of TMAs is

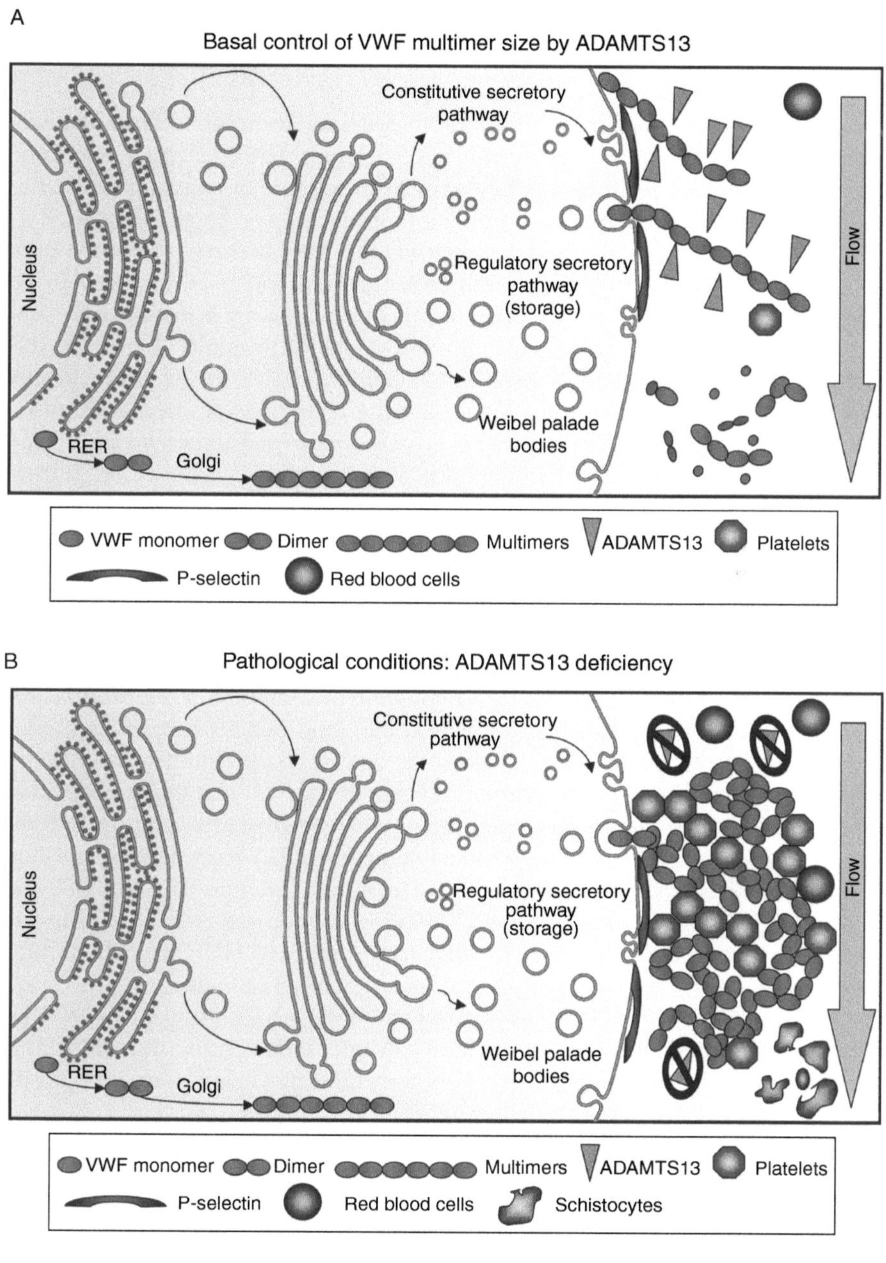

Fig. 5. Pathogenesis of TMA caused by ADAMTS13 deficiency. (A) Von Willebrand factor (VWF) multimers, produced and stored in the Weibel–Palade bodies of the endothelial cells, are secreted and adhere to endothelial cell membranes via GpIbα and P-selectin. Platelets adhere to VWF multimers through platelet membrane glycoprotein GPIbα. In flowing blood under high shear stress, VWF in the platelet-rich thrombus is in a stretched conformation and is trimmed by ADAMTS13, which limits thrombus growth. (B) If ADAMTS13 is absent or inhibited by autoantibodies, VWF-dependent platelet accumulation is uncontrolled and may cause microvascular thrombosis, formation of schistocytes and, ultimately, TMA.

constituted by different clinical settings referred to as Thrombotic Thrombocytopenic Purpura (TTP), hemolytic uremic syndrome (HUS), diarrhea-associated HUS, or atypical HUS. Unfortunately, any existing clinical or pathological classification of TMAs is based on assumptions that have never been validated. The greatest uncertainty has involved deciding whether certain cases represent examples of TTP or HUS. A rule of thumb has suggested that HUS may usually be distinguished from TTP because HUS occurs predominantly in individuals younger than 10 years, while TTP occurs predominantly in adults. However, this differentiation is not reliable, as either condition can occur in either group. Other clinical features aid in distinguishing the conditions at any age of onset. For instance, renal manifestations are usually more prominent in HUS than neurological ones, whereas neurological manifestations are usually more prominent in TTP than renal ones. Fever precedes TTP more commonly than it precedes HUS.[90] Despite these distinctions, continued recognition of borderline or atypical cases has generated doubts about the possibility that objective criteria other than age are able to distinguish "atypical" HUS from "atypical" TTP. This problem led to the application of the unsatisfactory term TTP–HUS to mean an indistinctly defined and clinically heterogenous collection of cases between classic TTP and classic HUS. The recognition of phenotypic instability in recurrent cases encouraged the use of this term. For example, one patient had five episodes manifesting the HUS phenotype before the age of 15 years and nine episodes manifesting the TTP phenotype after 20 years of age.[91] It should be noted that TTP and HUS share the fundamental pathologic feature of arteriolar thrombosis with vessel wall intimal swelling and fibrinoid necrosis. However, the composition of the thrombi differs histopathologically, at least in well-defined cases of TTP and HUS. Those of such well-defined TTP cases contain degranulated platelets and von Willebrand factor. Those of Shiga toxin-provoked HUS are rich in fibrin and thus arise from activation of the plasma coagulation cascade.[89]

Fortunately, recent advances in understanding the pathogenesis of TTP somewhat clarified the boundaries between microangiopathic disorders with renal or neurological manifestations, and they have produced useful diagnostic tests for some forms of clinically defined TTP.

A. Relationship Between ADAMTS13 and Occurrence of TMAs

Investigations have demonstrated a high degree of relevance in the relationship of ADAMTS13 to TTP. These investigations defined a heritable form of TTP with severe (<5%) ADAMTS13 activity deficiency and an acquired form due to the elaboration of antibodies directed at one or more ADAMTS13 epitopes.[92]

However, many thrombotic microangiopathies (TMAs) are not associated either with severe ADAMTS13 activity deficiency or with antibodies that block ADAMTS13 activity. This class of patients may represent > 30% of all TMA patients.[93] In some instances, the clinical syndrome is indistinguishable from typical TTP. At autopsy, widespread hyaline thrombi, accompanied by variable fibroblastic infiltration and endothelial overlay, are found in the terminal arterioles and capillaries of multiple organs. The thrombi are found most extensively in the heart, brain, kidney, pancreas, spleen, mesentery, and adrenal gland, and are composed primarily of platelets and von Willebrand factor.[94–96] A small amount of fibrin may be present surrounding the amorphous or granular materials. In older lesions, hyaline deposits may be seen in the subendothelial layers of capillaries and between the endothelium and muscular layers of arterioles. Preocclusive pseudoaneurysmal dilatation may also be present. Fibrinoid necrosis and vascular or perivascular inflammatory cell infiltration are characteristically absent or minimal. Some cases, especially those in adults, are associated with promoting factors that are associated with the development of typical hereditary or acquired TTP. Recent schemes have used the identification of such promoting factors to classify TTP-like thrombotic disorders without severe or acquired abnormalities of ADAMTS13 function, as just defined. These entities tend to occur in adults and sometimes manifest features that occur along a clinical spectrum between TTP and HUS. Many of these illnesses cannot be distinguished by using currently available laboratory tests, except when the underlying etiologic illnesses are symptomatic. These conditions share with TTP and HUS the fundamental finding of thrombocytopenic and hemolytic TMA on peripheral blood smear. In the following paragraphs, we will treat only the ADAMTS13-related forms of TMAs/TTP. These syndromes include (1) the congenital and (2) acquired deficiency of the metalloprotease. Finally, we will mention a recently discovered pathogenetic mechanism that can be responsible for accumulation of UL-VWF multimers and promote forms of TMAs in cardiovascular and metabolic disorders by perturbing the VWF–ADAMTS13 interaction.

B. Congenital ADAMTS13 Deficiency

Many studies in different ethnic populations have demonstrated the presence of ADAMTS13 mutations in patients with TTP.[7,24,45,48,60,71,97–120] Some aspects emerging from studies of ADAMTS13 congenital deficiency in mice could help to unravel the role of ADAMTS13 and UL-VWF multimers in the pathogenesis of TMAs. For instance, inactivation of the *ADAMTS13* gene in mice failed to generate the phenotype of TTP microvascular thrombosis until the ADAMTS13 null allele was transferred to a particular mouse strain, CASA/Rk, which has increased levels of VWF.[121,122] Nevertheless, cross-breeding studies showed that the development of TTP is independent of mouse plasma

VWF levels. In CASA/Rk mice with homozygous ADAMTS13 null alleles, spontaneous thrombosis and death occur in postneonatal life.[121] Only administration of shiga toxin is able to induce a massive secretion of UL-VWF multimers from endothelial cells.[123] From a clinical standpoint, there is no evidence of antecedent shiga toxin exposure in patients of TTP. Only a small fraction of TTP patients have elevated plasma VWF levels. Thus, the relevance of the shiga toxin-ADAMTS13-deficient mouse model to either TTP or shiga toxin-associated HUS remains uncertain. We can only speculate that the lack of a thrombotic phenotype in some mouse strains with severe deficiency of ADAMTS13 due to its gene inactivation suggests in these strains the presence of modifiers that affect the response of VWF to shear stress. To date, 77 mutations responsible for hereditary TTP have been identified in the *ADAMTS13* gene[7,24,45,48,60,71,97–120] (see Table I). Seven are splice mutations, ten frameshift deletions, four frameshift insertions, 11 nonsense mutations, and the remaining 45 mutations lead to codon changes. Moreover, numerous Single Nucleotide Polymorphisms (SNPs) have been recognized in recent years: eight of these SNPs are expressed and affect expression, secretion, and activity of the enzyme, whereas 18 are silent. The mutated sites in *ADAMTS13* are distributed across many exons and introns throughout the gene. The absence of clusters (hot-spots) of mutations within the metalloprotease domain implies the structural and functional importance of other regions besides the catalytic site. This finding is in-line with the observed relevance of exosites in ADAMTS13 in the molecular recognition and proteolytic processing of VWF[46,55,73] under both static and high shear rate conditions (see above). In patients with hereditary TTP, homozygous or compound heterozygous mutations of the *ADAMTS13* gene lead to severe ADAMTS13 deficiency. Globally, the affected residues span the entire spectrum of the *ADAMTS13* gene. Table I lists the principal mutations and ethnical associations discovered in the *ADAMTS13* gene and the predicted effects on enzyme sequence (see Fig. 6). Mutations of the *ADAMTS13* gene may cause impaired protein synthesis, secretion, or proteolytic activity, depending on its localization, as determined in numerous site-directed mutagenesis and expression studies. Heterozygous individuals have ADAMTS13 activity at 40–70% of normal values, while a TTP phenotype is present in more than 90% of the patients with double heterozygous or homozygous mutations. However, it may be predicted that variable phenotypic severity of TTP may arise from the various *ADAMTS13* mutations.[124] Only a few mutations have been described in more than one pedigree. A notable exception is 4143dupA, which has been described in multiple pedigrees of Northern and Central Europe and in Turkey. Haplotype analysis suggests that many, if not all, of the 4143dupA mutant alleles probably originated from a common ancestry.[101] Why this particular allele is much more frequent than other mutant alleles remains an unanswered question. Other

TABLE I
MUTATIONS OF ADAMTS13 GENE AND PREDICTED EFFECT ON MATURE ENZYME

Location	DNA	Predicted effect (protein or mRNA)	Domain	Ethnicity	Reference
Splice site					
Int 3	c.330 + 1G > A	r.[330_331ins105, 330_331ins27]		Japan/Korea	104
Int 4	c.414 + 1G > A	r.spl?		Japan	103
Int 6	c.686 + 1G > A			Japan	103
Int 6	c.687 − 2A > G			Turkey	48
Int 10	c.1244 + 2T > G	r.spl?		Japan	103
Int 11	c.1309 − ?G > A			France	98
Int 13	c.1584 + 5G > A				8
Deletion					
Ex 3	c.291_319del	p.E98PfsX31	Metalloprotease[a]	Turkey	118
Ex 7	c.718_724del	p.S240AfsX7	Metalloprotease[a]		100
Int.8_Ex.8	c.825-?_?del		Disintegrin	France	98
Ex 10	c.1095_1112del	p.W365_R370del	Disintegrin	USA	114
Ex 15	c.1783_1784del	p.L595GfsX19	Spacer[a]	USA (Yemen)	102
Ex 19	c.2279del	p.G760AfsX18	TSP1-3[a]		100
Ex 19	c.2376_2401del	p.A793PfsX43	TSP1-4[a]		8
Ex 20	c.2549_2550del	p.D850GfsX7	TSP1-5[a]		60
Ex 23	c.2930_2935del	p.C977_R979 delinsW	TSP1-6	Iran	111
Ex 25	c.3254_3255del	p.S1085CfsX12	TSP1-8[a]	France (Haiti)	98
Insertion					
Ex 1	c.82dupT	p.W28LfsX111	Metalloprotease[a]		109
Ex 4	c.372_373insGT	p.R125VfsX6	Metalloprotease		119
Ex 27	c.3770dupT	p.L1258VfsX36	CUB-1[a]		8
Ex 29	c.4143dupA	p.E1382RfsX6	CUB-2[a]	Germany	60
Nonsense					
Ex 2	c.130C > T	p.Q44X	Propeptide[a]		7
Ex 10	c.1169G > A	p.W390X	TSP1-1[a]		60
Ex 12	c.1345C > T	p.Q449X	Cysteine-rich[a]	Japan	24
Ex 21	c.2728C > T	p.R910X	TSP1-5[a]	Germany	60
Ex 22	c.2785C > T	p.Q929X	TSP1-5[a]		119
Ex 24	c.3047G > A	p.W1016X	TSP1-7[a]		109
Ex 24	c.3100A > T	p.R1034X	TSP1-7[a]	Germany	60
Ex 24	c.3170C > A	p.S1036X	TSP1-7[a]		166
Ex 26	c.3616C > T	p.R1206X	CUB-1[a]		113
Ex 27	c.3735G > A	p.W1245X	CUB-1[a]	Germany	105
Ex 28	c.3904C > T	p.Q1302X	CUB-2[a]		119
Missense					
Ex 1	c.19C > T	p.R7W[d]	Signal peptide		8
Ex 3	c.237C > G	p.I79M	Metalloprotease	France (Haiti)	98
Ex 3	c.262G > A	p.V88M	Metalloprotease	Italy	45
Ex 3	c.286C > G	p.H96D	Metalloprotease		8
Ex 3	c.304C > T	p.R102C	Metalloprotease		8

(*Continues*)

TABLE I (*Continued*)

Location	DNA	Predicted effect (protein or mRNA)	Domain	Ethnicity	Reference
Ex 4	c.356C > T	p.S119F	Metalloprotease	Tunisia	116
Ex 5	c.533T > C	p.I178T	Metalloprotease		119
Ex 6	c.577C > T	p.R193W	Metalloprotease	Japan	103
Ex 6	c.587C > T	p.T196I	Metalloprotease	Australia (Germany)	8
Ex 6	c.607T > C	p.S203P	Metalloprotease	France (Haiti)	98
Ex 7	c.695T > A	p.L232Q	Metalloprotease	Turkey	60
Ex 7	c.702C > A	p.H234Q	Metalloprotease		113
Ex 7	c.703G > C	p.D235H	Metalloprotease		100
Ex 7	c.749C > T	p.A250V	Metalloprotease	Japan	104
Ex 7	c.788C > G	p.S263C	Metalloprotease	Germany	60
Ex 7	c.803G > C	p.R268P	Metalloprotease	Japan/France (Haiti)	24
Ex 8	c.911A > G	p.Y304C	Disintegrin		119
Ex 8	c.932G > A	p.C311Y	Disintegrin		100
Ex 8	–[b]	–[c]	Disintegrin	Japan	117
Ex 9	c.1039T > A	p.C347S	Disintegrin	Croatia	101
Ex 9	c.1045C > T	p.R349C	Disintegrin		119
Ex 9	c.1058C > T	p.P353L	Disintegrin	Germany	60
Ex 10	c.1170G > C	p.W390C	TSP1-1	Germany	105
Ex 10	c.1193G > A	p.R398H	TSP1-1		8
Ex 12	c.1342C > G	p.Q448E[d]	Cysteine-rich		8
Ex 12	c.1370C > T	p.P457L	Cysteine-rich		100
Ex 12	c.1423C > T	p.P475S[d]	Cysteine-rich	Japan	24
Ex 13	c.1520G > A	p.R507Q	Cysteine-rich	France/Norway	98
Ex 13	c.1523G > A	p.C508Y	Cysteine-rich	Japan	24
Ex 13	c.1574G > A	p.G525D	Cysteine-rich		119
Ex 13	c.1582A > G	p.R528G	Cysteine-rich		8
Ex 16	c.1787C > T	p.A596V	Spacer	France (Haiti)	98
Ex 16	c.1816G > C	p.A606P	Spacer		119
Ex 16	c.1852C > G	p.P618A[d]	Spacer		8
Ex 16	c.1874G > A	p.R625H[d]	Spacer		8
Ex 17	c.2012C > T	p.P671L	Spacer	Sweden	101
Ex 17	c.2017A > T	p.I673F	Spacer		103
Ex 17	c.2074C > T	p.R692C	TSP1-2		8
Ex 18	c.2195T > C	p.A732V[d]	TSP1-2		8
Ex 19	c.2272T > C	p.C758R	TSP1-3	France	98
Ex 21	c.2699C > T	p.A900V[d]	TSP1-5		8
Ex 21	c.2723G > C	p.C908S	TSP1-5	France	98
Ex 21	c.2723G > A	p.C908Y	TSP1-5	Japan	103
Ex 22	c.2851T > G	p.C951G	TSP1-5		8
Ex 24	c.3070T > G	p.C1024G	TSP1-7		8
Ex 24	c.3097G > A	p.A1033T[d]	TSP1-7		8
Ex 24	c.3178C > T	p.R1060W	TSP1-7	USA/Italy/UK	114
Ex 25	c.3367C > T	p.R1123C	TSP1-8	Japan	103

(*Continues*)

TABLE I (*Continued*)

Location	DNA	Predicted effect (protein or mRNA)	Domain	Ethnicity	Reference
Ex 26	c.3638G > A	p.C1213Y	CUB-1		8
Ex 26	c.3650T > C	p.I1217T	CUB-1	Korea	167
Ex 26	c.3655C > T	p.R1219W	CUB-1		109
Ex 27	c.3716G > T	p.G1239V	CUB-1	Italy	107
Ex 28	c.4006C > T	p.R1336W	CUB-2		7

When available, the geographic origin of the patients carrying the mutations is indicated. The geographic origin of the parents of the patients with a given mutation is also indicated in brackets. Nucleotide numbering reflects cDNA numbering with +1 corresponding to the A of the ATG translation initiation codon in the reference sequence. The initiation codon is codon 1.

[a]The domain localization of these mutations refers to the localization of the stop codon.

[b]c.[964T > G (+) 968C > G (+) 969C > A (+) 970T > C].

[c]p.[C322G (+) T323R (+) F324L].

[d]Single Nucleotide Polymorphism (SNP) affecting expression, secretion, and activity of ADAMTS13.

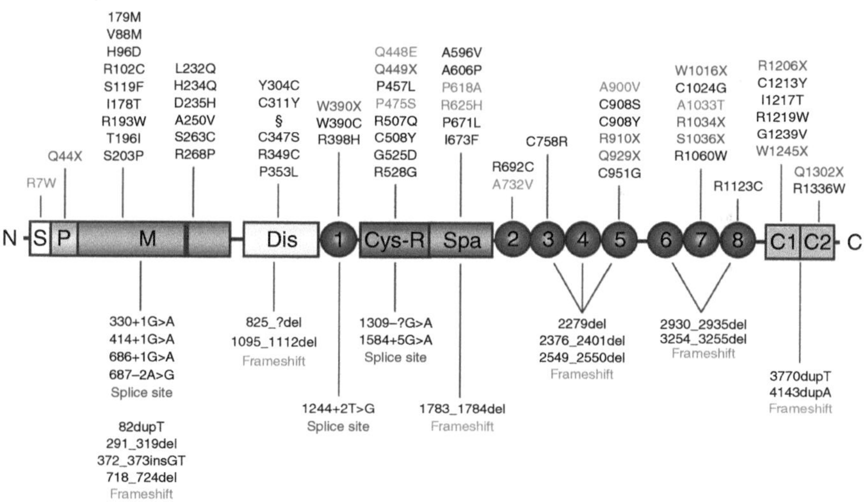

FIG. 6. Linear map of the location of ADAMTS13 mutations found in patients with congenital thrombotic thrombocytopenic purpura (TTP) (Upshaw–Schulman syndrome (USS)). The missense mutations, nonsense mutations (red), and single nucleotide polymorphisms (SNPs) (green) are shown above the domain structure of ADAMTS13. The mutations that result in alternative splicing of ADAMTS13 mRNA or frameshifts are listed under the domain structure of ADAMTS13. S indicates the signal peptide; P, propeptide; M, metalloprotease (location of zinc-binding motif shown in red); Dis, disintegrin domain; 1, first thrombospondin type 1 (TSP1) repeat; Cys-R, cysteine-rich domain; Spa, spacer domain; 2 through 8, the second to eighth TSP1 repeats; C1 and C2, two CUB domains (for complement C1r/C1s, Uegf, Bmp1 domain) §p.[C322G (+) T323R (+) F324L]. (See Color Insert.)

ethnical characteristics concern one ADAMTS13 variant allele, 1423C > T (P475S), found in Japanese (5.1%), Koreans (4%), and Chinese (0.5–1.7%) but not detected among Caucasians or African Americans.[24,125] This polymorphism had raised considerable interest, because in expression studies, this mutation markedly reduces the activity of ADAMTS13 to ~10% of control, raising the possibility that partial deficiency of ADAMTS13 may be quite common among Northeast Asians. Nevertheless, this prediction was not correct, as more recent investigations have shown that carriers of the P475S polymorphism have only a minor decrease (10%) of the ADAMTS13 activity and revealed that the previously reported low activity of the P475S variant resulted from the effect of high urea concentration used in the ADAMTS13 activity assay.[126] Thus, this mutant might have only an abnormal stability.

1. Splice Mutations

Among the splice mutations, the c.414 + 1G > A (intron 4), c.686 + 1G > A (intron 6), and c.1244 + 2T > G (intron 10) mutations cause inconsistencies with the GT–AG rules of mRNA splicing. However, the c.1584 + 5G > A (intron 13) mutation also alters the donor splice site from the consensus sequence, as experimentally verified. Total RNAs were isolated from patients' lymphocytes or whole blood cells and used as templates for reverse-transcriptase polymerase chain reaction (RT-PCR). The data indicated that little or no normally spliced products were indeed generated from the mutant alleles.[8,103]

2. Frameshift Mutations

In five frameshift mutations, c.1783_1784del, c.2376_2401del, c.2549_2550del, c.3770dupT, and c.4143dupA, ADAMTS13 is expressed as a truncated mutant with an aberrant C-terminal end. The c.1783_1784del mutation replaces L595-T1427 sequence in the spacer domain with the peptide sequence dGGEDRRALCRGWEDEHLP, the c.2376_2401del mutation in the TSP1-4 domain (A793-T1427) with the sequence PALPCQVGGVRAQLMHISWWSRPGLGERDLCARGRWPGGSSD, the c.2549_2550del mutation in the TSP1-5 domain (D850-T1427) with the GEAACP sequence, the c.3770dupT in the CUB-1 domain (L1258-T1427) with VGHDFQL QDQHAGGEAALRAARRWGAAAVWEPACS, and the c.4143dupA frameshift mutation in the CUB-2 domain (E1382-T1427) with the REQPG sequence. These aberrant mutants cannot be secreted, whereas others are secreted in sufficient amounts but are dysfunctional. However, it cannot be excluded that mRNAs with these frameshift mutations may be eliminated by gene expression quality control systems. Finally, it should be noted that the c.3254_3255del mutation, which is responsible for the congenital thrombotic thrombocytopenic purpura called Upshaw–Schulman syndrome, is often reported in the literature as the R1096X mutation.[98]

3. Missense Mutations

Several missense mutations have been characterized by site-directed mutagenesis of the *ADAMTS13* gene, followed by transfection into different expression cell systems. All the domains of the metalloprotease are affected by gene mutations, although the majority were found in the metalloprotease domain. The mutations involving the metalloprotease domain include residues I79 to R268. *In vitro* studies allowed assaying for biosynthesis, secretion, and activity in conditioned media. Several mutants, including V88M,[110] R193W,[103] A250V,[104] and R268P,[24] are characterized by low or no secretion and exhibit very low activity in the media of transfected cells. Recombinant ADAMTS13 V88M was found to be synthesized in normal amounts in cell lyzates, but its secretion was impaired ~40–50% compared to wild-type (WT) ADAMTS13.[110] Notably, this pattern was confirmed by immunofluorescence studies, showing that ADAMTS13 V88M was visualized in both the endoplasmic reticulum and Golgi compartments, but with less intensity than the WT recombinant protease.[110] The mutant protease had also reduced enzymatic activity in comparison with the WT protease. Analysis of the predicted secondary structure of the V88M mutant showed a rearrangement of the region between residues 61 and 70 and a slight increase of the sheet content over residues 80–90 compared to ADAMTS13 WT, possibly perturbing the Ca^{++}-binding site of the protease.[110] R193W[103] is close to the active site of the metalloprotease domain, whereas A250V is near the Met_{249} that is part of the S1′ catalytic subsite, as shown by recent molecular modeling studies (Pozzi *et al.*, submitted for publication). This finding may explain why the enzyme, secreted by transfected cells, shows a defective proteolytic activity.[104] On the whole, these expression studies corroborate the available knowledge on the key role of the metalloprotease domain on the secretion and enzymatic activity of the protease. The mutations found in the spacer, Cys-rich, TSP-1, and CUB domains usually (see Table I) show very low or absent activity, although some of those, as P475S, also have a defect in secretion. Several mutations consist of substitutions to/from cysteine residues, suggesting that disulfide bond formation is important for proper conformation and function. In particular, a proper conformation of the spacer and Cys-rich domain is required for a correct recognition of VWF. This requirement is also indirectly confirmed by the negative effect exerted by autoantibody against the Cys-rich domain and spacer domain on the VWF–ADAMTS13 interaction (see below).

4. Common Single-Nucleotide Polymorphisms (SNPs)

Several silent SNPs in both exons and introns were reported (see Table II). However, so far, no relationship between these SNPs and development of TTP has been suspected and/or reported. The allele frequencies of R7W, Q448E,

TABLE II
SINGLE NUCLEOTIDE POLYMORPHISMS (SNPs)[8,24]

Exon/intron	Nucleotide	Amino acid
ex1	19C > T	R7W[a]
ex4	354G > A	Silent
ex5	420T > C	Silent
ex6	582C > T	Silent
int6	686 + 4T > G	N/A
int8	987 + 11C > T	N/A
int8	987 + 69C > T	N/A
int9	1092 + 67G > A	N/A
int10	1245 − 32C > G	N/A
ex12	1342C > G	Q448E[a]
ex12	1423C > T	P475S[a]
int13	1584 + 106C > G	N/A
int13	1584 + 236T > C	N/A
ex15	1716G > A	Silent
int15	1787 − 26G > A	N/A
ex16	1852C > G	P618A[a]
ex16	1874G > A	R625H[a]
ex18	2195C > T	A732V[a]
ex19	2280T > C	Silent
ex21	2699C > T	A900V[a]
int22	2861 + 55C > T	N/A
ex23	2910C > T	Silent
ex24	3097G > A	A1033T[a]
ex24	3108G > A	Silent
int28	4077 + 32T > C	N/A
ex29	4221C > A	Silent

[a]SNP affecting expression, secretion, and activity of ADAMTS13.

P475S, and P618A SNPs have been reported. In Japan, the allele frequencies of the heterozygous and homozygous subjects with the P475S SNP were calculated to be 19% and 5%, respectively.[24] Expression studies with ADAMTS13 carrying the P475S SNP showed that this polymorphism causes a moderate but significant decrease (≈ 10%) of the protease activity toward VWF. Notably, a 43% frequency of Q448E was also calculated among 120 European alleles.[7] In the same study, the allele frequencies of R7W and P618A were calculated to be 10% and 9%, respectively. Due to the allele frequency of this SNP in the Japanese population, it may be inferred that ~ 10% of the Japanese population might have significantly reduced ADAMTS13 activity.[24]

C. Inhibitors of ADAMTS13

A strong deficiency of ADAMTS13 activity can also be associated to the development of autoantibodies against the protease. The formation of IgG or IgM anti-ADAMTS13 antibodies may be responsible for the onset of TMAs idiopathic or secondary to drugs, pregnancy, or diseases such as infections, cancers, and autoimmune diseases.[21,127] In patients with acquired TTP, deficiency of ADAMTS13 results from the autoimmune inhibitors of ADAMTS13, which either inhibit its catalytic activity or induce a rapid clearance from the circulation.[93,98,128–132] Similar to other autoimmune disorders, the etiologies of acquired TTP are unknown, and TTP patients often exhibit positive autoimmune reactions to different target antigens,[128] suggesting that defective immune regulation may contribute to the development of TTP. A defective regulation of T-reg and tolerogenic dendritic cells may be responsible for the occurrence of anti-ADAMTS13 antibodies, in analogy with what has been shown in other autoimmune coagulation inhibitors such as anti-FVIII antibodies.[133] HIV infection may also be a risk factor for TTP, although this association has not been confirmed by all authors.[134] The inhibitors are more frequently IgG, although occasional production of IgA and IgM antibodies has been described. In a recent study, IgG(4) was found to be the most prevalent IgG subclass (90%) in 58 patients with acquired TTP, followed by IgG(1) (52%), IgG(2) (50%), and IgG(3) (33%).[135] These studies also showed that IgG(4) may be found either alone (33%) or with other IgG subclasses (67%).[135] IgG(4) was not detected in 10% of the patients. Patients with high IgG(4) levels and undetectable IgG(1) are more prone to relapse than patients with low IgG(4) levels and detectable IgG(1).[135] Remarkably, a rising ADAMTS13 inhibitor level may be associated with switching of the IgG subclasses, suggesting that cytokine dysregulation may be responsible for the rising inhibitor levels observed in some cases of TTP.[136] Epitope mapping studies showed that the spacer domain,[48,137–140] specifically, residues T572-N579 and V657-G666,[140] comprise a common antigenic core region that is a relevant target for ADAMTS13 antibodies in TTP. Notably, the proteolytic activity of ADAMTS13 variants truncated upstream of the Cys-rich domain is not generally inhibited by the inhibitors of patients with TMAs. These noninhibited ADAMTS13 recombinant constructs may be used to overcome, at least in part, the difficult management of patients with high inhibitor levels. The levels of the ADAMTS13 inhibitors tend to be low (< 10 U/mL),[129,141] often receding to even lower or undetectable levels within weeks or months. Such characteristics of the ADAMTS13 inhibitors suggest that the immune response is induced by exposure to exogenous antigens with molecular mimicry to ADAMTS13. The level of anti-ADAMTS13 inhibitors determines the efficacy of therapeutic strategies, particularly plasma exchange, aimed at eliminating their pathologic effects. Usually, ADAMTS13 inhibitors, measured by the Bethesda assay in clinical laboratories,[142] have low titers (< 10 Bethesda units/mL) and self-limited course. However, if the level of anti-ADAMTS13 inhibitors is high, the treatment may fail.[136] Moreover, refractory TMA forms, characterized by persistent anti-ADAMTS13 inhibitors, have also been reported in patients requiring long-term plasma exchange treatment and immunosuppressive therapy with rituximab.[143]

VII. Are ADAMTS13 Changes Always Associated with Thrombotic Microangiopathies?

Although the onset of TMAs is associated with the production of anti-ADAMTS13 antibodies or congenital defects of the enzyme in the majority of cases, some clinical settings present with the same symptoms and laboratory signs of hematological abnormalities, although quantitative or qualitative deficiency of the metalloprotease is not present. These conditions are represented by a variety of pathological conditions, including all the hemolytic–uremic syndromes (HUS) arising from congenital complement dysregulation caused by factor H or I deficiency.[144] Moreover, a moderate decrease in ADAMTS13 levels was reported for miscellaneous conditions such as connective tissue disorders, sepsis, liver diseases, and *Plasmodium falciparum* infection.[31,145–151] However, in most cases, the decrease in ADAMTS13 level is unlikely to contribute to thrombosis, and some of the above observations are indeed the object of controversy, since a severe decrease of ADAMTS13 is the only condition considered to be responsible for TTP occurrence.[124,152] Likewise, measurements of ADAMTS13 activity have shown that pregnancy is characterized by a progressive decrease of ADAMTS13 during the various periods of gestation. The ADAMTS13 activity level is decreased to a mean of 52% (range 22–89%) of normal plasma by the end of pregnancy.[153] However, if gestation is complicated by the occurrence of the syndrome referred to as HELLP (High Enzyme Liver Low Platelet), ADAMTS13 levels are even lower (31%).[154] Thus, instead of considering pregnancy as a risk factor *per se* for TTP, this setting may simply precipitate or exacerbate a preexisting TTP/HUS (immune-mediated or congenital), since it decreases the ADAMTS13 level below the threshold necessary for preventing the VWF/platelet-mediated thrombosis.[155]

As anticipated above, some inorganic ions, such as chloride,[76,156] or biological macromolecules, such as thrombospondin,[76,156] free hemoglobin, or bilirubin,[48,157] are able to inhibit the interaction of VWF with ADAMTS13.[76,156] However, in all cases, the inhibitory activity was observed *in vitro*, and therefore, the role of these modulators *in vivo* remains to be determined. Finally, in inflammatory conditions, ADAMTS13 level decreases as a consequence of various mechanisms, such as (1) downregulation on a transcriptional level, (2) proteolytic degradation by other proteases such as thrombin or plasmin, (3) consumption or functional inability to sufficiently proteolyze a massive endothelial secretion of UL-VWF multimers causing a high substrate level, and (4) posttranslational modification of VWF, which inhibits the interaction with ADAMTS13. Notably, in all clinical settings characterized by a massive endothelial release of UL-VWF multimers, the enzymatic ability to cleave a sufficient amount of VWF due to high substrate level may cause a TTP-like syndrome (thrombocytopenia and haemolytic anemia) without altering the level of ADAMTS13. Furthermore, recent data have shown that if VWF multimers undergo an oxidative modification involving Met_{1606} with the formation of sulfoxy-methionine, the proteolysis by ADAMTS13 is severely inhibited.[158,159] However, although this oxidative modification is highly efficient *in vitro* in inducing a severe inhibition of ADAMTS13 proteolysis, its role in contributing to thrombotic microangiopathy in clinical settings with oxidative stress remains to be investigated.

In conclusion, much remains to be learned and investigated on how the various potential modifiers affect the phenotype of ADAMTS13 deficiency and elevation of UL-VWF levels with occurrence of TTP or TTP-like syndromes.

VIII. Issues Concerning Laboratory Methods Used to Measure the ADAMTS13 Activity in Clinical Samples

The laboratory assays employed to measure the activity of ADAMTS13 in plasma are based on the degradation by ADAMTS13 of purified, plasma-derived/recombinant VWF multimers or synthetic VWF peptides (A2 domain, 73aa- or 86aa-fragment) in patient plasma and the direct or indirect detection of the VWF cleavage products. The assays based on multimeric VWF are sensitive (3–6% of ADAMTS13 activity) and reproducible, but cumbersome, time-consuming (2–3 days), and performed in nonphysiological conditions, using denaturing agents that are required to promote the conformational changes to enhance the susceptibility of VWF multimers to cleavage, but that could also affect the enzymatic activity of ADAMTS13 and inhibit antibody–ADAMTS13 interactions. On the other hand, the assays based on VWF peptides are very sensitive (1–3% of ADAMTS13 activity), reproducible, easy, rapid (1–4 h), and performed in the absence of denaturing agents, but using nonphysiologic substrates. A number of variables may interfere with assay results. The results of some activity assays are influenced by high levels of VWF in test plasmas, hyperlipemia, plasma hemoglobinemia (acting as an inhibitor of ADAMTS13),[48] and hyperbilirubinemia (that may interfere with the fluorescence assays).[157,160] In addition, the presence of other proteases in plasma may interfere with VWF cleavage[161] or cause some degree of

ADAMTS13 degradation.[88] Because all these factors might influence the clinical usefulness of each assay, more efforts should be done to determine whether or not the results of these assays are comparable in different clinical settings. Accordingly, performance characteristics of these methods were recently investigated in the course of a multicenter study.[48,162–164]

IX. Future Directions

After the discovery that normal plasma contains a zinc protease able to specifically proteolyze VWF, the past decade has witnessed the most exciting advances in the history of studies on the pathogenesis of TMAs. However, many issues still need to be addressed. The knowledge of some mechanistic aspects of ADAMTS13 catalysis and its regulation, the development of sensitive and reliable assays in the clinical diagnostics of TMAs, and the nature of modifiers of ADAMTS13 activity on VWF multimers in patients affected by TMAs require further improvement. From a biotechnological standpoint, industrial production of partially deleted ADAMTS13, nonsuppressible by pathological autoantibodies, may circumvent the difficulties that replacement therapies with recombinant full-length ADAMTS13 may encounter in patients with acquired TTP. Finally, basic research to clarify the immunological mechanisms of generation of ADAMTS13 inhibitors[165] will aid in the discovery of new strategies able to improve the prevention, diagnosis, and management of TMAs.

References

1. Moake JL, Rudy CK, Troll JH, Weinstein MJ, Colannino NM, Azocar J, et al. Unusually large plasma factor VIII: von Willebrand factor multimers in chronic relapsing thrombotic thrombocytopenic purpura. *N Engl J Med* 1982;**307**:1432–5.
2. Sadler JE. Biochemistry and genetics of von Willebrand factor. *Annu Rev Biochem* 1998;**67**:395–424.
3. Ruggeri ZM. Von Willebrand factor, platelets and endothelial cell interactions. *J Thromb Haemost* 2003;**1**:1335–42.
4. Furlan M, Robles R, Lamie B. Partial purification and characterization of a protease from human plasma cleaving von Willebrand factor to fragments produced by in vivo proteolysis. *Blood* 1996;**87**:4223–34.
5. Tsai HM. Physiologic cleavage of von Willebrand factor by a plasma protease is dependent on its conformation and requires calcium ion. *Blood* 1996;**87**:4235–44.
6. Chung DW, Fujikawa K. Processing of von Willebrand factor by ADAMTS-13. *Biochemistry* 2002;**41**:11065–70.

7. Plaimauer B, Zimmermann K, Volkel D, Antoine G, Kerschbaumer R, Jenab P, et al. Cloning, expression, and functional characterization of the von Willebrand factor-cleaving protease (ADAMTS13). *Blood* 2002;**100**:3626–32.
8. Levy GG, Nichols WC, Lian EC, Foroud T, McClintick JN, McGee BM, et al. Mutations in a member of the ADAMTS gene family cause thrombotic thrombocytopenic purpura. *Nature* 2001;**413**:488–94.
9. Soejima K, Mimura N, Hirashima M, Maeda H, Hamamoto T, Nakagaki T, et al. A novel human metalloprotease synthesized in the liver and secreted into the blood: possibly, the von Willebrand factor-cleaving protease? *J Biochem* 2001;**130**:475–80.
10. Zheng X, Chung D, Takayama TK, Majerus EM, Sadler JE, Fujikawa K. Structure of von Willebrand factor-cleaving protease (ADAMTS13), a metalloprotease involved in thrombotic thrombocytopenic purpura. *J Biol Chem* 2001;**276**:41059–63.
11. Tsai HM. Von Willebrand factor, ADAMTS13, and thrombotic thrombocytopenic purpura. *J Mol Med* 2002;**80**:639–47.
12. Wagner DD, Marder VJ. Biosynthesis of von Willebrand protein by human endothelial cells. Identification of a large precursor polypeptide chain. *J Biol Chem* 1983;**258**:2065–7.
13. Huang RH, Wang Y, Roth R, Yu X, Purvis AR, Heuser JE, et al. Assembly of Weibel-Palade body-like tubules from N-terminal domains of von Willebrand factor. *Proc Natl Acad Sci USA* 2008;**105**:482–7.
14. Ruggeri ZM, Zimmerman TS. The complex multimeric composition of factor VIII/von Willebrand factor. *Blood* 1981;**57**:1140–3.
15. Tsai HM, Nagel RL, Hatcher VB, Sussman II. Multimeric composition of endothelial cell-derived von Willebrand factor. *Blood* 1989;**73**:2074–6.
16. Padilla A, Moake JL, Bernardo A, Ball C, Wang Y, Arya M, et al. P-selectin anchors newly released ultralarge von Willebrand factor multimers to the endothelial cell surface. *Blood* 2004;**103**:2150–6.
17. Sadler JE. von Willebrand factor. *J Biol Chem* 1991;**266**:22777–80.
18. Tsai HM, Sussman II, Nagel RL. Shear stress enhances the proteolysis of von Willebrand factor in normal plasma. *Blood* 1994;**83**:2171–9.
19. Lopez JA, Dong JF. Cleavage of von Willebrand factor by ADAMTS-13 on endothelial cells. *Semin Hematol* 2004;**41**:15–23.
20. Tsai HM, Sussman II, Ginsburg D, Lankhof H, Sixma JJ, Nagel RL. Proteolytic cleavage of recombinant type 2A von Willebrand factor mutants R834W and R834Q: inhibition by doxycycline and by monoclonal antibody VP-1. *Blood* 1997;**89**:1954–62.
21. Furlan M, Robles R, Galbusera M, Remuzzi G, Kyrle PA, Brenner B, et al. von Willebrand factor-cleaving protease in thrombotic thrombocytopenic purpura and the hemolytic-uremic syndrome. *N Engl J Med* 1998;**339**:1578–84.
22. Tsai HM, Lian EC. Antibodies to von Willebrand factor-cleaving protease in acute thrombotic thrombocytopenic purpura. *N Engl J Med* 1998;**339**:1585–94.
23. Hulstein JJ, de Groot PG, Silence K, Veyradier A, Fijnheer R, Lenting PJ. A novel nanobody that detects the gain-of-function phenotype of von Willebrand factor in ADAMTS13 deficiency and von Willebrand disease type 2B. *Blood* 2005;**106**:3035–42.
24. Kokame K, Matsumoto M, Soejima K, Yagi H, Ishizashi H, Funato M, et al. Mutations and common polymorphisms in ADAMTS13 gene responsible for von Willebrand factor-cleaving protease activity. *Proc Natl Acad Sci USA* 2002;**99**:11902–7.
25. Shang D, Zheng XW, Niiya M, Zheng XL. Apical sorting of ADAMTS13 in vascular endothelial cells and Madin-Darby canine kidney cells depends on the CUB domains and their association with lipid rafts. *Blood* 2006;**108**:2207–15.
26. Turner N, Nolasco L, Tao Z, Dong JF, Moake J. Human endothelial cells synthesize and release ADAMTS-13. *J Thromb Haemost* 2006;**4**:1396–404.

27. Liu L, Choi H, Bernardo A, Bergeron AL, Nolasco L, Ruan C, et al. Platelet-derived VWF-cleaving metalloprotease ADAMTS-13. *J Thromb Haemost* 2005;**3**:2536–44.
28. Suzuki M, Murata M, Matsubara Y, Uchida T, Ishihara H, Shibano T, et al. Detection of von Willebrand factor-cleaving protease (ADAMTS-13) in human platelets. *Biochem Biophys Res Commun* 2004;**313**:212–6.
29. Soejima K, Nakamura H, Hirashima M, Morikawa W, Nozaki C, Nakagaki T. Analysis on the molecular species and concentration of circulating ADAMTS13 in blood. *J Biochem* 2006;**139**:147–54.
30. Crawley JT, Lam JK, Rance JB, Mollica LR, O'Donnell JS, Lane DA. Proteolytic inactivation of ADAMTS13 by thrombin and plasmin. *Blood* 2005;**105**:1085–93.
31. Ono T, Mimuro J, Madoiwa S, Soejima K, Kashiwakura Y, Ishiwata A, et al. Severe secondary deficiency of von Willebrand factor-cleaving protease (ADAMTS13) in patients with sepsis-induced disseminated intravascular coagulation: its correlation with development of renal failure. *Blood* 2006;**107**:528–34.
32. Uemura M, Tatsumi K, Matsumoto M, Fujimoto M, Matsuyama T, Ishikawa M, et al. Localization of ADAMTS13 to the stellate cells of human liver. *Blood* 2005;**106**:922–4.
33. Zhou W, Inada M, Lee TP, Benten D, Lyubsky S, Bouhassira EE, et al. ADAMTS13 is expressed in hepatic stellate cells. *Lab Invest* 2005;**85**:780–8.
34. Niiya M, Uemura M, Zheng XW, Pollak ES, Dockal M, Scheiflinger F, et al. Increased ADAMTS-13 proteolytic activity in rat hepatic stellate cells upon activation in vitro and in vivo. *J Thromb Haemost* 2006;**4**:1063–70.
35. Cal S, Obaya AJ, Llamazares M, Garabaya C, Quesada V, Lopez-Otin C. Cloning, expression analysis, and structural characterization of seven novel human ADAMTSs, a family of metalloproteinases with disintegrin and thrombospondin-1 domains. *Gene* 2002;**283**:49–62.
36. Bork P, Beckmann G. The CUB domain. A widespread module in developmentally regulated proteins. *J Mol Biol* 1993;**231**:539–45.
37. Colige A, Sieron AL, Li SW, Schwarze U, Petty E, Wertelecki W, et al. Human Ehlers-Danlos syndrome type VII C and bovine dermatosparaxis are caused by mutations in the procollagen I N-proteinase gene. *Am J Hum Genet* 1999;**65**:308–17.
38. Dagoneau N, Benoist-Lasselin C, Huber C, Faivre L, Megarbane A, Alswaid A, et al. ADAMTS10 mutations in autosomal recessive Weill-Marchesani syndrome. *Am J Hum Genet* 2004;**75**:801–6.
39. Dietz HC, Cutting GR, Pyeritz RE, Maslen CL, Sakai LY, Corson GM, et al. Marfan syndrome caused by a recurrent de novo missense mutation in the fibrillin gene. *Nature* 1991;**352**:337–9.
40. Hurskainen TL, Hirohata S, Seldin MF, Apte SS. ADAM-TS5, ADAM-TS6, and ADAM-TS7, novel members of a new family of zinc metalloproteases. General features and genomic distribution of the ADAM-TS family. *J Biol Chem* 1999;**274**:25555–63.
41. Shindo T, Kurihara H, Kuno K, Yokoyama H, Wada T, Kurihara Y, et al. ADAMTS-1: a metalloproteinase-disintegrin essential for normal growth, fertility, and organ morphology and function. *J Clin Invest* 2000;**105**:1345–52.
42. Tortorella MD, Burn TC, Pratta MA, Abbaszade I, Hollis JM, Liu R, et al. Purification and cloning of aggrecanase-1: a member of the ADAMTS family of proteins. *Science* 1999;**284**:1664–6.
43. Chauhan AK, Kisucka J, Brill A, Walsh MT, Scheiflinger F, Wagner DD. ADAMTS13: a new link between thrombosis and inflammation. *J Exp Med* 2008;**205**:2065–74.
44. Akiyama M, Takeda S, Kokame K, Takagi J, Miyata T. Crystal structures of the noncatalytic domains of ADAMTS13 reveal multiple discontinuous exosites for von Willebrand factor. *Proc Natl Acad Sci USA* 2009;**106**:19274–9.

45. Bestetti G, Stellari A, Lattuada A, Corbellino M, Parravicini C, Calzarossa C, et al. ADAMTS 13 genotype and vWF protease activity in an Italian family with TTP. *Thromb Haemost* 2003;**90**:955–6.
46. Majerus EM, Anderson PJ, Sadler JE. Binding of ADAMTS13 to von Willebrand factor. *J Biol Chem* 2005;**280**:21773–8.
47. Van Wart HE, Birkedal-Hansen H. The cysteine switch: a principle of regulation of metalloproteinase activity with potential applicability to the entire matrix metalloproteinase gene family. *Proc Natl Acad Sci USA* 1990;**87**:5578–82.
48. Klaus C, Plaimauer B, Studt JD, Dorner F, Lammle B, Mannucci PM, et al. Epitope mapping of ADAMTS13 autoantibodies in acquired thrombotic thrombocytopenic purpura. *Blood* 2004;**103**:4514–9.
49. Yiallouros I, Grosse Berkhoff E, Stocker W. The roles of Glu93 and Tyr149 in astacin-like zinc peptidases. *FEBS Lett* 2000;**484**:224–8.
50. Bode W, Fernandez-Catalan C, Tschesche H, Grams F, Nagase H, Maskos K. Structural properties of matrix metalloproteinases. *Cell Mol Life Sci* 1999;**55**:639–52.
51. Bode W, Gomis-Ruth FX, Stockler W. Astacins, serralysins, snake venom and matrix metalloproteinases exhibit identical zinc-binding environments (HEXXHXXGXXH and Met-turn) and topologies and should be grouped into a common family, the 'metzincins'. *FEBS Lett* 1993;**331**:134–40.
52. Rawlings ND, Barrett AJ. Evolutionary families of metallopeptidases. *Methods Enzymol* 1995;**248**:183–228.
53. Stocker W, Grams F, Baumann U, Reinemer P, Gomis-Ruth FX, McKay DB, et al. The metzincins–topological and sequential relations between the astacins, adamalysins, serralysins, and matrixins (collagenases) define a superfamily of zinc-peptidases. *Protein Sci* 1995;**4**:823–40.
54. Soejima K, Matsumoto M, Kokame K, Yagi H, Ishizashi H, Maeda H, et al. ADAMTS-13 cysteine-rich/spacer domains are functionally essential for von Willebrand factor cleavage. *Blood* 2003;**102**:3232–7.
55. Zheng X, Nishio K, Majerus EM, Sadler JE. Cleavage of von Willebrand factor requires the spacer domain of the metalloprotease ADAMTS13. *J Biol Chem* 2003;**278**:30136–41.
56. Ai J, Smith P, Wang S, Zhang P, Zheng XL. The proximal carboxyl-terminal domains of ADAMTS13 determine substrate specificity and are all required for cleavage of von Willebrand factor. *J Biol Chem* 2005;**280**:29428–34.
57. Gardner MD, Chion CK, de Groot R, Shah A, Crawley JT, Lane DA. A functional calcium-binding site in the metalloprotease domain of ADAMTS13. *Blood* 2009;**113**:1149–57.
58. Gerhardt S, Hassall G, Hawtin P, McCall E, Flavell L, Minshull C, et al. Crystal structures of human ADAMTS-1 reveal a conserved catalytic domain and a disintegrin-like domain with a fold homologous to cysteine-rich domains. *J Mol Biol* 2007;**373**:891–902.
59. Huizinga EG, Tsuji S, Romijn RA, Schiphorst ME, de Groot PG, Sixma JJ, et al. Structures of glycoprotein Ibalpha and its complex with von Willebrand factor A1 domain. *Science* 2002;**297**:1176–9.
60. Schneppenheim R, Budde U, Oyen F, Angerhaus D, Aumann V, Drewke E, et al. von Willebrand factor cleaving protease and ADAMTS13 mutations in childhood TTP. *Blood* 2003;**101**:1845–50.
61. Bornstein P. Thrombospondins: structure and regulation of expression. *FASEB J* 1992;**6**:3290–9.
62. Guo NH, Krutzsch HC, Negre E, Zabrenetzky VS, Roberts DD. Heparin-binding peptides from the type I repeats of thrombospondin. Structural requirements for heparin binding and promotion of melanoma cell adhesion and chemotaxis. *J Biol Chem* 1992;**267**:19349–55.

63. Schultz DR, Arnold PI, Jy W, Valant PA, Gruber J, Ahn YS, et al. Anti-CD36 autoantibodies in thrombotic thrombocytopenic purpura and other thrombotic disorders: identification of an 85 kD form of CD36 as a target antigen. *Br J Haematol* 1998;**103**:849–57.
64. Tandon NN, Rock G, Jamieson GA. Anti-CD36 antibodies in thrombotic thrombocytopenic purpura. *Br J Haematol* 1994;**88**:816–25.
65. Bornstein P. Thrombospondins as matricellular modulators of cell function. *J Clin Invest* 2001;**107**:929–34.
66. McLane MA, Marcinkiewicz C, Vijay-Kumar S, Wierzbicka-Patynowski I, Niewiarowski S. Viper venom disintegrins and related molecules. *Proc Soc Exp Biol Med* 1998;**219**:109–19.
67. Gao W, Anderson PJ, Majerus EM, Tuley EA, Sadler JE. Exosite interactions contribute to tension-induced cleavage of von Willebrand factor by the antithrombotic ADAMTS13 metalloprotease. *Proc Natl Acad Sci USA* 2006;**103**:19099–104.
68. Di Stasio E, Lancellotti S, Peyvandi F, Palla R, Mannucci PM, De Cristofaro R. Mechanistic studies on ADAMTS13 catalysis. *Biophys J* 2008;**95**:2450–61.
69. Zhou Z, Jing H, Tao Z, Choi H, Aboulfatova K, Moake J, et al. Effects of naturally occurring mutations in CUB-1 domain on synthesis, stability, and activity of ADAMTS-13. *Thromb Res* 2009;**124**:323–7.
70. Tao Z, Wang Y, Choi H, Bernardo A, Nishio K, Sadler JE, et al. Cleavage of ultralarge multimers of von Willebrand factor by C-terminal-truncated mutants of ADAMTS-13 under flow. *Blood* 2005;**106**:141–3.
71. Pimanda JE, Maekawa A, Wind T, Paxton J, Chesterman CN, Hogg PJ. Congenital thrombotic thrombocytopenic purpura in association with a mutation in the second CUB domain of ADAMTS13. *Blood* 2004;**103**:627–9.
72. Shelat SG, Ai J, Zheng XL. Molecular biology of ADAMTS13 and diagnostic utility of ADAMTS13 proteolytic activity and inhibitor assays. *Semin Thromb Hemost* 2005;**31**:659–72.
73. Tao Z, Peng Y, Nolasco L, Cal S, Lopez-Otin C, Li R, et al. Recombinant CUB-1 domain polypeptide inhibits the cleavage of ULVWF strings by ADAMTS13 under flow conditions. *Blood* 2005;**106**:4139–45.
74. Zhang P, Pan W, Rux AH, Sachais BS, Zheng XL. The cooperative activity between the carboxyl-terminal TSP1 repeats and the CUB domains of ADAMTS13 is crucial for recognition of von Willebrand factor under flow. *Blood* 2007;**110**:1887–94.
75. Tsai HM. Shear stress and von Willebrand factor in health and disease. *Semin Thromb Hemost* 2003;**29**:479–88.
76. De Cristofaro R, Peyvandi F, Palla R, Lavoretano S, Lombardi R, Merati G, et al. Role of chloride ions in modulation of the interaction between von Willebrand factor and ADAMTS-13. *J Biol Chem* 2005;**280**:23295–302.
77. Dong JF, Moake JL, Nolasco L, Bernardo A, Arceneaux W, Shrimpton CN, et al. ADAMTS-13 rapidly cleaves newly secreted ultralarge von Willebrand factor multimers on the endothelial surface under flowing conditions. *Blood* 2002;**100**:4033–9.
78. Dong JF. Cleavage of ultra-large von Willebrand factor by ADAMTS-13 under flow conditions. *J Thromb Haemost* 2005;**3**:1710–6.
79. Auton M, Cruz MA, Moake J. Conformational stability and domain unfolding of the Von Willebrand factor A domains. *J Mol Biol* 2007;**366**:986–1000.
80. McGrath RT, McRae E, Smith OP, O'Donnell JS. Platelet von Willebrand factor–structure, function and biological importance. *Br J Haematol* 2010;**148**:834–43.
81. Moake JL. Thrombotic microangiopathies. *N Engl J Med* 2002;**347**:589–600.
82. Hassenpflug WA, Budde U, Obser T, Angerhaus D, Drewke E, Schneppenheim S, et al. Impact of mutations in the von Willebrand factor A2 domain on ADAMTS13-dependent proteolysis. *Blood* 2006;**107**:2339–45.

83. Sutherland JJ, O'Brien LA, Lillicrap D, Weaver DF. Molecular modeling of the von Willebrand factor A2 Domain and the effects of associated type 2A von Willebrand disease mutations. *J Mol Model* 2004;**10**:259–70.

84. Zhang Q, Zhou YF, Zhang CZ, Zhang X, Lu C, Springer TA. Structural specializations of A2, a force-sensing domain in the ultralarge vascular protein von Willebrand factor. *Proc Natl Acad Sci USA* 2009;**106**:9226–31.

85. Shankaran H, Neelamegham S. Hydrodynamic forces applied on intercellular bonds, soluble molecules, and cell-surface receptors. *Biophys J* 2004;**86**:576–88.

86. Nishio K, Anderson PJ, Zheng XL, Sadler JE. Binding of platelet glycoprotein Ibalpha to von Willebrand factor domain A1 stimulates the cleavage of the adjacent domain A2 by ADAMTS13. *Proc Natl Acad Sci USA* 2004;**101**:10578–83.

87. Bernardo A, Ball C, Nolasco L, Moake JF, Dong JF. Effects of inflammatory cytokines on the release and cleavage of the endothelial cell-derived ultralarge von Willebrand factor multimers under flow. *Blood* 2004;**104**:100–6.

88. Lam JK, Chion CK, Zanardelli S, Lane DA, Crawley JT. Further characterization of ADAMTS-13 inactivation by thrombin. *J Thromb Haemost* 2007;**5**:1010–8.

89. George JN. Clinical practice. Thrombotic thrombocytopenic purpura. *N Engl J Med* 2006;**354**:1927–35.

90. Silverstein A. Thrombotic thrombocytopenic purpura. The initial neurologic manifestations. *Arch Neurol* 1968;**18**:358–62.

91. Ruggenenti P, Remuzzi G. Thrombotic microangiopathies. *Crit Rev Oncol Hematol* 1991;**11**:243–65.

92. McLeod BC. Thrombotic microangiopathies in bone marrow and organ transplant patients. *J Clin Apher* 2002;**17**:118–23.

93. Peyvandi F, Ferrari S, Lavoretano S, Canciani MT, Mannucci PM. von Willebrand factor cleaving protease (ADAMTS-13) and ADAMTS-13 neutralizing autoantibodies in 100 patients with thrombotic thrombocytopenic purpura. *Br J Haematol* 2004;**127**:433–9.

94. Asada Y, Sumiyoshi A, Hayashi T, Suzumiya J, Kaketani K. Immunohistochemistry of vascular lesion in thrombotic thrombocytopenic purpura, with special reference to factor VIII related antigen. *Thromb Res* 1985;**38**:469–79.

95. Tsai HM, Chandler WL, Sarode R, Hoffman R, Jelacic S, Habeeb RL, et al. von Willebrand factor and von Willebrand factor-cleaving metalloprotease activity in Escherichia coli O157: H7-associated hemolytic uremic syndrome. *Pediatr Res* 2001;**49**:653–9.

96. Hosler GA, Cusumano AM, Hutchins GM. Thrombotic thrombocytopenic purpura and hemolytic uremic syndrome are distinct pathologic entities. A review of 56 autopsy cases. *Arch Pathol Lab Med* 2003;**127**:834–9.

97. Camilleri RS, Cohen H, Mackie IJ, Scully M, Starke RD, Crawley JT, et al. Prevalence of the ADAMTS-13 missense mutation R1060W in late onset adult thrombotic thrombocytopenic purpura. *J Thromb Haemost* 2008;**6**:331–8.

98. Mahdian R, Rayes J, Girma JP, Houllier A, Obert B, Meyer D, et al. Comparison of FRETS-VWF73 to full-length VWF as a substrate for ADAMTS13 activity measurement in human plasma samples. *Thromb Haemost* 2006;**95**:1049–51.

99. Kentouche K, Budde U, Furlan M, Scharfe V, Schneppenheim R, Zintl F. Remission of thrombotic thrombocytopenic purpura in a patient with compound heterozygous deficiency of von Willebrand factor-cleaving protease by infusion of solvent/detergent plasma. *Acta Paediatr* 2002;**91**:1056–9.

100. Assink K, Schiphorst R, Allford S, Karpman D, Etzioni A, Brichard B, et al. Mutation analysis and clinical implications of von Willebrand factor-cleaving protease deficiency. *Kidney Int* 2003;**63**:1995–9.

101. Schneppenheim R, Kremer Hovinga JA, Becker T, Budde U, Karpman D, Brockhaus W, et al. A common origin of the 4143insA ADAMTS13 mutation. *Thromb Haemost* 2006;**96**:3–6.

102. Savasan S, Lee SK, Ginsburg D, Tsai HM. ADAMTS13 gene mutation in congenital thrombotic thrombocytopenic purpura with previously reported normal VWF cleaving protease activity. *Blood* 2003;**101**:4449–51.

103. Matsumoto M, Kokame K, Soejima K, Miura M, Hayashi S, Fujii Y, et al. Molecular characterization of ADAMTS13 gene mutations in Japanese patients with Upshaw-Schulman syndrome. *Blood* 2004;**103**:1305–10.

104. Uchida T, Wada H, Mizutani M, Iwashita M, Ishihara H, Shibano T, et al. Identification of novel mutations in ADAMTS13 in an adult patient with congenital thrombotic thrombocytopenic purpura. *Blood* 2004;**104**:2081–3.

105. Licht C, Stapenhorst L, Simon T, Budde U, Schneppenheim R, Hoppe B. Two novel ADAMTS13 gene mutations in thrombotic thrombocytopenic purpura/hemolytic-uremic syndrome (TTP/HUS). *Kidney Int* 2004;**66**:955–8.

106. Snider CE, Moore JC, Warkentin TE, Finch CN, Hayward CP, Kelton JG. Dissociation between the level of von Willebrand factor-cleaving protease activity and disease in a patient with congenital thrombotic thrombocytopenic purpura. *Am J Hematol* 2004;**77**:387–90.

107. Noris M, Bucchioni S, Galbusera M, Donadelli R, Bresin E, Castelletti F, et al. Complement factor H mutation in familial thrombotic thrombocytopenic purpura with ADAMTS13 deficiency and renal involvement. *J Am Soc Nephrol* 2005;**16**:1177–83.

108. Liu F, Jin J, Dong NZ, Wang YG, Ruan CG. Identification of two novel mutations in ADAMTS13 gene in a patient with hereditary thrombotic thrombocytopenic purpura. *Zhonghua Xue Ye Xue Za Zhi* 2005;**26**:521–4.

109. Donadelli R, Banterla F, Galbusera M, Capoferri C, Bucchioni S, Gastoldi S, et al. In-vitro and in-vivo consequences of mutations in the von Willebrand factor cleaving protease ADAMTS13 in thrombotic thrombocytopenic purpura. *Thromb Haemost* 2006;**96**:454–64.

110. Peyvandi F, Lavoretano S, Palla R, Valsecchi C, Merati G, De Cristofaro R, et al. Mechanisms of the interaction between two ADAMTS13 gene mutations leading to severe deficiency of enzymatic activity. *Hum Mutat* 2006;**27**:330–6.

111. Palla R, Lavoretano S, Lombardi R, Garagiola I, Karimi M, Afrasiabi A, et al. The first deletion mutation in the TSP1-6 repeat domain of ADAMTS13 in a family with inherited thrombotic thrombocytopenic purpura. *Haematologica* 2009;**94**:289–93.

112. Plaimauer B, Fuhrmann J, Mohr G, Wernhart W, Bruno K, Ferrari S, et al. Modulation of ADAMTS13 secretion and specific activity by a combination of common amino acid polymorphisms and a missense mutation. *Blood* 2006;**107**:118–25.

113. Shibagaki Y, Matsumoto M, Kokame K, Ohba S, Miyata T, Fujimura Y, et al. Novel compound heterozygote mutations (H234Q/R1206X) of the ADAMTS13 gene in an adult patient with Upshaw-Schulman syndrome showing predominant episodes of repeated acute renal failure. *Nephrol Dial Transplant* 2006;**21**:1289–92.

114. Tao Z, Anthony K, Peng Y, Choi H, Nolasco L, Rice L, et al. Novel ADAMTS-13 mutations in an adult with delayed onset thrombotic thrombocytopenic purpura. *J Thromb Haemost* 2006;**4**:1931–5.

115. Hommais A, Rayes J, Houllier A, Obert B, Legendre P, Veyradier A, et al. Molecular characterization of four ADAMTS13 mutations responsible for congenital thrombotic thrombocytopenic purpura (Upshaw-Schulman syndrome). *Thromb Haemost* 2007;**98**:593–9.

116. Meyer SC, Jeddi R, Meddeb B, Gouider E, Lammle B, Kremer Hovinga JA. A first case of congenital TTP on the African continent due to a new homozygous mutation in the catalytic domain of ADAMTS13. *Ann Hematol* 2008;**87**:663–6.

117. Kokame K, Aoyama Y, Matsumoto M, Fujimura Y, Miyata T. Inherited and de novo mutations of ADAMTS13 in a patient with Upshaw-Schulman syndrome. *J Thromb Haemost* 2008;**6**:213–5.

118. Garagiola I, Valsecchi C, Lavoretano S, Oren H, Bohm M, Peyvandi F. Nonsense-mediated mRNA decay in the ADAMTS13 gene caused by a 29-nucleotide deletion. *Haematologica* 2008;**93**:1678–85.

119. Fujimura Y, Matsumoto M, Kokame K, Isonishi A, Soejima K, Akiyama N, et al. Pregnancy-induced thrombocytopenia and TTP, and the risk of fetal death, in Upshaw-Schulman syndrome: a series of 15 pregnancies in 9 genotyped patients. *Br J Haematol* 2009;**144**:742–54.

120. Manea M, Kristoffersson A, Tsai HM, Zhou W, Winqvist I, Oldaeus G, et al. ADAMTS13 phenotype in plasma from normal individuals and patients with thrombotic thrombocytopenic purpura. *Eur J Pediatr* 2007;**166**:249–57.

121. Motto DG, Chauhan AK, Zhu G, Homeister J, Lamb CB, Desch KC, et al. Shigatoxin triggers thrombotic thrombocytopenic purpura in genetically susceptible ADAMTS13-deficient mice. *J Clin Invest* 2005;**115**:2752–61.

122. Banno F, Kokame K, Okuda T, Honda S, Miyata S, Kato H, et al. Complete deficiency in ADAMTS13 is prothrombotic, but it alone is not sufficient to cause thrombotic thrombocytopenic purpura. *Blood* 2006;**107**:3161–6.

123. Nolasco LH, Turner NA, Bernardo A, Tao Z, Cleary TG, Dong JF, et al. Hemolytic uremic syndrome-associated Shiga toxins promote endothelial-cell secretion and impair ADAMTS13 cleavage of unusually large von Willebrand factor multimers. *Blood* 2005;**106**:4199–209.

124. Tsai HM. Pathophysiology of thrombotic thrombocytopenic purpura. *Int J Hematol* 2010;**91**:1–19.

125. Jang MJ, Kim NK, Chong SY, Kim HJ, Lee SJ, Kang MS, et al. Frequency of Pro475Ser polymorphism of ADAMTS13 gene and its association with ADAMTS-13 activity in the Korean population. *Yonsei Med J* 2008;**49**:405–8.

126. Akiyama M, Kokame K, Miyata T. ADAMTS13 P475S polymorphism causes a lowered enzymatic activity and urea lability in vitro. *J Thromb Haemost* 2008;**6**:1830–2.

127. Sadler JE, Moake JL, Miyata T, George JN. Recent advances in thrombotic thrombocytopenic purpura. *Hematology Am Soc Hematol Educ Program* 2004;407–23.

128. Coppo P, Bengoufa D, Veyradier A, Wolf M, Bussel A, Millot GA, et al. Severe ADAMTS13 deficiency in adult idiopathic thrombotic microangiopathies defines a subset of patients characterized by various autoimmune manifestations, lower platelet count, and mild renal involvement. *Medicine (Baltimore)* 2004;**83**:233–44.

129. Zheng XL, Kaufman RM, Goodnough LT, Sadler JE. Effect of plasma exchange on plasma ADAMTS13 metalloprotease activity, inhibitor level, and clinical outcome in patients with idiopathic and nonidiopathic thrombotic thrombocytopenic purpura. *Blood* 2004;**103**:4043–9.

130. Rieger M, Mannucci PM, Kremer Hovinga JA, Herzog A, Gerstenbauer G, Konetschny C, et al. ADAMTS13 autoantibodies in patients with thrombotic microangiopathies and other immunomediated diseases. *Blood* 2005;**106**:1262–7.

131. Mannucci PM. Thrombotic thromboytopenic purpura: another example of immunomediated thrombosis. *Pathophysiol Haemost Thromb* 2006;**35**:89–97.

132. Mannucci PM, Bohm M, Scharrer I, Scheiflinger F. Patterns of changes of anti-ADAMTS13 after plasma exchange. *J Thromb Haemost* 2006;**4**:1405–6.

133. Waters B, Qadura M, Burnett E, Chegeni R, Labelle A, Thompson P, et al. Anti-CD3 prevents factor VIII inhibitor development in hemophilia A mice by a regulatory CD4+CD25+-dependent mechanism and by shifting cytokine production to favor a Th1 response. *Blood* 2009;**113**:193–203.

134. Terrell DR, Williams LA, Vesely SK, Lammle B, Hovinga JA, George JN. The incidence of thrombotic thrombocytopenic purpura-hemolytic uremic syndrome: all patients, idiopathic patients, and patients with severe ADAMTS-13 deficiency. *J Thromb Haemost* 2005;**3**:1432–6.

135. Ferrari S, Mudde GC, Rieger M, Veyradier A, Kremer Hovinga JA, Scheiflinger F. IgG subclass distribution of anti-ADAMTS13 antibodies in patients with acquired thrombotic thrombocytopenic purpura. *J Thromb Haemost* 2009;**7**:1703–10.

136. Dong L, Chandrasekaran V, Zhou W, Tsai HM. Evolution of ADAMTS13 antibodies in a fatal case of thrombotic thrombocytopenic purpura. *Am J Hematol* 2008;**83**:815–7.

137. Zhou W, Dong L, Ginsburg D, Bouhassira EE, Tsai HM. Enzymatically active ADAMTS13 variants are not inhibited by anti-ADAMTS13 autoantibodies: a novel therapeutic strategy? *J Biol Chem* 2005;**280**:39934–41.

138. Luken BM, Turenhout EA, Hulstein JJ, Van Mourik JA, Fijnheer R, Voorberg J. The spacer domain of ADAMTS13 contains a major binding site for antibodies in patients with thrombotic thrombocytopenic purpura. *Thromb Haemost* 2005;**93**:267–74.

139. Luken BM, Kaijen PH, Turenhout EA, Kremer Hovinga JA, van Mourik JA, Fijnheer R, et al. Multiple B-cell clones producing antibodies directed to the spacer and disintegrin/thrombospondin type-1 repeat 1 (TSP1) of ADAMTS13 in a patient with acquired thrombotic thrombocytopenic purpura. *J Thromb Haemost* 2006;**4**:2355–64.

140. Luken BM, Turenhout EA, Kaijen PH, Greuter MJ, Pos W, van Mourik JA, et al. Amino acid regions 572-579 and 657-666 of the spacer domain of ADAMTS13 provide a common antigenic core required for binding of antibodies in patients with acquired TTP. *Thromb Haemost* 2006;**96**:295–301.

141. Tsai HM, Li A, Rock G. Inhibitors of von Willebrand factor-cleaving protease in thrombotic thrombocytopenic purpura. *Clin Lab* 2001;**47**:387–92.

142. Favaloro EJ, Bonar R, Kershaw G, Duncan E, Sioufi J, Marsden K. Investigations from external quality assurance programs reveal a high degree of variation in the laboratory identification of coagulation factor inhibitors. *Semin Thromb Hemost* 2009;**35**:794–805.

143. Yomtovian R, Niklinski W, Silver B, Sarode R, Tsai HM. Rituximab for chronic recurring thrombotic thrombocytopenic purpura: a case report and review of the literature. *Br J Haematol* 2004;**124**:787–95.

144. Ruiz-Torres MP, Casiraghi F, Galbusera M, Macconi D, Gastoldi S, Todeschini M, et al. Complement activation: the missing link between ADAMTS-13 deficiency and microvascular thrombosis of thrombotic microangiopathies. *Thromb Haemost* 2005;**93**:443–52.

145. Mannucci PM, Vanoli M, Forza I, Canciani MT, Scorza R. Von Willebrand factor cleaving protease (ADAMTS-13) in 123 patients with connective tissue diseases (systemic lupus erythematosus and systemic sclerosis). *Haematologica* 2003;**88**:914–8.

146. Uemura M, Matsuyama T, Ishikawa M, Fujimoto M, Kojima H, Sakurai S, et al. Decreased activity of plasma ADAMTS13 may contribute to the development of liver disturbance and multiorgan failure in patients with alcoholic hepatitis. *Alcohol Clin Exp Res* 2005;**29**:264S–NaN.

147. Martin K, Borgel D, Lerolle N, Feys HB, Trinquart L, Vanhoorelbeke K, et al. Decreased ADAMTS-13 (A disintegrin-like and metalloprotease with thrombospondin type 1 repeats) is associated with a poor prognosis in sepsis-induced organ failure. *Crit Care Med* 2007;**35**:2375–82.

148. Larkin D, de Laat B, Jenkins PV, Bunn J, Craig AG, Terraube V, et al. Severe Plasmodium falciparum malaria is associated with circulating ultra-large von Willebrand multimers and ADAMTS13 inhibition. *PLoS Pathog* 2009;**5**:e1000349.

149. Uemura M, Fujimura Y, Matsumoto M, Ishizashi H, Kato S, Matsuyama T, et al. Comprehensive analysis of ADAMTS13 in patients with liver cirrhosis. *Thromb Haemost* 2008;**99**:1019–29.
150. Bridges DJ, Bunn J, van Mourik JA, Grau G, Preston RJ, Molyneux M, et al. Rapid activation of endothelial cells enables Plasmodium falciparum adhesion to platelet-decorated von Willebrand factor strings. *Blood* 2010;**115**:1472–4.
151. Lowenberg EC, Charunwatthana P, Cohen S, van den Born BJ, Meijers JC, Yunus EB, et al. Severe malaria is associated with a deficiency of von Willebrand factor cleaving protease, ADAMTS13. *Thromb Haemost* 2010;**103**:181–7.
152. Bianchi V, Robles R, Alberio L, Furlan M, Lammle B. Von Willebrand factor-cleaving protease (ADAMTS13) in thrombocytopenic disorders: a severely deficient activity is specific for thrombotic thrombocytopenic purpura. *Blood* 2002;**100**:710–3.
153. Sanchez-Luceros A, Farias CE, Amaral MM, Kempfer AC, Votta R, Marchese C, et al. von Willebrand factor-cleaving protease (ADAMTS13) activity in normal non-pregnant women, pregnant and post-delivery women. *Thromb Haemost* 2004;**92**:1320–6.
154. Lattuada A, Rossi E, Calzarossa C, Candolfi R, Mannucci PM. Mild to moderate reduction of a von Willebrand factor cleaving protease (ADAMTS-13) in pregnant women with HELLP microangiopathic syndrome. *Haematologica* 2003;**88**:1029–34.
155. George JN. The association of pregnancy with thrombotic thrombocytopenic purpura-hemolytic uremic syndrome. *Curr Opin Hematol* 2003;**10**:339–44.
156. Bonnefoy A, Daenens K, Feys HB, De Vos R, Vandervoort P, Vermylen J, et al. Thrombospondin-1 controls vascular platelet recruitment and thrombus adherence in mice by protecting (sub)endothelial VWF from cleavage by ADAMTS13. *Blood* 2006;**107**:955–64.
157. Eckmann CM, De Laaf RT, Van Keulen JM, Van Mourik JA, De Laat B. Bilirubin oxidase as a solution for the interference of hyperbilirubinemia with ADAMTS-13 activity measurement by FRETS-VWF73 assay. *J Thromb Haemost* 2007;**5**:1330–1.
158. Chen J, Fu X, Wang Y, Ling M, McMullen B, Kulman J, et al. Oxidative modification of von Willebrand factor by neutrophil oxidants inhibits its cleavage by ADAMTS13. *Blood* 2010;**115**:706–12.
159. Lancellotti S, De Filippis V, Pozzi N, Peyvandi F, Palla R, Rocca B, et al. Formation of methionine sulfoxide by peroxynitrite at position 1606 of von Willebrand factor inhibits its cleavage by ADAMTS-13: a new prothrombotic mechanism in diseases associated with oxidative stress. *Free Radic Biol Med* 2010;**48**:446–56.
160. Meyer SC, Sulzer I, Lammle B, Kremer Hovinga JA. Hyperbilirubinemia interferes with ADAMTS-13 activity measurement by FRETS-VWF73 assay: diagnostic relevance in patients suffering from acute thrombotic microangiopathies. *J Thromb Haemost* 2007;**5**:866–7.
161. Raife TJ, Cao W, Atkinson BS, Bedell B, Montgomery RR, Lentz SR, et al. Leukocyte proteases cleave von Willebrand factor at or near the ADAMTS13 cleavage site. *Blood* 2009;**114**:1666–74.
162. Hovinga JA, Vesely SK, Terrell DR, Lammle B, George JN. Survival and relapse in patients with thrombotic thrombocytopenic purpura. *Blood* 2010;**115**:1500–11, quiz 1662.
163. Tripodi A, Chantarangkul V, Bohm M, Budde U, Dong JF, Friedman KD, et al. Measurement of von Willebrand factor cleaving protease (ADAMTS-13): results of an international collaborative study involving 11 methods testing the same set of coded plasmas. *J Thromb Haemost* 2004;**2**:1601–9.
164. Tripodi A, Peyvandi F, Chantarangkul V, Palla R, Afrasiabi A, Canciani MT, et al. Second international collaborative study evaluating performance characteristics of methods measuring the von Willebrand factor cleaving protease (ADAMTS-13). *J Thromb Haemost* 2008;**6**:1534–41.

165. Wada H, Kaneko T, Ohiwa M, Tanigawa M, Tamaki S, Minami N, et al. Plasma cytokine levels in thrombotic thrombocytopenic purpura. *Am J Hematol* 1992;**40**:167–70.
166. Klukowska A, Niewiadomska E, Budde U, Oyen F, Schneppenheim R. Difficulties in diagnosing congenital thrombotic thrombocytopenic purpura. *J Pediatr Hematol Oncol* 2010;**32**:103–7.
167. Park HW, Oh D, Kim N, Cho HY, Moon KC, Chae JH, et al. Congenital thrombotic thrombocytopenic purpura associated with unilateral moyamoya disease. *Pediatr Nephrol* 2008;**23**:1555–8.

Thrombin as an Anticoagulant

Enrico Di Cera

Department of Biochemistry and Molecular Biology, Saint Louis University School of Medicine, St. Louis, Missouri, USA

I. Preamble 146
II. Thrombin Interactions 147
III. Thrombin Structure 148
IV. Thrombin is an Allosteric Enzyme 151
V. Structures of E*, E, and E:Na$^+$ 153
VI. Dissociating Procoagulant and Anticoagulant Activities 158
VII. WE: A Prototypic Anticoagulant Thrombin 163
VIII. Molecular Mechanism of Anticoagulant Activity of Thrombin Mutants 165
IX. Beyond WE 170
X. Conclusions 175
References 175

Thrombosis is the most prevalent cause of fatal diseases in developed countries. An antithrombotic agent that can be administered to patients with severe acute thrombotic diseases without the risk of causing hemorrhage, as experienced with antithrombotic/thrombolytic therapy in the treatment of acute ischemic stroke or systemic anticoagulants like heparin, would likely revolutionize the treatment of cardiovascular and cerebrovascular diseases. Thrombin remains at the forefront of cardiovascular medicine and a major target of antithrombotic and anticoagulant therapies, due to its involvement in thrombotic deaths. Heparins and direct thrombin inhibitors currently used in the treatment of acute thrombotic complications, especially in the venous circulation, are plagued by complications related to dosage and bleeding. A new strategy of intervention has been proposed in recent years aiming at modulating, rather than inhibiting, thrombin function. Specifically, efforts have been directed toward finding ways of exploiting the anticoagulant function of thrombin unleashed by the activation of protein C, either using small modulators or protein engineering. The ability of thrombin to activate protein C coexists with its procoagulant and prothrombotic functions, mediated respectively by cleavage of fibrinogen and the protease-activated receptor 1 (PAR1). A strategy that inhibits thrombin at the active site abrogates the procoagulant and prothrombotic functions, but also shuts down activity toward the anticoagulant protein C. On the other hand, a strategy that selectively compromises fibrinogen and PAR1 recognition may take advantage of the anticoagulant and cytoprotective functions of activated protein C and prove

Progress in Molecular Biology
and Translational Science, Vol. 99
DOI: 10.1016/S1877-1173(11)99004-8

1877-1173/11 $35.00

of interest for *in vivo* applications. This chapter summarizes current protein engineering efforts to convert thrombin into a potent and safe anticoagulant for *in vivo* applications.

I. Preamble

Although substantial progress has been made in the prevention and treatment of cardiovascular disease and its major risk factors, it has been predicted that thrombotic complications will remain the leading cause of death and disability and will represent a major burden to productivity worldwide well into the year 2020.[1] Indeed, thrombosis is the most prevalent cause of fatal diseases in developed countries. More than 80% of stroke cases are of thrombotic origin, and stroke is among the three leading causes of mortality and severe chronic disability in the US.[2] One of the causal treatment options for acute ischemic stroke is early antithrombotic/thrombolytic therapy. A critical problem with such therapy is that the currently available potent antithrombotic agents, most notably tissue-type plasminogen activator, are not specific for thrombosis. They disable the hemostatic system at their most efficacious doses and, for this reason, cannot be administered at their fully effective doses.[3] Systemic anticoagulants like heparin prevent comorbidity from deep vein thrombosis and reduce the progression of acute cerebral thrombosis. However, their bleeding side effects in the acute phase outweigh the benefits.[4] An antithrombotic agent that can be administered to patients with severe acute thrombotic diseases, such as heart attack and stroke, without the risk of causing hemorrhage, would be so significant as to revolutionize the treatment of cardiovascular and cerebrovascular diseases. Progress in the developments of new anticoagulants and antithrombotics, therefore, remains a top medical and social priority and has the potential to impact the lifestyle and life expectancy of millions of people worldwide.

Due to its involvement in thrombotic deaths, thrombin remains at the forefront of cardiovascular medicine and a major target of antithrombotic and anticoagulant therapies.[5] Because heparin and direct thrombin inhibitors are plagued with complications related to dosage and bleeding,[4,5] an entirely new strategy of intervention has been proposed in recent years to take advantage of the so-called protein C anticoagulant pathway.[6] Protein C requires thrombin for activation and activated protein C acts as a potent anticoagulant and cytoprotective agent.[6,7] A thrombin mutant engineered for exclusive activity toward protein C and devoid of activity toward fibrinogen and the platelet receptor PAR1 could represent an innovative and potentially powerful tool to achieve anticoagulation without disruption of the hemostatic balance.

Proof-of-principle that such a strategy could benefit the treatment of thrombotic diseases has come from *in vivo* studies in nonhuman primates.[8–11] Specifically, the anticoagulant thrombin mutant, W215A/E217A (WE), has been shown to elicit an antithrombotic effect that is more efficacious than the direct administration of activated protein C and safer than the administration of low molecular weight heparins.[10] Substantial room for optimization remains. The procoagulant activity of the mutant can be abrogated by further protein engineering to eliminate any possible side effects due to fibrinogen clotting or platelet aggregation. This problem presents a challenging task for basic science with tremendous relevance to the medical community. Developments toward the goal of dissociating thrombin functions and optimizing its anticoagulant activity toward protein C are the main focus of this chapter.

II. Thrombin Interactions

Blood coagulation is initiated by the exposure of tissue factor that forms a complex with factor VIIa and results in the generation of small quantities of factors IXa and Xa.[12,13] The small quantities of Xa generate minute concentrations of thrombin that result in the activation of factor XI and the cofactors VIII and V. At this point, the VIIIa–IXa complex generates sufficient quantities of Xa to form the prothrombinase complex, composed of factors Va, Xa, Ca^{2+}, and phospholipids, which leads to the explosive generation of thrombin from prothrombin.[14] Thrombin is a trypsin-like protease endowed with important physiological functions that are mediated and regulated by interaction with numerous macromolecular substrates, receptors, and inhibitors.[15–18] Activity of the enzyme toward synthetic and physiological substrates is enhanced allosterically by the binding of Na^+ to a site located > 15 Å away from residues of the catalytic triad H57, D102, and S195[19,20] (Fig. 1). Na^+ activation is present in all vitamin K-dependent clotting enzymes and many complement factors.[21] The effect of Na^+ is seen not only on cleavage of fibrinogen and PAR1,[22–24] the primary procoagulant and prothrombotic substrates,[16,25,26] but also on PAR3 and PAR4[15,27] and activation of factors V,[28] VIII,[29] and XI[30] that ensure the build-up of coagulation factors responsible for the explosive generation of thrombin from prothrombin.[14,31] Importantly, Na^+ binding has no effect on protein C activation in the absence or presence of thrombomodulin,[23,24] which makes Na^+ an exquisite procoagulant/prothrombotic cofactor of thrombin. Evidence that Na^+ plays an important role in blood coagulation through its specific interaction with thrombin comes from the observation that several naturally occurring mutations, like prothrombin Frankfurt (E146A),[32] Salakta (E146A),[33] Greenville (R187Q),[34] Scranton (K224T),[35] Copenhagen (A190V),[36] and Saint Denis (D221E),[37] affect residues linked to Na^+ binding[20]

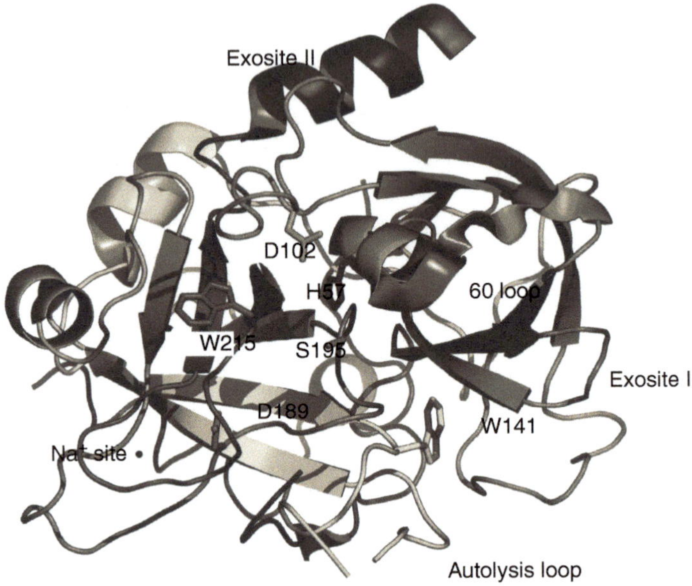

FIG. 1. The structure of thrombin in the E:Na^+ form (PDB ID code 1SG8[20]) rendered as a ribbon in spectrum color. The enzyme is composed of two polypeptide chains of 36 (A chain) and 259 (B chain) residues that are covalently linked through a disulfide bond.[17,44] In this standard orientation,[44] the A chain (blue) runs in the back of the molecule, opposite to the front hemisphere of the catalytic B chain. Catalytic residues (H57, D102, S195) and the S1 site (D189) are labeled, along with the two major fluorophores, W215 and W141. The bound Na^+ (red ball) is 15 Å away from the residues of the catalytic triad and within 5 Å from the side chain of D189 in the S1 site, nestled between the 220- (red) and 186- (orange) loops. Exosite I is on the opposite side of the B chain relative to the Na^+ site and comprises residues of the 30- (cyan) and 70- (green) loops. Other important regions of the enzyme are noted. (For interpretation of the references to color in this figure legend, the reader is referred to the Web version of this chapter.)

and are associated with bleeding. Furthermore, anticoagulant thrombin mutants have been engineered rationally by perturbing the Na^+ site.[8,9,11,24,38,39] The remarkable success in preclinical studies[9,10,40,41] indicates that such mutants may soon offer viable alternatives to existing anticoagulants like heparin.[42]

III. Thrombin Structure

The prothrombinase complex converts prothrombin to the mature enzyme along two pathways by cleaving sequentially at R271 and R320 (prothrombin numbering). Initial cleavage at R320 (R15 in the chrymotrypsin numbering) between the A and B chains is the preferred pathway under physiological conditions and generates the active intermediate, meizothrombin, by triggering formation of the I16–D194 ion-pair, and structuring of the active site and oxyanion hole.[43] The alternative initial cleavage at R271 sheds the Gla domain and the two kringles and generates the inactive precursor, prethrombin-2, with the R15–I16 peptide bond intact. Thrombin is composed of two polypeptide chains of 36 (A chain) and 259 (B chain) residues that are covalently linked through a disulfide bond between residues C1 and C122.[17,44] The standard orientation[44] puts the A chain at the back of the molecule, opposite to the front hemisphere of the B chain that hosts the entrance to the active site and all known functional epitopes of the enzyme (Fig. 1).[20] The A chain has received little attention in thrombin studies and is considered an appendage of the activation process from prothrombin. Previous studies have suggested that the A chain may be dispensable for function.[45,46] However, several naturally occurring mutations of prothrombin involve residues of the A chain[47–50] and are associated with severe bleeding. The functional defects in prothrombins Denver (E8K and E14cK),[47] Segovia (G14mR),[49] and San Antonio (R15H)[50] have been attributed to perturbation of the zymogen → enzyme conversion and processing by factor Xa, resulting in severe bleeding. Such explanation is obvious for the G14mR and R15H mutations that affect the P1 (R15) and P2 (G14m) sites of recognition by factor Xa, but not for the E8K and E14cK mutations of prothrombin Denver. Other naturally occurring mutations, like deletion of K9 or K10,[48] are also associated with severe bleeding. Interestingly, the defect causes impaired fibrinogen and PAR1 cleavage, reduced response to Na^+ activation,[51,52] and long-range perturbation of active site residues.[52] Recent mutagenesis of the A chain has explained the phenotype observed in naturally occurring mutations and pointed out the importance of this domain in thrombin function.[53]

The B chain contains all residues responsible for catalytic activity, substrate recognition, and allosteric regulation. Trypsin-like specificity for Arg residues at P1[54] is conferred to thrombin by the presence of D189 in the S1 site occupying the bottom of the catalytic pocket. Unlike trypsin, however, thrombin can efficiently cleave chymotrypsin-specific substrates carrying Phe at the P1 position,[55] as documented more than 40 years ago from studies on ester substrates.[56–58] Thrombin has a preference for small and hydrophobic side chains at P2 that pack tightly against the hydrophobic wall of the S2 site defined by residues Y60a-P60b-P60c-W60d of the 60-loop. Residues at P3 point away from the thrombin surface, whereas aromatic and hydrophobic residues at P4 tend to fold back on the thrombin surface[59] and engage the aryl binding site defined by L99, I174, and W215.[44] An important exception has been identified recently from the structure of thrombin in complex with the uncleaved extracellular fragment of PAR1, where the acidic residue at the P3 position of substrate makes an H-bonding interaction with the backbone N atom of G219 instead of pointing away from the thrombin surface.[60] The autolysis loop shapes the lower rim of access to the active site and contributes to recognition of fibrinogen[61] and the intrinsic allosteric properties of the enzyme.[39] The loop centered on K70 defines exosite I and is homologous to the Ca^{2+} binding loop of trypsin and chymotrypsin.[62] In these proteases, Ca^{2+} stabilizes the fold and confers increased resistance to proteolytic digestion. In thrombin, the need for Ca^{2+} is eliminated by insertion of K70 in the cavity available for binding this cation. Thrombin does not bind Ca^{2+} up to mM concentrations.[23,63] Exosite I contains several positively charged residues that give rise to an intense electrostatic field. The field provides steering and optimal preorientation for fibrinogen, thrombomodulin, the natural inhibitor, hirudin, and PAR1 to facilitate formation of a productive complex upon binding. Structural and site-directed mutagenesis data support exosite I as a binding epitope for fibrinogen,[22,64,65] fibrin,[22,66] thrombomodulin,[64,67–70] and the thrombin receptors, PAR1[22,60,71,72] and PAR3.[22,73] On the side of the enzyme opposite to exosite I, a C-terminal helix and its neighbor domains host a number of positively charged residues and define exosite II. This site is the locale for interaction with polyanionic ligands, like heparin and glucosaminoglycans,[74–78] and fragment 2 in thrombin precursors.[79,80] Heparin enhances inhibition of thrombin by antithrombin via a template mechanism in which a high affinity heparin–antithrombin complex is first formed and then docked into exosite II and the thrombin active site by electrostatic coupling.[77,81–83] Exosite II is also the locale for thrombin interaction with the platelet receptor GpIb,[84–87] the acidic moiety of the fibrinogen γ' chain,[88] and has been involved in the binding of autoantibodies.[89]

The first X-ray structure of thrombin was solved in 1992 and revealed relevant information on the overall fold of the enzyme and especially on the arrangement of loops involved in macromolecular substrate recognition.[44] However, this and many subsequent structures of thrombin overlooked the bound Na^+, although crystals were solved at high resolution and in the presence of high concentrations of Na^+. The Na^+ binding site of thrombin was first identified crystallographically in 1995 from Rb^+ replacement.[19] Notably, this was also the first Na^+ binding site identified in the large family of M^+-activated enzymes to which thrombin belongs.[90] Na^+ binds 16–20 Å away from residues of the catalytic triad and within 5 Å from D189 in the S1 site, nestled between the 220- and 186- loops, and coordinated octahedrally by two carbonyl O atoms from the protein (residues R221a and K224) and four buried water molecules anchored to the side chains of D189, D221, and the backbone atoms of G223 and Y184a. The site is highly specific for Na^+ that binds with significantly ($>$ 10-fold) higher affinity compared with Li^+, K^+, or Rb^+.[91] Elucidation of the architecture of the Na^+ binding site of thrombin facilitated the subsequent identification of the analogous Na^+ binding sites in factor Xa,[92,93] factor VIIa,[94] and activated protein C.[95] Identification of the Na^+ binding site of factor IXa remains controversial.[96] Several groups have shown that Na^+ has a significant effect on the activity of factors VIIa,[21,97] IXa,[98,99] Xa,[100–106] and activated protein C,[107–111] and interest continues to grow on the physiological role of Na^+ in these proteases and blood coagulation in general.[112,113]

Structural biology has revealed important information on how thrombin utilizes both the active site and exosites for interaction with substrates, inhibitors, and effectors,[17,114,115] a basic strategy that is also exploited by factor Xa and other clotting proteases.[115–117] Information on how thrombin recognizes substrate at the active site has come from the structure of the enzyme in complex with the irreversible active site inhibitor, H-D-Phe-Pro-Arg-CH_2Cl (PPACK).[44] Arg at P1 ion-pairs to D189 in the S1 site, Pro at P2 fits snugly against P60b, P60c, and W60d in the S2 site and Phe at P3, in the D-enantiomer, makes an edge-to-face interaction with W215 in the aryl binding site. The PPACK-inhibited structure reveals interactions that are relevant to recognition of natural substrates and confirms the key role played by the H-bonding network found within the active site of all trypsin-like enzymes bound to substrate.[118–120] Crucial components of this network are the bidentate ion-pair between D189 and the guanidinium group of Arg at P1, the H-bonds of the carbonyl O atom of the P1 residue with the N atoms of G193 and S195 forming the oxyanion hole, the H-bond between the N atom of the P1 residue and the carbonyl O atom of S214, and the H-bonds between the backbone O and N atoms of the P3 residue with the N and O atoms of G216. This important arrangement of H-bonds has been documented in the structures of thrombin bound to fragments of the natural substrates fibrinogen,[121] PAR1,[60] PAR4,[122] and factor XIII.[123] The structure of thrombin in complex with the potent natural inhibitor, hirudin, has revealed how thrombin recognizes ligands at exosite I.[124] Hirudin blocks access to the active site of thrombin using its compact N-terminal domain and binds to exosite I via its extended, acidic C-terminal domain. The mode of interaction of the C-terminal domain of hirudin has later been documented in the structures of thrombin bound to hirugen,[125] fibrinogen,[65,66,126] PAR1,[60,127] PAR3,[122] thrombomodulin,[68] and heparin cofactor II.[128] Finally, the role of exosite II has been documented in the structures of thrombin bound to heparin,[77,78,83] the fibrinogen γ' peptide,[88] and GpIb.[85,86]

IV. Thrombin is an Allosteric Enzyme

The most striking feature of thrombin is its ability to assume different conformations.[15] This property has been revealed by a series of functional and structural studies aimed at delineating the interaction of thrombin with Na^+ and its effect on procoagulant (fibrinogen), prothrombotic (PARs), and anticoagulant (protein C) substrates.[23,24,129] Na^+ activation of thrombin involves the transition of the enzyme between two active forms,[130] leading to an increase in k_{cat} and a decrease in K_m. These forms are E (Na^+-free) with low activity and E:Na^+ (Na^+-bound) with high activity, originally defined as the slow and fast forms.[130] The E and E:Na^+ forms are significantly (2:3 ratio) populated under physiological conditions because the K_d for Na^+ binding is 110 mM at 37 °C[91,130–133] and the physiological [NaCl] = 140 mM is not sufficient for saturation. The role of the two active species in equilibrium, E and E:Na^+, can be cast mathematically in terms of the well-known Botts–Morales scheme for the action of a "modifier" on enzyme function.[134] In this scheme, binding of Na^+ to thrombin is expected to produce a single kinetic phase reflecting the conversion of E into E:Na^+. Recent findings, however, indicate that this is not the case. Na^+ binding to thrombin gives rise to two kinetic phases, one fast (in the μs time scale) due to Na^+ binding to E to produce E:Na^+[133,135] and another considerably slower (in the ms time scale) that reflects a pre-equilibrium between E and another form, E*, that is unable to bind Na^+ or substrate to the active site.[133,136] The three forms of thrombin lead to the kinetic scheme:

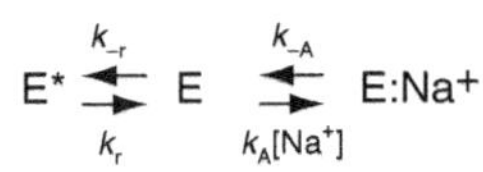

SCHEME 1.

Two Na^+-free forms, E* and E, interconvert with kinetic rate constants, k_r and k_{-r}, and E interacts with Na^+ with a rate constant, k_A, to populate E:Na^+ that may dissociate into the parent components with a rate constant, k_{-A}. The presence of E* redefines the slow form[130] as a mixture of two conformations in equilibrium, E* and E, which requires extension of the Botts–Morales scheme[134] in terms of the E*, E, and E:Na^+ forms in Scheme 1, as indicated in Scheme 2[137]:

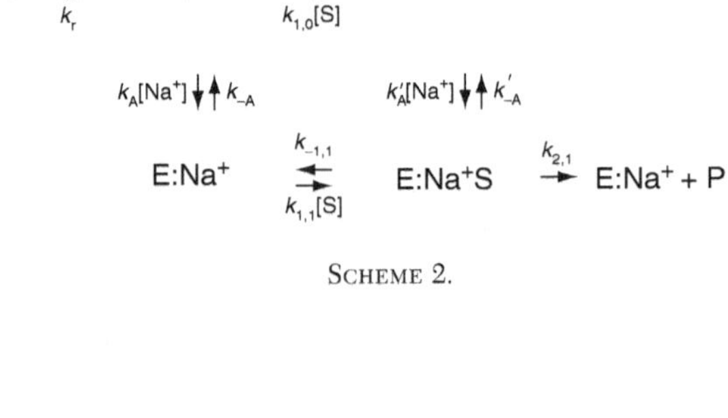

SCHEME 2.

The active forms, E and E:Na^+, feature different values of kinetic rate constants for binding ($k_{1,0}$, $k_{1,1}$) and dissociation ($k_{-1,0}$, $k_{-1,1}$) of substrate S, as well as rates of substrate conversion into product ($k_{2,0}$, $k_{2,1}$). The parameters $K_A = k_A/k_{-A}$ and $K_A' = k_A'/k_{-A}'$ are the equilibrium association constants for Na^+ binding to E and ES, respectively. The ratio $r = k_{-r}/k_r = [E^*]/[E]$ is the equilibrium constant for the E*–E interconversion and removal of E* ($r = 0$) converts Scheme 2 into the original Botts–Morales scheme. To derive the expressions for the independent parameters $s = k_{cat}/K_m$ and k_{cat}, Scheme 2 must be solved for the velocity of product formation at steady state, v, per unit active enzyme concentration, e_{tot}. The solution is known in the case of the Botts–Morales scheme.[15,134,138] Hence, solution of Scheme 2 can be obtained in terms of the known solution of the Botts–Morales scheme perturbed by the presence of E*.[137]

V. Structures of E*, E, and E:Na^+

In 2002, the thrombin mutant R77aA, devoid of the autoproteolytic site in exosite I, was crystallized free of Na^+ and inhibitors and yielded the first structure of thrombin in the Na^+-free form, E.[139] This structure was followed in 2004 by higher resolution structures of the E and E:Na^+ forms, free or bound to the active site inhibitor, PPACK,[20] that have revealed some of the changes caused by Na^+ binding. Structures of E and E:Na^+ are highly similar, with r.m.s. deviations of the Cα traces of only 0.38 Å. PPACK-bound forms of thrombin are practically identical to each other (r.m.s. = 0.19 Å), except for the obvious absence of Na^+ in the PPACK-inhibited E form. A small (1 Å) upward shift is observed in the 60-loop in the structure of the E form relative to the E:Na^+ form that could explain the involvement of W60d in the fluorescence change linked to E to E:Na^+ transition.[133] There are five main structural differences between the E and E:Na^+ forms of thrombin (Fig. 2): (1) the R187–D222 ion-pair; (2) orientation of D189 in the primary specificity pocket; (3) conformation of E192 at the entrance of the active site; (4). position of the catalytic S195; and (5) architecture of the water network spanning > 20 Å from the Na^+ site to the active site. The R187–D222 ion-pair connects the 220- and 186- loops that define the Na^+ site. D222 belongs to the allosteric core and the mutant D222A has drastically impaired Na^+ binding, a property mirrored by the R187A mutant.[20] In the E:Na^+ form, the guanidinium N atoms of R187 are 2.7 and 3.1 Å from the carboxylate O atoms of D222. In the slow form, these distances become 3.3 and 4.8 Å, respectively. Breakage of the ion-pair was observed in the low-resolution structure of the E form[139] and is consistent with the properties of the R187A and D222A mutants. Notably, the broken ion-pair shifts the backbone O atom of R221a that directly coordinates Na^+ in the

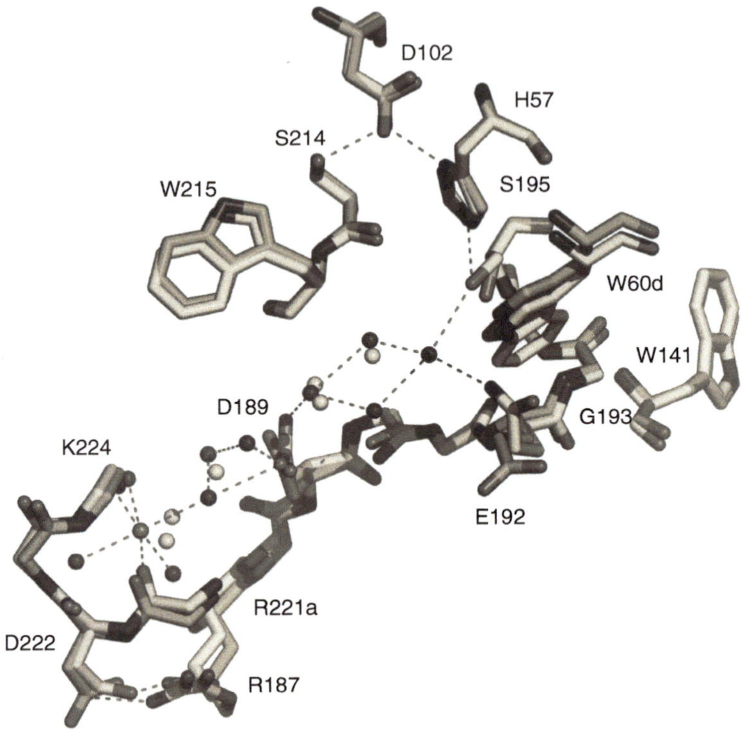

FIG. 2. Changes induced by Na^+ binding to thrombin revealed by the structures of the E (PDB ID code 1SGI, blue) and E:Na^+ (PDB ID code 1SG8, yellow) forms. The main changes induced by Na^+ (magenta ball) binding are formation of the R187:D222 ion-pair that causes a shift in the backbone O atom of R221a, reorientation of D189 that accounts for the change in substrate binding (lower K_m), shift of the side chain of E192, and shift in the position of the Oγ atom of S195 that engages H57 in an H-bond required for efficient catalysis (higher k_{cat}). A network of water molecules in the E:Na^+ form (red balls) connects Na^+ to the side chain of D189 and continues on to reach the Oγ atom of S195. A critical link in the network is provided by a water molecule that H-bonds to S195 and E192. This water molecule is removed in the E form, causing a reorientation of E192 and S195. The connectivity of water molecules in the E form (cyan balls) is compromised by the lack of Na^+ and proper anchoring of the side chain of D189. H-bonds (broken lines) refer to the E:Na^+ form. No change is observed for W215 and W141, but W60d shifts slightly in the E form. (See Color Insert.)

E:Na^+ form and moves it into an orientation that is incompatible with Na^+ binding. The structure of the thrombin mutant D221A/D222K also assumes signatures characteristic of the E form.[140] It should be pointed out that residue 222 is Lys in murine thrombin and that K222 confers this enzyme high activity without Na^+ activation due to molecular mimicry of the bound Na^+.[55,141]

Perturbation of the primary specificity pocket is evidenced in the E → E:Na^+ transition of thrombin. D189 in the E:Na^+ form is optimally oriented for electrostatic coupling with the P1 Arg residue of substrate. In the E form, the carboxylate of D189 experiences a 30° rotation that moves the Oδ1 atom up to 1.1 Å away from its optimal coupling with the guanidinium group of Arg at the P1 position of substrate. Rearrangement of D189 upon Na^+ binding enhances substrate specificity by improving the K_m. The structure of the E form supports a key role for D189 in both Na^+ coordination and allosteric transduction of Na^+ binding into enhanced catalytic activity[142] consistent with results from mutagenesis data. Further conformational changes are observed at the level of the side chain of E192 that moves away from the active site region in the E form. Such movement could minimize the potential electrostatic clash with the P3 and P3′ acidic residues of protein C and could explain why the E form retains high activity toward this anticoagulant substrate.[23,143,144] Subtle changes at the enzyme active site also include conformational alteration of the catalytic triad. In the E form of thrombin, the nucleophilic S195 side chain rotates about 35° in the E form relative to the E:Na^+ form and breaks the critical H-bond with the catalytic H57, which itself also shifts slightly away from S195. Integrity of the H-bond is important for catalysis,[145] and the unfavorable position of S195 in the E form may explain the lower k_{cat} observed in the absence of Na^+.

Current structures of the E and E:Na^+ forms of thrombin fail to document a number of important changes supported by functional data. Mutant N143P demonstrates that the oxyanion hole is disrupted by a flipped E192–G193 peptide bond in the E form and assumes the correct architecture upon transition to the E:Na^+ form.[137] W215 and W141 involved in the spectral changes linked to Na^+ binding[133] do not undergo significant movement in the E and E:Na^+ forms. A recent structure of thrombin in the free form and in the absence of Na^+ shows a 3.0 Å shift in the position of E192 that occludes access to the active site and flips the E192–G193 peptide bond,[146] which would validate the predictions drawn from the recent results on the N143P mutant.[137] The structure also features a 190° flip in the side chain of W215, which would support involvement of this residue as a fluorophore in the E–E:Na^+ transition. However, this structure is biased by the presence of extensive crystal packing involving the active site and Na^+ site and should be confirmed under other crystallization conditions and packing before it could be considered a genuine representation of the E form of thrombin.

A major advance to our understanding of the molecular basis of thrombin allostery has come from the structural identification of the E* form from a number of thrombin mutants.[39,137,147,148] The structure of E* portrays a conformation of thrombin that is unable to bind Na^+ and ligands to the active site. The overall fold of E* has a backbone rms deviation of 0.77 Å compared to the

E:Na^+ form[20] and shows changes in the oxyanion hole, primary specificity pocket, and in the Na^+ binding site that provide a plausible interpretation of the peculiar properties of this inactive form. The oxyanion hole is disrupted by a flip of the E192–G193 peptide bond. The 215–219 segment collapses into the primary specificity pocket (Fig. 3). W215 shifts 130° at the Cβ atom and causes the indole ring to pack against the hydrophobic pocket in the active site formed by W60d, Y60a, H57, and L99. In this conformation, the indole of W215 occupies the same position as the Pro ring of the active site inhibitor PPACK in the PPACK-inhibited structure of thrombin, suggesting that the collapse of W215 into the active site is driven by the favorable hydrophobic environment of the S2 site near the catalytic H57.[44] Downstream from the 215–219 segment, the twist in the backbone caused by the collapse of W215 moves the entire

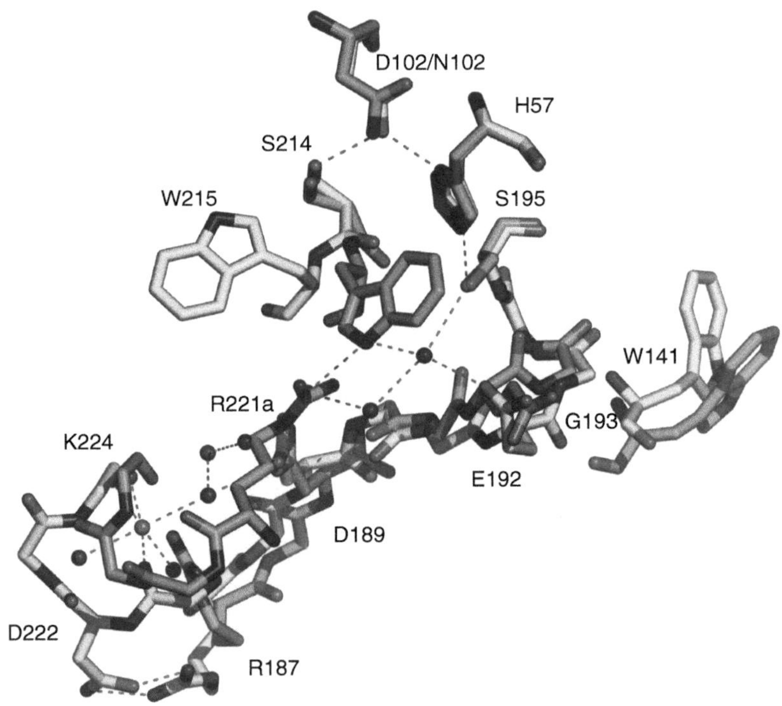

FIG. 3. The structure of E* (PDB ID code 3BEI, green) compared to E:Na^+ (yellow as in Fig. 2). H-bonds (broken lines) refer to the E:Na^+ form. W215 collapses into the active site, R221a penetrates the protein core and engages the side chain of D189 in a strong ion-pair interaction, and R187 obliterates the Na^+ (magenta ball) site by positioning its guanidinium group within 1 Å from the bound cation. Significant movement is also observed at the level of W141. (See Color Insert.)

220-loop upward and changes the orientation of several residues. The side chain of R221a, located > 10 Å away from the site of mutation, rotates 95° and brings the guanidinium group in contact with D189 in the primary specificity pocket. In this conformation, the guanidinium group of R221a occupies a position analogous to that of the guanidinium group of the Arg of PPACK, or of an incoming substrate molecule. Altogether, the drastic shifts of W215 and R221a produce a conformation of thrombin that is self-inhibited by the hydrophobic engagement of the 60-loop and active site H57 by Trp-215, and of the acidic moiety of D189 in the primary specificity pocket by R221a. Residues D221 and D222 flanking R221a shift upward, together with the entire 220-loop, and free the side chain of R187, which is ion-paired to them in the E:Na^+ form. The acquired mobility of R187 and the upward shift of the 220-loop cause the side chain of R187 to penetrate the protein core and the Na^+ site, with the guanidinium group positioned within 1 Å from where Na^+ would bind.

The drastic structural changes in the E* form involving W215 and the oxyanion hole occur with the ion-pair I16–D194 intact, suggesting that E* is not equivalent to the zymogen form of the protease and that the E*–E equilibrium is established after the conversion from the zymogen form has taken place. Indeed, existing structures of the zymogen forms of trypsin,[149] chymotrypsin,[150] chymase,[151] and thrombin itself[125,152] feature a broken I16–D194 ion-pair, but no collapse of the 215–217 segment. Stopped-flow experiments show that the E*–E conversion takes place on a time scale < 10 ms,[133] as opposed to the much longer (100–1000 ms) time scale required for the zymogen–protease conversion.[153,154]

The E* form is not a peculiarity of thrombin. A collapsed conformation of the 215–217 segment is also seen in the structures of αI-tryptase,[155] the high-temperature-requirement-like protease,[156] complement factor D,[157] granzyme K,[158] hepatocyte growth factor activator,[159] prostate kallikrein,[160] and prostasin.[161,162] Examples of a perturbed conformation of the oxyanion, as seen in the E* form of thrombin,[39,137,147,148] have been reported for complement factor B,[163] the arterivirus nsp4,[164] the epidermolyic toxin A,[165,166] and clotting factor VIIa.[94] The physiological relevance of the E*–E equilibrium is illustrated by numerous examples. Stabilization of E* is particularly important in maintaining the resting state of complement factor D that does not have a zymogen form or known natural inhibitors.[157] In this case, binding of substrate induces the transition to the active E form. In the case of complement factor B,[163] hepatocyte growth factor activator[159] or clotting factor VIIa,[94] stabilization of E* is achieved after the zymogen → protease conversion has taken place, and transition to the active E form relies on the binding of substrate and/or cofactors. Hence, the E*–E equilibrium is a basic property of the trypsin fold that fine-tunes activity and specificity once the zymogen → protease conversion has taken place.[167]

VI. Dissociating Procoagulant and Anticoagulant Activities

The allosteric nature of thrombin has a direct bearing on its multifunctional roles in the blood and provides context and strategies toward dissociating procoagulant and anticoagulant activities. Thrombin mutants with anticoagulant activity help rationalize the phenotypes of several naturally occurring mutations and could eventually provide new tools for pharmacological intervention.[42] The group of Sadler first reported that mutation of R75 to Glu in exosite I has normal fibrinogen clotting activity but only 7% activity toward protein C in the presence of thrombomodulin, whereas mutation of K60f to Glu in the 60-loop had 2.5-fold increased protein C activation and 17% activity toward fibrinogen.[168] More pronounced anticoagulant effects were reported by the group of Leung with the E217A and E217K mutations that significantly shift thrombin specificity toward protein C relative to wild-type.[8,11] This group also provided the first proof-of-principle on the efficacy of such anticoagulant thrombin mutants *in vivo*[8,11] and pioneered systematic Ala scanning mutagenesis of the epitopes recognizing fibrinogen and protein C.

Figure 4 and Table I summarize the results of an Ala scan of 97 residues of thrombin covering $> 53\%$ of the accessible surface area of the enzyme in terms of the activity toward the procoagulant substrate fibrinogen, the prothrombotic substrate PAR1, and the anticoagulant substrate protein C in the presence of the cofactor, thrombomodulin, and 5 mM $CaCl_2$.[169] Residues targeted by mutagenesis are highlighted in Fig. 5 and belong to critical regions of the B chain of thrombin defining the active site, the 60-loop, the autolysis loop, exosites I and II, the Na^+ site, and the A chain. In the case of wild-type, the value of k_{cat}/K_m for cleavage of fibrinogen or activation of PAR1 is > 100-fold higher than that for activation of protein C in the presence of thrombomodulin (Table I). A general trend gleaned from the plot is that an Ala mutation of the thrombin scaffold tends to perturb recognition of fibrinogen and PAR1 more than protein C. Residues important for fibrinogen recognition are distributed over the entire surface of contact between the enzyme and the substrate,[126] and involve the 60-loop, exosite I, the primary specificity pocket, the aryl binding site, and the Na^+ binding site.[129] There is substantial overlap between this epitope and the one recognizing PAR1. The surface of recognition between thrombin and protein C changes significantly upon thrombomodulin binding and is reduced mainly to the primary specificity pocket and portions of the 60-loop in the presence of cofactor.[70] This change makes it possible to identify residues that are significantly more important for fibrinogen and PAR1 recognition than for protein C activation and to engineer thrombin mutants with enhanced anticoagulant activity.

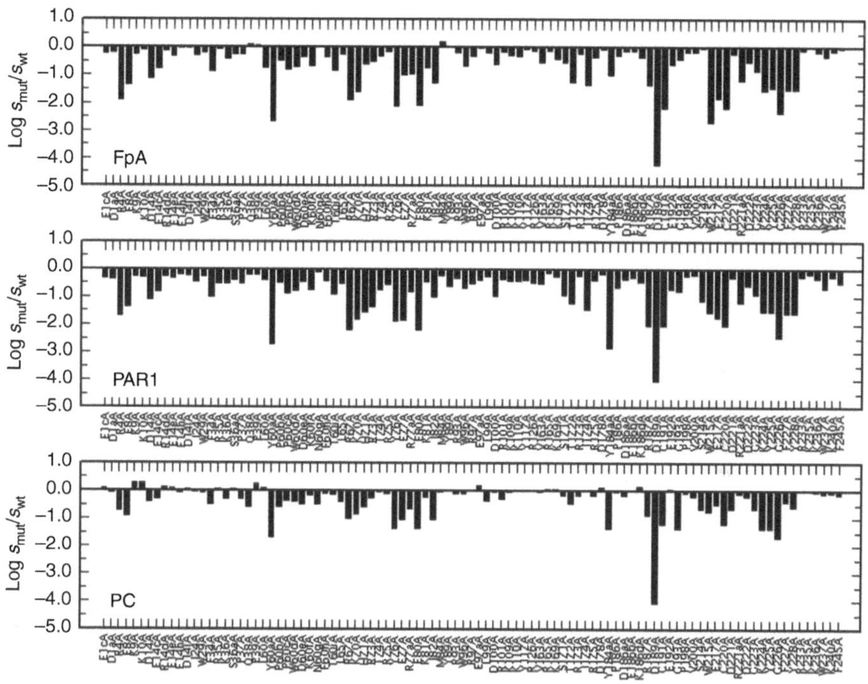

FIG. 4. The effect of Ala substitution of 97 residues of thrombin on the cleavage of fibrinogen (FpA) and activation of PAR1 or protein C (PC) in the presence of 50-nM thrombomodulin and 5-mM $CaCl_2$. Shown is the change in the specificity constant $s = k_{cat}/K_m$ due to the mutation, expressed as log s_{mut}/s_{wt}, under experimental conditions of 5-mM Tris, 0.1% PEG, 145-mM NaCl, pH 7.4 at 37 °C. Note the similarity in the profiles for fibrinogen and PAR1 hydrolysis that are distinct from that for activation of protein C. The values of s for wild-type are $17 \pm 1\ \mu M^{-1}s^{-1}$ (FpA), $30 \pm 1\ \mu M^{-1}s^{-1}$ (PAR1), and $0.22 \pm 0.02\ \mu M^{-1}s^{-1}$ (PC). The values of s for each mutant are listed in Table I, along with the relative differences for each pair of the three substrates. This research was originally published in Ref. 169. © The American Society for Biochemistry and Molecular Biology.

The double mutant, W215A/E217A (WE), was constructed by combining the two single mutations, W215A and E217A,[38] that stand out for their significant perturbation of fibrinogen and PAR1 cleavage over protein C activation (Fig. 4). WE shows anticoagulant/antithrombotic activity both *in vitro* and *in vivo*.[9,10,38,40,41,170,171] Its antithrombotic effect in nonhuman primates is more efficacious than the direct administration of activated protein C and safer than the administration of low molecular weight heparins.[10] Activated protein C generated *in situ* with the mutant WE offers cytoprotective advantages over activated protein C administered to the circulation.[171] Furthermore, WE acts as a potent and safe antithrombotic by blocking the interaction of von Willebrand factor with the platelet receptor, GpIb.[10,40]

TABLE I
PROPERTIES OF ALA MUTANTS OF THROMBIN TOWARD FIBRINOGEN (FPA), PAR1, AND PROTEIN C (PC) IN THE PRESENCE OF CA^{2+} AND THROMBOMODULIN

Mutant	FpA	PAR1	PC	FpA−PAR1	PAR1−PC	FpA−PC
wt	7.23	7.48	5.34	− 0.25	2.14	1.89
E1cA	7.00	7.15	5.43	− 0.15	1.73	1.57
D1aA	7.04	7.13	5.28	− 0.09	1.85	1.76
R4A	5.36	5.80	4.64	− 0.44	1.16	0.72
E8A	5.89	6.13	4.44	− 0.24	1.68	1.44
K9A	6.97	7.20	5.62	− 0.23	1.58	1.35
K10A	7.15	7.18	5.62	− 0.03	1.56	1.52
D14A	6.11	6.37	4.94	− 0.26	1.43	1.18
E14cA	6.46	6.68	5.04	− 0.22	1.64	1.42
R14dA	7.11	7.22	5.47	− 0.11	1.75	1.64
E14eA	6.93	7.15	5.44	− 0.22	1.71	1.49
E14hA	7.20	7.29	5.28	− 0.08	2.01	1.93
D14lA	7.20	7.25	5.40	− 0.04	1.85	1.81
I24A	6.96	7.04	5.30	− 0.08	1.74	1.66
W29A	7.04	7.20	5.28	− 0.16	1.93	1.76
F34A	6.38	6.47	4.84	− 0.09	1.63	1.54
R35A	7.18	6.97	5.41	0.21	1.55	1.76
K36A	6.82	6.97	5.04	− 0.15	1.93	1.78
S36aA	6.99	7.10	5.40	− 0.11	1.70	1.59
P37A	6.98	6.97	5.04	0.01	1.93	1.94
Q38A	7.34	7.27	4.78	0.08	2.49	2.57
E39A	7.30	7.29	5.59	0.02	1.70	1.71
L60A	6.49	7.10	5.43	− 0.61	1.67	1.06
Y60aA	4.59	4.76	3.68	−0.17	1.08	0.91
P60bA	6.75	7.00	4.78	− 0.26	2.23	1.97
P60cA	6.43	6.60	5.00	− 0.17	1.61	1.43
W60dA	6.53	6.71	4.95	− 0.18	1.76	1.58
D60eA	6.88	7.04	4.84	− 0.16	2.19	2.04
K60fA	6.57	6.74	5.17	− 0.17	1.56	1.39
N60gA	7.23	7.39	4.84	− 0.16	2.54	2.39
F60hA	6.89	7.07	5.25	− 0.18	1.82	1.63
T60iA	6.41	6.59	5.20	− 0.17	1.38	1.21
L65A	6.97	6.97	4.95	0.00	2.01	2.02
R67A	5.36	5.29	4.34	0.08	0.95	1.02
K70A	5.62	5.68	4.52	− 0.06	1.16	1.11
H71A	6.62	5.92	4.78	0.70	1.15	1.85
R73A	6.72	6.13	5.08	0.60	1.05	1.65
T74A	6.91	6.75	5.30	0.16	1.45	1.61
R75A	7.08	6.92	5.25	0.16	1.67	1.83
Y76A	5.11	5.62	4.00	− 0.51	1.63	1.12
E77A	6.23	5.63	4.30	0.60	1.33	1.93
R77aA	6.28	6.68	4.70	− 0.40	1.98	1.58
E80A	5.15	5.29	4.00	− 0.14	1.29	1.15
K81A	6.50	7.04	5.11	− 0.53	1.93	1.39

(*Continues*)

TABLE I (*Continued*)

Mutant	FpA	PAR1	PC	FpA−PAR1	PAR1−PC	FpA−PC
I82A	5.96	6.48	4.30	−0.52	2.18	1.66
M84A	7.43	7.25	5.32	0.18	1.93	2.11
Y89A	7.25	6.86	5.38	0.39	1.49	1.88
R93A	7.06	7.15	5.23	−0.09	1.93	1.83
W96A	6.58	6.79	5.23	−0.21	1.56	1.35
R97A	6.92	6.97	5.38	−0.04	1.59	1.55
E97aA	7.20	7.10	5.52	0.10	1.58	1.68
L99A	7.04	7.20	5.00	−0.16	2.20	2.04
D100A	6.62	6.50	5.30	0.12	1.20	1.32
R101A	7.08	7.13	5.04	−0.05	2.09	2.04
K109A	6.96	7.07	5.32	−0.11	1.75	1.64
K110A	6.92	7.07	5.36	−0.15	1.71	1.56
Y117A	7.18	7.10	5.36	0.08	1.74	1.82
R126A	7.11	7.00	5.36	0.11	1.64	1.75
V163A	6.69	6.97	5.30	−0.28	1.67	1.39
R165A	7.12	7.34	5.40	−0.22	1.94	1.72
K169A	6.81	7.22	5.41	−0.42	1.81	1.40
S171A	6.71	6.53	5.17	0.18	1.35	1.53
T172A	5.98	6.25	4.90	−0.27	1.35	1.07
R173A	7.01	7.25	5.17	−0.23	2.07	1.84
I174A	5.88	6.04	5.34	−0.16	0.70	0.54
R175A	6.89	7.10	5.17	−0.21	1.93	1.72
D178A	7.19	7.30	5.47	−0.11	1.83	1.72
Y184aA	6.25	4.66	4.00	1.60	0.66	2.26
P186A	6.95	6.82	5.32	0.14	1.50	1.63
D186aA	7.11	7.13	5.17	−0.01	1.95	1.94
E186bA	7.11	7.18	5.38	−0.07	1.80	1.74
K186dA	6.89	7.04	5.50	−0.15	1.53	1.38
R187A	5.90	5.46	4.47	0.44	0.98	1.42
D189A	3.00	3.47	1.28	−0.47	2.19	1.72
C191A	5.04	5.44	4.14	−0.40	1.30	0.90
E192A	6.63	6.76	5.41	−0.12	1.34	1.22
G193A	6.84	6.70	4.00	0.14	2.70	2.84
P198A	7.08	7.25	5.28	−0.17	1.97	1.80
V200A	7.09	7.27	5.14	−0.17	2.12	1.95
S214A	7.20	6.34	4.70	0.87	1.64	2.51
W215A	4.53	5.92	4.60	−1.39	1.32	−0.07
E217A	5.41	5.74	4.84	−0.33	0.90	0.57
C220A	5.04	5.44	4.14	−0.40	1.30	0.90
D221A	7.00	7.18	4.70	−0.18	2.48	2.30
R221aA	6.04	6.29	5.23	−0.24	1.06	0.81
D222A	6.74	6.91	5.14	−0.17	1.77	1.60
G223A	6.40	6.57	4.70	−0.17	1.87	1.70
K224A	5.69	5.97	4.00	−0.28	1.97	1.69
Y225A	5.78	5.97	4.00	−0.18	1.97	1.79
G226A	4.88	5.04	3.66	−0.16	1.38	1.22

(*Continues*)

TABLE I (*Continued*)

Mutant	FpA	PAR1	PC	FpA–PAR1	PAR1–PC	FpA–PC
F227A	5.74	5.91	4.95	−0.17	0.96	0.79
Y228A	5.73	5.90	4.78	−0.17	1.13	0.96
R233A	7.14	7.20	5.34	−0.06	1.86	1.80
K235A	7.24	7.30	5.34	−0.06	1.96	1.90
K236A	7.08	7.15	5.30	−0.08	1.86	1.78
W237A	6.92	6.79	5.25	0.13	1.54	1.67
K240A	7.10	7.22	5.28	−0.12	1.95	1.82
F245A	7.22	7.00	5.20	0.22	1.80	2.02

Shown are the values of log k_{cat}/K_m in $M^{-1}s^{-1}$ and the difference in such values between the three pairs of substrates. Errors are < 0.04 log units.

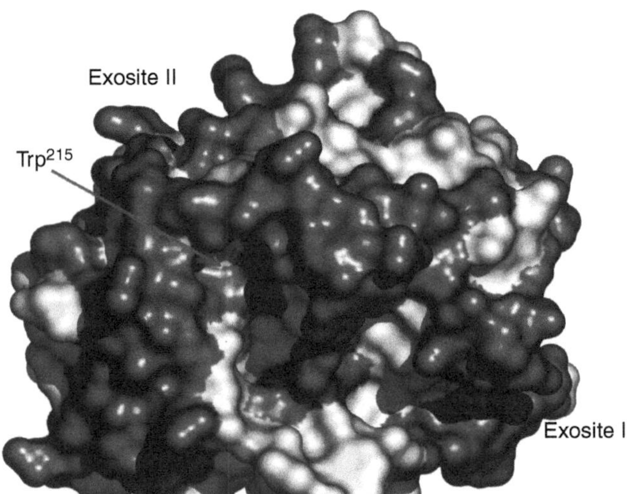

FIG. 5. Accessible surface area of thrombin in the standard orientation with the active site at the center and other regulatory sites as indicated. The position of Trp[215] within the active site is noted by a red arrow. Residues of thrombin subject to Ala scanning mutagenesis are colored in cyan. These residues cover 53% of the thrombin surface area. Not seen in this orientation are residues of the A chain at the back of the molecule. This research was originally published in Ref. 169. © The American Society for Biochemistry and Molecular Biology. (For interpretation of the references to color in this figure legend, the reader is referred to the Web version of this chapter.)

VII. WE: A Prototypic Anticoagulant Thrombin

Systemic administration of activated protein C (APC) has antithrombotic and anti-inflammatory effects[172–174] that are now utilized in the treatment of severe sepsis.[175] Since infused thrombin activates protein C and activated protein C is antithrombotic, thrombin infusion could act, in theory, as an antithrombotic agent. The thrombin analog, WE, was tested in a wide dose range for safety and efficacy in a baboon model of acute vascular graft thrombosis.[9,10] In this model, a permanent, surgically implanted arteriovenous shunt is temporarily extended with a thrombogenic vascular graft segment. Acute thrombus formation is visualized and quantified in real time using gamma camera imaging of the deposition of radiolabeled platelets in the graft. WE was initially administered as an intravenous bolus in order to evaluate pharmacokinetics. The lowest bolus dose of WE tested, 11 μg/kg, reduced platelet accumulation by 80% 1 h after the beginning of thrombosis, and was at least as effective as the direct administration of 40-fold more (0.45 mg/kg bolus) APC. Baboons treated with WE at doses as high as 200 μg/kg showed no clinical or laboratory signs of thrombosis, hemorrhage, or organ failure. No procoagulant activity could be detected for up to 1 week in baboon plasma obtained following bolus WE administration. Meanwhile, rapid systemic anticoagulation was observed, which dissipated with the biological half-life of circulating APC, as determined in previous experiments in baboons. Higher doses of WE (> 20 μg/kg) had pronounced anticoagulant effect, triggered by a burst in the levels of circulating endogenous APC following injection of WE.[41] High-dose WE infusion exhausted the cofactors of protein C activation before consumption of the substrate reserve, and the process was rapidly downregulated > 90% following a WE overdose. The exact molecular or cellular mechanism of this self-limiting pharmacological process has not been known yet, but failure of protein C activation upon WE overdose can be overcome by cofactor supplementation using soluble recombinant thrombomodulin.[41] These results suggest that the thrombin–thrombomodulin complex is efficiently and rapidly inhibited *in vivo* and that thrombomodulin does not recycle rapidly once WE (and possibly thrombin) is bound to it. The true significance of this surprising finding is that pharmacological protein C activation by WE appears to be intrinsically safe compared to all other accepted methods of anticoagulation that are ultimately fatal when overdosed. Studies with low-dose WE infusions revealed that pharmacological levels of APC and marginal systemic anticoagulation could be maintained by continuous WE infusion for at least 5 h without a substantial decrease in protein C levels. Since WE has potent antithrombotic effects, even at very low doses that do not induce systemic anticoagulation,[10] natural protein C reserves and production could keep up with the consumption of protein C during sustained antithrombotic WE treatment.

The hemostatic safety of continuous WE infusion in comparison to equiefficacious doses of low molecular weight heparin infusion for the prevention of acute vascular graft thrombus propagation in the baboon model were also evaluated. Based on previously established anticoagulant effect and antithrombotic efficacy of circulating APC in the baboon model,[172,173] we originally predicted that, in systemic blood samples, an APTT prolongation of at least 1.5-fold over baseline and APC levels at least 20-fold over baseline, sustained for at least 40 min, would result in a significant antithrombotic benefit. The results exceeded our expectations and we found that WE had a potent antithrombotic effect even at a dose (2.1 μg/kg/70 min) when the APTT was not demonstrably affected.[10] This finding indicated that WE was a very potent antithrombotic agent in primates.[10] Low doses of WE (2.1, 4.2, or 8.3 μg/kg/70 min) outperformed higher doses of exogenous APC (28 and 222 μg/kg/70 min) and were as efficient as interventional doses of intravenous enoxaparin (325–2600 μg/kg/70 min) in preventing the propagation of thrombi. The lowest dose of WE tested still increased circulating APC levels by approximately fivefold (to about 20 ng/mL), and we do not know yet whether even lower doses of WE that may not detectably increase APC levels would also be efficacious. The comparably effective plasma concentration of exogenous (recombinant) APC exceeded the endogenous APC level following low-dose WE infusion by several-fold. The fact that WE is very effective at such small doses strongly argues for limited receptor-mediated mechanism. The considerably higher antithrombotic efficacy of WE compared to exogenous APC suggests that APC generated by WE on the surface of endothelial cells may remain transiently bound to the receptor[176,177] and elicit additional effects by signaling via PAR1.[10,171] Although *ex vivo* data suggest that the signaling effect is dependent on the generation of APC on the cell surface, one may speculate that WE could exhibit direct (PAR1-mediated) cytoprotective signaling in the presence of EPCR-bound protein C.[178,179] WE may be retained on the cell surface in the vicinity of PAR1 by receptors other than thrombomodulin, similar to the case of its interaction with platelet GpIb, which appears to be active only under shear flow conditions.[40]

The molar ratio of the equiefficacious doses of WE and enoxaparin exceeded 1:1000. However, this extraordinary potency would be ultimately and clinically irrelevant if the safety profile of WE were not substantially different from other anticoagulants. The striking finding was that these potent antithrombotic doses of WE infusion did not impair primary hemostasis, an outcome never before seen in baboons that received comparably effective doses of commercially available antithrombotic agents.[10] Interestingly, the comparably antithrombotic and reasonably low doses of wild-type thrombin, 40 μg/kg/h,[180] and WE, about 2 μg/kg/h,[10] appear to be at least one order of magnitude apart to the advantage of WE, despite the 90% reduced activity of

WE toward protein C. One of the possible explanations for this discrepancy could be the propensity of wild-type thrombin to bind to and activate abundant prothrombotic substrates (e.g., fibrinogen, PAR1) in the blood flow, leaving only a fraction of the dose to diffuse across the boundary layer of flow to surfaces. Meanwhile, binding and interaction of WE with fibrinogen in the blood is dramatically reduced and the rate of such interactions as fibrinogen cleavage and inhibition by antithrombin are delayed by several orders of magnitude. The delay leaves more time for the administered enzyme to reach surfaces and become available for interaction with or consumption by transmembrane cofactors and receptors such as thrombomodulin and PAR1. This interaction can thus lead to an effective activation of the protein C-dependent pathways on solid surfaces, such as the endothelium. Altogether, these properties make WE thrombin an agent that targets thrombosis and help explain the very high antithrombotic efficacy of this enzyme.

VIII. Molecular Mechanism of Anticoagulant Activity of Thrombin Mutants

The E*–E equilibrium is a basic feature of the trypsin fold, affording a simple mechanism of allosteric regulation of protease activity after the irreversible zymogen → protease conversion has taken place.[167] Collapse of the 215–217 segment into the active site and disruption of the oxyanion hole due to a flip of the 192–193 peptide bond are distinctive features of the E* form of thrombin and occur with the I16–D194 ion-pair intact.[39,137,147,148] The E*–E equilibrium also provides a context to interpret the effect of mutations associated with the loss of biological activity in highly active proteases. In some cases, as documented by thrombin,[15,53] the molecular origin of the effect is unclear because the mutation does not affect residues in direct contact with substrate. Stabilization of E* through molecular conduits not necessarily involved in substrate recognition may offer a plausible explanation.

The allosteric E*–E equilibrium has far-reaching implications for protein engineering. Stabilization of E* by selected mutations, coupled with a transition to E triggered by suitable cofactors, may result in expression of protease activity on demand in a biological context. Thrombin exists predominantly in the active E and E:Na^+ forms in rapid equilibrium because of the extremely fast rates of binding and dissociation of Na^+ with the enzyme.[135] However, a thrombin mutant stabilized in the E* form and converting to E upon interaction with thrombomodulin and protein C would be an anticoagulant of potential clinical relevance. Such a mutant would show little or no activity toward fibrinogen and PAR1, but would retain activity toward protein C. A number of thrombin mutants have been reported with an altered specificity that favors protein C activation over fibrinogen cleavage.[8,11,24,38,61,168,181] Among these mutants, E217K and WE are effective as anticoagulants and antithrombotics in nonhuman primates[8–11] and have been crystallized at 2.5–2.8 Å resolution.[182,183] The structures show a partial collapse of the 215–217 segment and disruption of the oxyanion hole that resemble the conformation of E*.[148] The collapse is similar to, but less pronounced than, that observed in the structure of the inactive E* form of thrombin, where W215 relinquishes its hydrophobic interaction with F227 to engage the catalytic H57 and residues of the 60-loop after a 10-Å shift in its position.[148] These more substantial changes have been observed recently in the structure of the anticoagulant mutant, Δ146-149e.[39]

The mutant, Δ146-149e, carries a deletion of the nine residues, ^{146}ETWTANVGK^{149e}, in the autolysis loop and was originally constructed to assess the role of this highly flexible domain in thrombin function.[61] The sequence, ^{149a}ANVGK^{149e}, is not present in trypsin and chymotrypsin and can be deleted without functional consequences.[61] Also, swapping the entire ^{146}ETWTANVGK^{149e} sequence of thrombin with the ^{146}SSGT^{149} sequence of trypsin[184] or deleting the ^{146}ETW^{148} sequence[185] produces perturbations of function that are recapitulated by the single mutations, E146A and R221aA.[20,24] Residue E146 makes an important ion-pair interaction with R221a in the adjacent 220-loop in the E form, but not in the E* form.[20,147,148] The entire autolysis loop likely participates in the long-range communication seen in the E*–E equilibrium[147] between the 220-loop, the active site, and exosite I where thrombomodulin binds.[15] Deletion of the entire ^{146}ETWTANVGK^{149e} sequence in the Δ146-149e mutant causes a significant loss of activity toward chromogenic and physiological substrates, but binding of thrombomodulin almost completely restores activity toward the anticoagulant protein C.[39] Importantly, thrombomodulin has only a modest effect on the hydrolysis of a chromogenic substrate, as already documented for other anticoagulant thrombin mutants[183] and wild-type.[186] These properties suggest that the Δ146-149e mutation shifts the E*–E equilibrium of thrombin in favor of E*, and thrombomodulin in complex with protein C switches the mutant back into the active conformation E. Therefore, the thrombin mutant, Δ146-149e, likely functions as an allosteric switch stabilized into the inactive form E* until the combined binding of thrombomodulin and protein C shifts E* to E and restores activity. Consistent with functional data, the mutant assumes a collapsed conformation that is practically identical (rmsd 0.154 Å) to the E* form identified recently from the structure of the thrombin mutant, D102N.[147,148]

It would be simple to assume that both E217K and WE, like Δ146-149e, are stabilized in the E* form. However, unlike Δ146-149e, both E217K and WE carry substitutions in the critical 215–217 segment that could result in additional functional effects overlapping with or mimicking a perturbation of the E*–E equilibrium. A significant concern is that both structures suffer from crystal packing interactions that may have biased the conformation of side chains and loops near the active site.[148] The collapsed structures of E217K and WE may be artifactual unless validated by structural studies documenting different crystal packing. No such controls have so far been carried out for E217K. In the case of WE, six crystal structures of the human and murine enzyme have been solved, four in the absence of any ligands and two bound to the extracellular fragment of human PAR1.[187] The six models vary in the number of molecules per asymmetric unit, space group, and crystallization conditions including pH, as well as in a number of critical residues that confer murine thrombin high catalytic activity in the absence of Na^+ activation.[55] Yet, the structures present a common active site architecture that differs significantly from the wild-type in the E form.[20] Each of the 11 WE monomers shows collapse of the 215–217 peptide segment into the active site, and the majority show disruption of the oxyanion hole, as seen in the E* form (Fig. 6). Binding of PAR1 to exosite I does not correct the collapsed conformation. It is likely that the collapsed conformation of WE is indeed the E* form biased by the effect of removing the indole of W215. The side chain of W215 is critical in maintaining the open conformation of the active site in the E form through a strong hydrophobic interaction with F227.[20] Likewise, in the E* form, the hydrophobic interaction of W215 with H57 and Y60a in the 60-loop produces the collapsed conformation that obliterates substrate access to the active site.[148] When the side chain of W215 is eliminated with the W215A substitution, the 215–217 β-strand loses a key structural element to stabilize either E or E*. Hence, it is not surprising that the conformation of the 215–217 β-strand in the WE mutant occupies a position that is intermediate to the two limiting positions in the E and E* forms. The collapse is sufficient to compromise substrate binding to the active site, even though it does not move the 215–217 peptide segment fully to the position seen in the E* form.

The crystal structures of human and murine thrombin mutant WE bound to a fragment of PAR1 in exosite I show no conversion to the E form and no change in the collapsed 215–217 peptide segment observed in the free form. These observations seem to contradict recent crystallographic evidence that the D102N mutant assumes the E* form when free and the E form when bound to a fragment of PAR1 at exosite I.[147] However, attention to linkage principles shows that the results are not in contradiction. An allosteric effector can only shift the E*–E equilibrium in favor of E by an amount equal to the ratio of affinities of the two forms. Because exosite I, unlike the active site, is similarly accessible in the E* and E form,[15,20,147,148] binding of thrombomodulin, hirugen, or fragments of PAR1 to the two forms is unlikely to result in extreme perturbations of the E*–E equilibrium. In fact, binding of hirugen to exosite I in the WE mutant takes place with an affinity that is fourfold lower

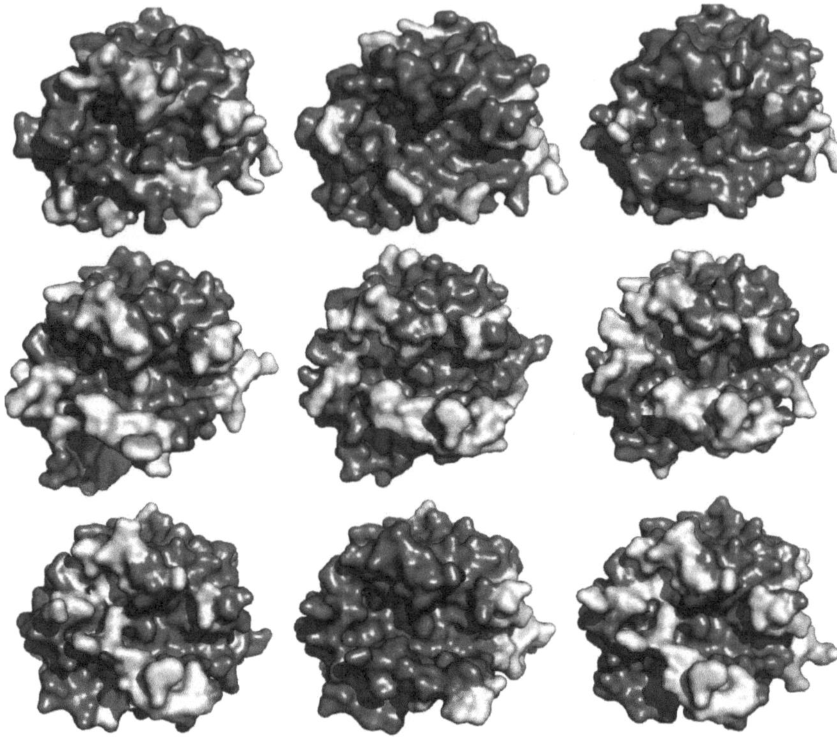

FIG. 6. Surface representation of the nine free monomers of human and murine WE (wheat). The wide variety of crystal contacts < 4 Å (light blue) documented in the datasets prove that the collapse of the 215–217 segment (arrow) is not an artifact of crystal packing. Top row: PDB ID 3EE0, PDB ID 1TQ0 molecule A, PDB ID 1TQ0 molecule B. Middle row: PDB ID 3HK3, PDB ID 3HK6 molecule A, PDB ID 3HK6 molecule B. Bottom row: PDB ID 3EDX molecule A, B and C. This research was originally published in Ref. 187. © The American Society for Biochemistry and Molecular Biology. (For interpretation of the references to color in this figure legend, the reader is referred to the Web version of this chapter.)

than that of wild-type,[187] confirming that the collapsed conformation of the mutant that compromises access to the active site has little effect on the architecture of exosite I. The available structures of D102N do not imply an all-or-none distribution of E* and E in solution. Kinetic studies where the E*–E equilibrium distribution can be measured directly[133] prove that $r = 1.1$ for D102N free in solution.[148] Therefore, it would have been equally likely for this mutant to crystallize in the E form when free, but E* predominated under the crystallization conditions so far explored.[147,148] If binding to exosite I takes place with fourfold higher affinity in the E form, then a value of $r = 0.28$ is obtained for the mutant D102N bound to exosite I. Under these conditions, the fraction of D102N in the active E form would be 78% of the total. It is not surprising, then, that the structure of D102N bound to a fragment of PAR1 to exosite I has revealed the active E form.[147] However, future crystallization studies may trap the D102N–PAR1 complex with the enzyme in the E* form because of the significant fraction (22%) of this inactive conformation still present in solution. The value of r for the WE mutant cannot be estimated directly from kinetic studies because W215 is a major fluorophore reporting the E*–E interconversion.[133] However, a lower bound of $r = 200$ can be inferred from the functional properties of the mutant Δ146-149e that crystallizes in the E* form when free[39] and is a less potent anticoagulant compared to WE. Binding of ligands to exosite I of WE should give a value of $r = 50$, which translates to 2% of the enzyme in the E form when bound. Under these conditions, it would be very difficult to trap the WE–PAR1 complex with the enzyme in E form, and the E* form-documented crystallographically reflects the conformation of 98% of the molecules in solution.

The foregoing argument also explains why thrombomodulin and hirugen have little effect on the hydrolysis of chromogenic substrates by the mutants WE[183] or Δ146-149e,[39] as well as why physiological substrates like fibrinogen and PAR1 are not cleaved efficiently by the mutants WE and Δ146-149e.[38,39] Assuming that these substrates bind to exosite I before contacting the active site, the interaction with exosite I would be insufficient to energetically move the E*–E equilibrium completely in favor of the active E form. On the other hand, when protein C and thrombomodulin are present in solution, then most of the activity of the mutant WE is restored. The thrombomodulin–protein C complex acting as a substrate may have a more profound effect on the E*–E equilibrium by accessing additional regions of the thrombin surface beyond exosite I.[70] In that case, the change in affinity between the E* and E forms may be substantial and could ensure significant repopulation of the E form that is not possible with either protein C or thrombomodulin alone.

We conclude that the mutant WE is stabilized in the E* form, but the complete collapse of the 215–217 peptide segment is impeded by the absence of the side chain of W215. Binding to exosite I fails to substantially shift the E*–E equilibrium in favor of the E form, unless protein C and thrombomodulin act in combination. The result of this mechanism is a mutant that acts efficiently in the anticoagulant pathway and retains minimal activity toward the procoagulant substrate fibrinogen and the prothrombotic substrate, PAR1. The intriguing functional properties of the mutant WE,[40,171] its well-established potency as an anticoagulant in nonhuman primates,[9,10] and the current elucidation of the role of E* in switching thrombin into an anticoagulant[39] greatly facilitate engineering a thrombin mutant with exclusive activity toward protein C for therapeutic applications.

IX. Beyond WE

The intriguing properties of the mutant WE provide proof-of-principle that a thrombin mutant with exclusive activity toward protein C would be a compelling anticoagulant/antithrombotic agent *in vivo*. Further mutagenesis of thrombin was therefore undertaken to remove any residual activity toward fibrinogen and PAR1 in preparation for clinical applications. Previous saturation mutagenesis studies of residue 217 yielded an incomplete set of constructs and identified E217K as the most potent anticoagulant mutant.[11] Although the properties of E217K provide a slight improvement over those of E217A, they are comparable to those of W215A[38,170] and make residue 217 of little interest as a target for further mutagenesis. Indeed, the W215A/E217K double mutant features a less pronounced anticoagulant/antithrombotic profile compared to WE. Results of saturation mutagenesis of residue 215 are summarized in Fig. 7.[169] All constructs could be expressed to homogeneity and characterized in terms of their interaction with key physiological substrates. W215 is highly conserved in trypsin-like proteases[188] and is located within a hydrophobic patch (Fig. 5) essential for recognition of the P4 residue of substrate.[20,44] The role of W215 in fibrinogen binding is illustrated directly by the crystal structure[121] and its edge-to-face interaction with residue F8 of the Aα chain. Strong hydrophobic interactions involve W215 in recognition of the procoagulant substrates factor XIII,[123,189] PAR1[60] and PAR4.[22,122] W215 also plays a key role in the conformational transition of thrombin from the inactive E* to the active E form[15,133] by relocating more than 10 Å within the active site.

Mutation of W215 has a profound effect on thrombin specificity toward macromolecular substrates. Several mutations of W215 result in > 10,000-fold loss of activity toward fibrinogen and/or PAR1. In some cases (D, E, I, L, R, V), the loss of activity toward fibrinogen approaches or exceeds five orders of magnitude and is more pronounced than that observed with the potent anticoagulant mutant WE. For example, the mutant W215I features a k_{cat}/K_m value for cleavage of fibrinogen that is > 100,000-fold lower than that of wild-type. The k_{cat}/K_m value for PAR1 activation is > 10,000-fold lower compared to wild-type, but the k_{cat}/K_m value for activation of protein C in the presence of thrombomodulin is perturbed < 100-fold. Perturbation produced by introduction of a charge in the hydrophobic aryl binding site explains the drastic loss in specificity toward fibrinogen and PAR1 for the W215D, W215E, and W215R mutants. However, the effects seen for the W215I, W215L, and W215V mutants are somewhat unexpected and support the need for an aromatic side chain or ring structure at this position. In fact, wild-type and W215F have the highest activity toward fibrinogen and PAR1,[170] with W215Y, W215P, and W215H not far behind. Activity toward PAR1 correlates ($r = 0.91$) with that of fibrinogen, underscoring the basic similarity of interaction of the two

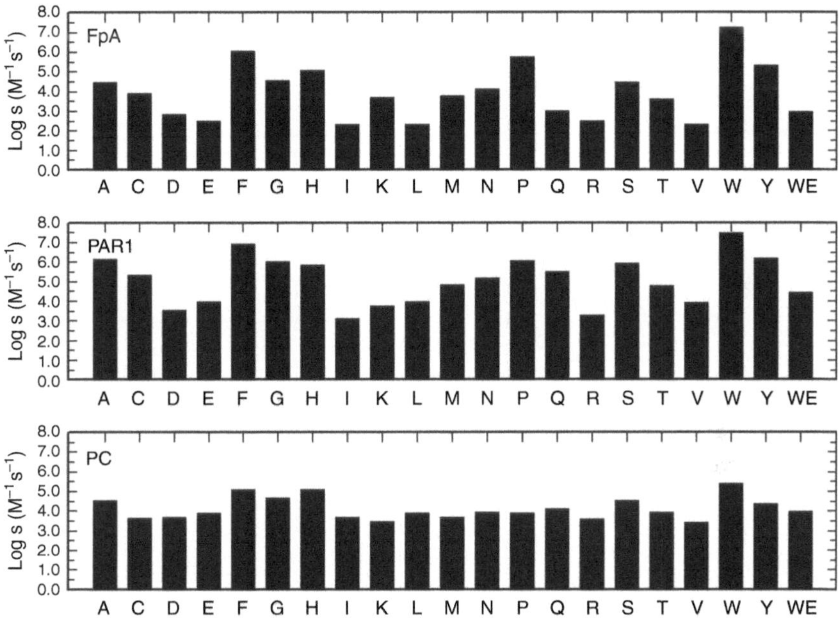

FIG. 7. Activity of the 20 possible amino acid substitutions at position 215 of thrombin toward fibrinogen (FpA), PAR1, or protein C (PC) in the presence of 50-nM thrombomodulin and 5-mM $CaCl_2$. W is the wild-type residue. Shown is the log of the specificity constant $s = k_{cat}/K_m$ under experimental conditions of 5-mM Tris, 0.1% PEG, 145-mM NaCl, and pH 7.4 at 37 °C. The properties of the anticoagulant mutant, WE, are also shown for comparison. The average values of log s for the 20 substitutions of residue 215 are 4.0 (FpA), 5.1 (PAR1), and 4.1 (PC), reflecting an average loss in log units relative to the wild-type of 3.2 (FpA), 2.4 (PAR1), and 1.2 (PC). A strong correlation ($r = 0.91$) is observed between the values for fibrinogen and PAR1 cleavage. Correlation between activation of protein C and fibrinogen or PAR1 cleavage is significantly weaker ($r = 0.79$ and $r = 0.84$, respectively). This research was originally published in Ref. 169. © The American Society for Biochemistry and Molecular Biology.

substrates with thrombin[121,147] and consistent with previous mutagenesis studies.[22] Interestingly, activity toward fibrinogen or PAR1 correlates with activity toward the chromogenic substrate, FPR ($r = 0.88$ and $r = 0.92$, respectively, data not shown), suggesting that perturbation of the active site moiety due to replacement of W215 is the dominant factor controlling hydrolysis of the procoagulant and prothrombotic substrates of thrombin. On the other hand, activity toward protein C correlates to a lower extent with activity toward FPR ($r = 0.77$), fibrinogen ($r = 0.79$), or PAR1 ($r = 0.84$). The concerted action of thrombomodulin and protein C on thrombin tends to correct the functional deficit associated with a mutation of the enzyme, as demonstrated recently in the case of the double mutant, WE.[187]

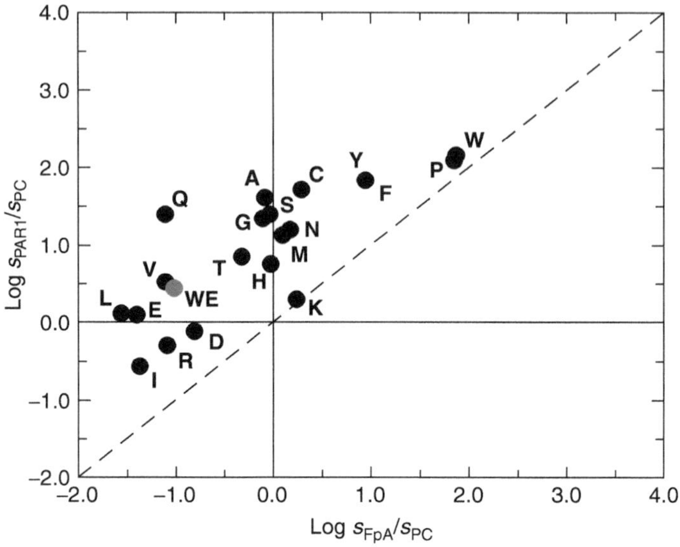

Fig. 8. Functional properties of the thrombin mutants of residue 215 (W corresponds to wild-type). Shown are the values of $s = k_{cat}/K_m$ for the hydrolysis of fibrinogen (s_{FpA}), PAR1 (s_{PAR1}), or protein C (s_{PC}) in the presence of 50-nM thrombomodulin and 5-mM Ca^{2+} plotted as the logarithm of the ratios s_{PAR1}/s_{PC} versus s_{FpA}/s_{PC}. Experimental conditions are 5-mM Tris, 0.1% PEG8000, 145-mM NaCl, pH 7.4 at 37 °C. The properties of the anticoagulant mutant WE are also shown for comparison (gray circle). Note how mutagenesis of residue 215 produces constructs (W215V) with properties equivalent to those of WE. In some cases (W215D, W215E, W215I, W215L, W215R), the anticoagulant/antithrombotic profile of WE is significantly improved with a single amino acid substitution. This research was originally published in Ref. 169. © The American Society for Biochemistry and Molecular Biology.

Shift in the specificity of the enzyme is best appreciated in the plot shown in Fig. 8. The plot contains many elements of interest as it incorporates information on three different substrates. The value of s_{PAR1} for the hydrolysis of PAR1 by thrombin is plotted versus the analogous value for fibrinogen cleavage, s_{FpA}, and both values are expressed in units of s_{PC}, that is, the k_{cat}/K_m value for the hydrolysis of protein C in the presence of thrombomodulin and Ca^{2+} (see also Fig. 7). The origin of the axes divides the plot into four regions, with the two of them further divided into two regions by the diagonal dotted line, reflecting a regime where perturbation of PAR1 cleavage is the same as that of fibrinogen cleavage. The region to the right of the vertical axis denotes activity toward fibrinogen that exceeds that toward protein C and the reverse is seen to the left of the vertical axis. Likewise, the region above the horizontal axis denotes activity toward PAR1 that exceeds that toward protein C and the reverse is seen below the horizontal axis. Interestingly, all points reflecting the properties of the 20 possible amino acid substitutions of residue 215 map on or above the dotted line, indicating that mutation of W215 fails to produce a construct for which the activity toward fibrinogen exceeds that toward PAR1. Distance from the diagonal dotted line quantifies the extent of preferential interaction toward fibrinogen and PAR1. The mutant W215Q emerges as the most specific toward PAR1 relative to fibrinogen, improving on the analogous property seen with the W215A[170] and WE mutants.[38] Wild-type maps in the region reflecting procoagulant/prothrombotic propensity almost on the diagonal dotted line, underscoring the similarity of PAR1 and fibrinogen cleavage that both exceed the activity toward protein C by two orders of magnitude. The mutant WE maps in the anticoagulant region of the plot but close to the horizontal line, indicating that although the activity toward protein C exceeds that toward fibrinogen, PAR1 is cleaved with a specificity constant $s = k_{cat}/K_m$ higher than that of protein C. This leaves considerable room for optimization by directly interfering with the prothrombotic activity of the mutant. The quadrant in the lower left corner of the plot defines a region where activity toward protein C exceeds that toward fibrinogen and PAR1 and represents a target for protein engineering studies aimed at optimizing the anticoagulant activity of the enzyme at the expense of the procoagulant and prothrombotic activities. A replot of the data shown in Fig. 4 and Table I reveals that none of the 97 Ala mutants of thrombin maps into the target region, nor does the mutant E217K. Mutant W215V, bearing a single amino acid substitution, recapitulates the properties of WE in the plot. Five mutants of W215 improve on the anticoagulant/antithrombotic profile of WE and map into the target region (D, I, R) or close by (E, L). These mutations make a compelling case for further mutagenesis.

Having explored the properties of residue 215 by saturation mutagenesis, we turned our attention to the autolysis loop of thrombin that decorates the lower rim of the active site and is strategically positioned between the Na^+ site and exosite I (Fig. 1). After being neglected for years because of its intrinsic disorder in crystal structures, the autolysis loop has recently gained attention because deletion of the nine residues ^{146}ETWTANVGK^{149e} produces a potent anticoagulant due to stabilization of the inactive E* form.[39,61] Because the loop is not in contact with the region around residue 215, we surmised that a combined mutation of residue 215 and deletion of ^{146}ETWTANVGK^{149e} in the autolysis loop could afford an additive effect and result in an improved anticoagulant/antithrombotic profile relative to the individual mutations. The mutant W215G was initially selected as a basis for the double mutation in view of its properties similar to those of W215A (Figs. 7 and 8). The double mutant, W215G/Δ146-149e, maps into the target region and features almost complete additivity of the individual W215G and Δ146-149e mutations (Fig. 9). Encouraged by these results, we used W215E as a basis for the combined mutation

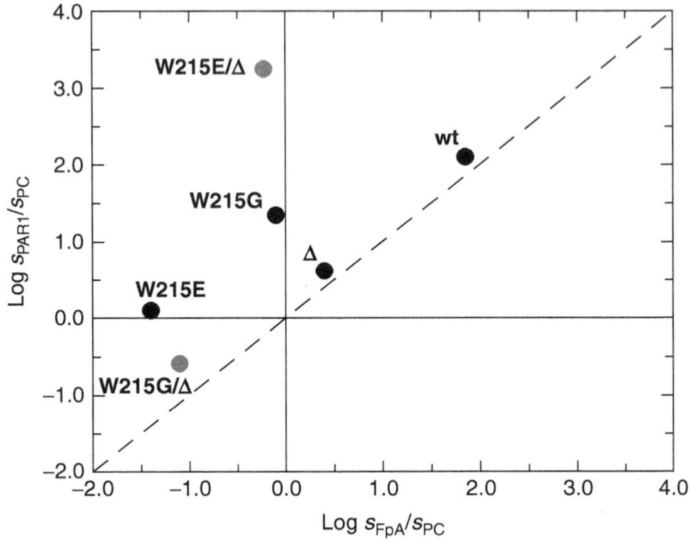

FIG. 9. Functional properties of the double mutants, W215G/Δ146-149e (W215G/Δ) and W215E/Δ146-149e (W215E/Δ) (gray circles), obtained by combining the single mutants, W215G and W215E (see Figs. 3 and 4), with deletion of residues 146-149e in the autolysis loop (Δ) that produces a thrombin mutant with an anticoagulant/antithrombotic profile.[39,61] Note how the double mutant, W215G/Δ146-149e, has additive properties relative to the single mutations, W215G and Δ146-149e, resulting in a construct with preferential activity toward protein C. On the other hand, the double mutant, W215E/Δ146-149e, has properties that cannot be predicted from additivity of the single mutations, W215E and Δ146-149e, resulting in a construct with almost exclusive activity toward PAR1. Experimental conditions are 5-mM Tris, 0.1% PEG8000, 145-mM NaCl, and pH 7.4 at 37 °C. The values of s for the two double mutants are (W215G/Δ146-149e) $0.43 \pm 0.02\ mM^{-1}s^{-1}$ (FpA), $1.4 \pm 0.1\ mM^{-1}s^{-1}$ (PAR1), and $5.3 \pm 0.2\ mM^{-1}s^{-1}$ (PC); (W215E/Δ146-149e) $0.30 \pm 0.01\ mM^{-1}s^{-1}$ (FpA), $890 \pm 10\ mM^{-1}s^{-1}$ (PAR1), and $0.50 \pm 0.02\ mM^{-1}s^{-1}$ (PC). This research was originally published in Ref. 169. © The American Society for Biochemistry and Molecular Biology.

with Δ146-149e. Our expectation was that the mutant W215E/Δ146-149e would feature no measurable activity toward fibrinogen and PAR1, but would retain appreciable activity toward the anticoagulant protein C. Contrary to this expectation, the mutant W215E/Δ146-149e features a k_{cat}/K_m for PAR1 cleavage > 1,000-fold higher than that of fibrinogen or protein C and behaves as an "exclusive" PAR1 agonist. An important lesson in protein engineering is learned from these constructs. Although the W215G substitution has similar effects on macromolecular substrate recognition on the wild-type or Δ146-149e scaffolds, the W215E substitution has profoundly different functional consequences on the two scaffolds. On the wild-type scaffold, it favors protein C activation over fibrinogen cleavage and equalizes PAR1 and protein C activation. On the Δ146-149e scaffold, it produces an almost exclusive agonist of PAR1. Hence, mutation of the same residue in two different thrombin scaffolds returns completely different specificity toward macromolecular substrates and generates a mutant with properties not seen in any of the individual constructs used in the mutagenesis.

X. Conclusions

These findings illustrate the power of a combined approach that links protein engineering and structural biology. The basic knowledge garnered on the allosteric properties of thrombin and the E*–E equilibrium can be brought to practical fruition with the construction of thrombin variants with selective activity toward macromolecular substrates. Anticoagulant thrombin mutants can be engineered rationally and should afford interesting and powerful new reagents for clinical applications.

Acknowledgments

This work was supported in part by NIH research grants HL49413, HL58141, HL73813, and HL95315.

References

1. Gerszten RE, Wang TJ. The search for new cardiovascular biomarkers. *Nature* 2008;**451**:949–52.
2. Anderson RN, Smith BL. Deaths: leading causes for 2002. *Natl Vital Stat Rep* 2005;**53**:1–89.
3. Padma V, Fisher M, Moonis M. Thrombolytic therapy for acute ischemic stroke: 3 h and beyond. *Expert Rev Neurother* 2005;**5**:223–33.
4. Busch M, Masuhr F. Thromboprophylaxis and early antithrombotic therapy in patients with acute ischemic stroke and cerebral venous and sinus thrombosis. *Eur J Med Res* 2004;**9**:199–206.
5. Mackman N. Triggers, targets and treatments for thrombosis. *Nature* 2008;**451**:914–8.
6. Esmon CT. The protein C pathway. *Chest* 2003;**124**:26S–32S.
7. Riewald M, Petrovan RJ, Donner A, Mueller BM, Ruf W. Activation of endothelial cell protease activated receptor 1 by the protein C pathway. *Science* 2002;**296**:1880–2.
8. Gibbs CS, Coutre SE, Tsiang M, Li WX, Jain AK, Dunn KE, et al. Conversion of thrombin into an anticoagulant by protein engineering. *Nature* 1995;**378**:413–6.
9. Gruber A, Cantwell AM, Di Cera E, Hanson SR. The thrombin mutant W215A/E217A shows safe and potent anticoagulant and antithrombotic effects in vivo. *J Biol Chem* 2002;**277**:27581–4.
10. Gruber A, Marzec UM, Bush L, Di Cera E, Fernandez JA, Berny MA, et al. Relative antithrombotic and antihemostatic effects of protein C activator versus low molecular weight heparin in primates. *Blood* 2007;**109**:3733–40.

11. Tsiang M, Paborsky LR, Li WX, Jain AK, Mao CT, Dunn KE, et al. Protein engineering thrombin for optimal specificity and potency of anticoagulant activity in vivo. *Biochemistry* 1996;**35**:16449–57.
12. Gailani D, Broze Jr. GJ. Factor XI activation in a revised model of blood coagulation. *Science* 1991;**253**:909–12.
13. Girard TJ, MacPhail LA, Likert KM, Novotny WF, Miletich JP, Broze Jr. GJ. Inhibition of factor VIIa-tissue factor coagulation activity by a hybrid protein. *Science* 1990;**248**:1421–4.
14. Mann KG, Butenas S, Brummel K. The dynamics of thrombin formation. *Arterioscler Thromb Vasc Biol* 2003;**23**:17–25.
15. Di Cera E. Thrombin. *Mol Aspects Med* 2008;**29**:203–54.
16. Davie EW, Kulman JD. An overview of the structure and function of thrombin. *Semin Thromb Hemost* 2006;**32**(Suppl. 1):3–15.
17. Bode W. Structure and interaction modes of thrombin. *Blood Cells Mol Dis* 2006;**36**:122–30.
18. Lane DA, Philippou H, Huntington JA. Directing thrombin. *Blood* 2005;**106**:2605–12.
19. Di Cera E, Guinto ER, Vindigni A, Dang QD, Ayala YM, Wuyi M, et al. The Na+ binding site of thrombin. *J Biol Chem* 1995;**270**:22089–92.
20. Pineda AO, Carrell CJ, Bush LA, Prasad S, Caccia S, Chen ZW, et al. Molecular dissection of Na+ binding to thrombin. *J Biol Chem* 2004;**279**:31842–53.
21. Dang QD, Di Cera E. Residue 225 determines the Na(+)-induced allosteric regulation of catalytic activity in serine proteases. *Proc Natl Acad Sci USA* 1996;**93**:10653–6.
22. Ayala YM, Cantwell AM, Rose T, Bush LA, Arosio D, Di Cera E. Molecular mapping of thrombin-receptor interactions. *Proteins* 2001;**45**:107–16.
23. Dang OD, Vindigni A, Di Cera E. An allosteric switch controls the procoagulant and anticoagulant activities of thrombin. *Proc Natl Acad Sci USA* 1995;**92**:5977–81.
24. Dang QD, Guinto ER, Di Cera E. Rational engineering of activity and specificity in a serine protease. *Nat Biotechnol* 1997;**15**:146–9.
25. Coughlin SR. Thrombin signalling and protease-activated receptors. *Nature* 2000; **407**:258–64.
26. Coughlin SR. Protease-activated receptors in hemostasis, thrombosis and vascular biology. *J Thromb Haemost* 2005;**3**:1800–14.
27. Di Cera E. Thrombin as procoagulant and anticoagulant. *J Thromb Haemost* 2007;**5**:196–202.
28. Myles T, Yun TH, Hall SW, Leung LL. An extensive interaction interface between thrombin and factor V is required for factor V activation. *J Biol Chem* 2001;**276**:25143–9.
29. Nogami K, Zhou Q, Myles T, Leung LL, Wakabayashi H, Fay PJ. Exosite-interactive regions in the A1 and A2 domains of factor VIII facilitate thrombin-catalyzed cleavage of heavy chain. *J Biol Chem* 2005;**280**:18476–87.
30. Yun TH, Baglia FA, Myles T, Navaneetham D, Lopez JA, Walsh PN, et al. Thrombin activation of factor XI on activated platelets requires the interaction of factor XI and platelet glycoprotein Ib alpha with thrombin anion-binding exosites I and II, respectively. *J Biol Chem* 2003;**278**:48112–9.
31. Mann KG. Thrombin formation. *Chest* 2003;**124**:4S–10S.
32. Degen SJ, McDowell SA, Sparks LM, Scharrer I. Prothrombin Frankfurt: a dysfunctional prothrombin characterized by substitution of Glu-466 by Ala. *Thromb Haemost* 1995;**73**:203–9.
33. Miyata T, Aruga R, Umeyama H, Bezeaud A, Guillin MC, Iwanaga S. Prothrombin Salakta: substitution of glutamic acid-466 by alanine reduces the fibrinogen clotting activity and the esterase activity. *Biochemistry* 1992;**31**:7457–62.
34. Henriksen RA, Dunham CK, Miller LD, Casey JT, Menke JB, Knupp CL, et al. Prothrombin Greenville, Arg517->Gln, identified in an individual heterozygous for dysprothrombinemia. *Blood* 1998;**91**:2026–31.

35. Sun WY, Smirnow D, Jenkins ML, Degen SJ. Prothrombin Scranton: substitution of an amino acid residue involved in the binding of Na+ (LYS-556 to THR) leads to dysprothrombinemia. *Thromb Haemost* 2001;**85**:651–4.
36. Stanchev H, Philips M, Villoutreix BO, Aksglaede L, Lethagen S, Thorsen S. Prothrombin deficiency caused by compound heterozigosity for two novel mutations in the prothrombin gene associated with a bleeding tendency. *Thromb Haemost* 2006;**95**:195–8.
37. Rouy S, Vidaud D, Alessandri JL, Dautzenberg MD, Venisse L, Guillin MC, et al. Prothrombin Saint-Denis: a natural variant with a point mutation resulting in Asp to Glu substitution at position 552 in prothrombin. *Br J Haematol* 2006;**132**:770–3.
38. Cantwell AM, Di Cera E. Rational design of a potent anticoagulant thrombin. *J Biol Chem* 2000;**275**:39827–30.
39. Bah A, Carrell CJ, Chen Z, Gandhi PS, Di Cera E. Stabilization of the E* form turns thrombin into an anticoagulant. *J Biol Chem* 2009;**284**:20034–40.
40. Berny MA, White TC, Tucker EI, Bush-Pelc LA, Di Cera E, Gruber A, et al. Thrombin mutant W215A/E217A acts as a platelet GpIb antagonist. *Arterioscler Thromb Vasc Biol* 2008;**18**:329–34.
41. Gruber A, Fernandez JA, Bush L, Marzec U, Griffin JH, Hanson SR, et al. Limited generation of activated protein C during infusion of the protein C activator thrombin analog W215A/E217A in primates. *J Thromb Haemost* 2006;**4**:392–7.
42. Bates SM, Weitz JI. The status of new anticoagulants. *Br J Haematol* 2006;**134**:3–19.
43. Doyle MF, Mann KG. Multiple active forms of thrombin. IV. Relative activities of meizothrombins. *J Biol Chem* 1990;**265**:10693–701.
44. Bode W, Turk D, Karshikov A. The refined 1.9-A X-ray crystal structure of D-Phe-Pro-Arg chloromethylketone-inhibited human alpha-thrombin: structure analysis, overall structure, electrostatic properties, detailed active-site geometry, and structure-function relationships. *Protein Sci* 1992;**1**:426–71.
45. Hageman TC, Endres GF, Scheraga HA. Mechanism of action of thrombin on fibrinogen. On the role of the A chain of bovine thrombin in specificity and in differentiating between thrombin and trypsin. *Arch Biochem Biophys* 1975;**171**:327–36.
46. DiBella EE, Maurer MC, Scheraga HA. Expression and folding of recombinant bovine prethrombin-2 and its activation to thrombin. *J Biol Chem* 1995;**270**:163–9.
47. Lefkowitz JB, Haver T, Clarke S, Jacobson L, Weller A, Nuss R, et al. The prothrombin Denver patient has two different prothrombin point mutations resulting in Glu-300–>Lys and Glu-309–>Lys substitutions. *Br J Haematol* 2000;**108**:182–7.
48. Akhavan S, Mannucci PM, Lak M, Mancuso G, Mazzucconi MG, Rocino A, et al. Identification and three-dimensional structural analysis of nine novel mutations in patients with prothrombin deficiency. *Thromb Haemost* 2000;**84**:989–97.
49. Akhavan S, Rocha E, Zeinali S, Mannucci PM. Gly319 –> arg substitution in the dysfunctional prothrombin Segovia. *Br J Haematol* 1999;**105**:667–9.
50. Sun WY, Burkart MC, Holahan JR, Degen SJ. Prothrombin San Antonio: a single amino acid substitution at a factor Xa activation site (Arg320 to His) results in dysprothrombinemia. *Blood* 2000;**95**:711–4.
51. De Cristofaro R, Akhavan S, Altomare C, Carotti A, Peyvandi F, Mannucci PM. A natural prothrombin mutant reveals an unexpected influence of a-chain structure on the activity of human {alpha}-thrombin. *J Biol Chem* 2004;**279**:13035–43.
52. De Cristofaro R, Carotti A, Akhavan S, Palla R, Peyvandi F, Altomare C, et al. The natural mutation by deletion of Lys9 in the thrombin A-chain affects the pKa value of catalytic residues, the overall enzyme's stability and conformational transitions linked to Na+ binding. *FEBS J* 2006;**273**:159–69.

53. Papaconstantinou ME, Bah A, Di Cera E. Role of the A chain in thrombin function. *Cell Mol Life Sci* 2008;**65**:1943–7.
54. Schechter I, Berger A. On the size of the active site in proteases. I. Papain. *Biochem Biophys Res Commun* 1967;**27**:157–62.
55. Bush LA, Nelson RW, Di Cera E. Murine thrombin lacks Na+ activation but retains high catalytic activity. *J Biol Chem* 2006;**281**:7183–8.
56. Martin CJ, Golubow J, Axelrod AE. The hydrolysis of carbobenzoxy-L-tyrosine p-nitrophenyl ester by various enzymes. *J Biol Chem* 1959;**234**:1718–25.
57. Lorand L, Brannen Jr. WT, Rule NG. Thrombin-catalyzed hydrolysis of p-nitrophenyl esters. *Arch Biochem Biophys* 1962;**96**:147–51.
58. Kezdy FJ, Lorand L, Miller KD. Titration of active centers in thrombin solutions. Standardization of the enzyme. *Biochemistry* 1965;**4**:2302–8.
59. Cleary DB, Trumbo TA, Maurer MC. Protease-activated receptor 4-like peptides bind to thrombin through an optimized interaction with the enzyme active site surface. *Arch Biochem Biophys* 2002;**403**:179–88.
60. Gandhi PS, Chen Z, Di Cera E. Crystal structure of thrombin bound to the uncleaved extracellular fragment of PAR1. *J Biol Chem* 2010;**285**:15393–8.
61. Dang QD, Sabetta M, Di Cera E. Selective loss of fibrinogen clotting in a loop-less thrombin. *J Biol Chem* 1997;**272**:19649–51.
62. Bartunik HD, Summers LJ, Bartsch HH. Crystal structure of bovine beta-trypsin at 1.5 A resolution in a crystal form with low molecular packing density. Active site geometry, ion pairs and solvent structure. *J Mol Biol* 1989;**210**:813–28.
63. Yang L, Prasad S, Di Cera E, Rezaie AR. The conformation of the activation peptide of protein C is influenced by Ca2+ and Na+ binding. *J Biol Chem* 2004;**279**:38519–24.
64. Tsiang M, Jain AK, Dunn KE, Rojas ME, Leung LL, Gibbs CS. Functional mapping of the surface residues of human thrombin. *J Biol Chem* 1995;**270**:16854–63.
65. Pechik I, Madrazo J, Mosesson MW, Hernandez I, Gilliland GL, Medved L. Crystal structure of the complex between thrombin and the central "E" region of fibrin. *Proc Natl Acad Sci USA* 2004;**101**:2718–23.
66. Pechik I, Yakovlev S, Mosesson MW, Gilliland GL, Medved L. Structural basis for sequential cleavage of fibrinopeptides upon fibrin assembly. *Biochemistry* 2006;**45**:3588–97.
67. Hall SW, Nagashima M, Zhao L, Morser J, Leung LL. Thrombin interacts with thrombomodulin, protein C, and thrombin-activatable fibrinolysis inhibitor via specific and distinct domains. *J Biol Chem* 1999;**274**:25510–6.
68. Fuentes-Prior P, Iwanaga Y, Huber R, Pagila R, Rumennik G, Seto M, et al. Structural basis for the anticoagulant activity of the thrombin-thrombomodulin complex. *Nature* 2000;**404**:518–25.
69. Pineda AO, Cantwell AM, Bush LA, Rose T, Di Cera E. The thrombin epitope recognizing thrombomodulin is a highly cooperative hot spot in exosite I. *J Biol Chem* 2002;**277**:32015–9.
70. Xu H, Bush LA, Pineda AO, Caccia S, Di Cera E. Thrombomodulin changes the molecular surface of interaction and the rate of complex formation between thrombin and protein C. *J Biol Chem* 2005;**280**:7956–61.
71. Vu TK, Wheaton VI, Hung DT, Charo I, Coughlin SR. Domains specifying thrombin-receptor interaction. *Nature* 1991;**353**:674–7.
72. Myles T, Le Bonniec BF, Stone SR. The dual role of thrombin's anion-binding exosite-I in the recognition and cleavage of the protease-activated receptor 1. *Eur J Biochem* 2001;**268**:70–7.
73. Ishihara H, Connolly AJ, Zeng D, Kahn ML, Zheng YW, Timmons C, et al. Protease-activated receptor 3 is a second thrombin receptor in humans. *Nature* 1997;**386**:502–6.

74. Gan ZR, Li Y, Chen Z, Lewis SD, Shafer JA. Identification of basic amino acid residues in thrombin essential for heparin-catalyzed inactivation by antithrombin III. *J Biol Chem* 1994;**269**:1301–5.
75. Sheehan JP, Sadler JE. Molecular mapping of the heparin-binding exosite of thrombin. *Proc Natl Acad Sci USA* 1994;**91**:5518–22.
76. Tsiang M, Jain AK, Gibbs CS. Functional requirements for inhibition of thrombin by antithrombin III in the presence and absence of heparin. *J Biol Chem* 1997;**272**:12024–9.
77. Li W, Johnson DJ, Esmon CT, Huntington JA. Structure of the antithrombin-thrombin-heparin ternary complex reveals the antithrombotic mechanism of heparin. *Nat Struct Mol Biol* 2004;**11**:857–62.
78. Carter WJ, Cama E, Huntington JA. Crystal structure of thrombin bound to heparin. *J Biol Chem* 2005;**280**:2745–9.
79. Martin PD, Malkowski MG, Box J, Esmon CT, Edwards BF. New insights into the regulation of the blood clotting cascade derived from the X-ray crystal structure of bovine meizothrombin des F1 in complex with PPACK. *Structure* 1997;**5**:1681–93.
80. Papaconstantinou ME, Gandhi PS, Chen Z, Bah A, Di Cera E. Na(+) binding to meizothrombin desF1. *Cell Mol Life Sci* 2008;**65**:3688–97.
81. Olson ST, Chuang YJ. Heparin activates antithrombin anticoagulant function by generating new interaction sites (exosites) for blood clotting proteinases. *Trends Cardiovasc Med* 2002;**12**:331–8.
82. Gettins PG. Serpin structure, mechanism, and function. *Chem Rev* 2002;**102**:4751–804.
83. Dementiev A, Petitou M, Herbert JM, Gettins PG. The ternary complex of antithrombin-anhydrothrombin-heparin reveals the basis of inhibitor specificity. *Nat Struct Mol Biol* 2004;**11**:863–7.
84. De Cristofaro R, De Candia E, Rutella S, Weitz JI. The Asp(272)-Glu(282) region of platelet glycoprotein Ibalpha interacts with the heparin-binding site of alpha-thrombin and protects the enzyme from the heparin-catalyzed inhibition by antithrombin III. *J Biol Chem* 2000;**275**:3887–95.
85. Celikel R, McClintock RA, Roberts JR, Mendolicchio GL, Ware J, Varughese KI, et al. Modulation of alpha-thrombin function by distinct interactions with platelet glycoprotein Ibalpha. *Science* 2003;**301**:218–21.
86. Dumas JJ, Kumar R, Seehra J, Somers WS, Mosyak L. Crystal structure of the GpIbalpha-thrombin complex essential for platelet aggregation. *Science* 2003;**301**:222–6.
87. Ramakrishnan V, DeGuzman F, Bao M, Hall SW, Leung LL, Phillips DR. A thrombin receptor function for platelet glycoprotein Ib-IX unmasked by cleavage of glycoprotein V. *Proc Natl Acad Sci USA* 2001;**98**:1823–8.
88. Pineda AO, Chen ZW, Marino F, Mathews FS, Mosesson MW, Di Cera E. Crystal structure of thrombin in complex with fibrinogen gamma' peptide. *Biophys Chem* 2007;**125**:556–9.
89. Colwell NS, Blinder MA, Tsiang M, Gibbs CS, Bock PE, Tollefsen DM. Allosteric effects of a monoclonal antibody against thrombin exosite II. *Biochemistry* 1998;**37**:15057–65.
90. Di Cera E. A structural perspective on enzymes activated by monovalent cations. *J Biol Chem* 2006;**281**:1305–8.
91. Prasad S, Wright KJ, Roy DB, Bush LA, Cantwell AM, Di Cera E. Redesigning the monovalent cation specificity of an enzyme. *Proc Natl Acad Sci USA* 2003;**100**:13785–90.
92. Zhang E, Tulinsky A. The molecular environment of the Na+ binding site of thrombin. *Biophys Chem* 1997;**63**:185–200.
93. Scharer K, Morgenthaler M, Paulini R, Obst-Sander U, Banner DW, Schlatter D, et al. Quantification of cation-pi interactions in protein-ligand complexes: crystal-structure analysis of Factor Xa bound to a quaternary ammonium ion ligand. *Angew Chem Int Ed Engl* 2005;**44**:4400–4.

94. Bajaj SP, Schmidt AE, Agah S, Bajaj MS, Padmanabhan K. High resolution structures of p-Aminobenzamidine- and Benzamidine-VIIa/soluble tissue factor: unpredicted conformation of the 192-193 peptide bond and mapping of Ca^{2+}, Mg^{2+}, Na^{+}, and Zn^{2+} sites in factor VIIa. *J Biol Chem* 2006;**281**:24873–88.
95. Schmidt AE, Padmanabhan K, Underwood MC, Bode W, Mather T, Bajaj SP. Thermodynamic linkage between the S1 site, the Na^{+} site, and the Ca^{2+} site in the protease domain of human activated protein C (APC). Sodium ion in the APC crystal structure is coordinated to four carbonyl groups from two separate loops. *J Biol Chem* 2002;**277**:28987–95.
96. Zogg T, Brandstetter H. Structural basis of the cofactor- and substrate-assisted activation of human coagulation factor IXa. *Structure* 2009;**17**:1669–78.
97. Petrovan RJ, Ruf W. Role of residue Phe225 in the cofactor-mediated, allosteric regulation of the serine protease coagulation factor VIIa. *Biochemistry* 2000;**39**:14457–63.
98. Schmidt AE, Stewart JE, Mathur A, Krishnaswamy S, Bajaj SP. Na+ site in blood coagulation factor IXa: effect on catalysis and factor VIIIa binding. *J Mol Biol* 2005;**350**:78–91.
99. Gopalakrishna K, Rezaie AR. The influence of sodium ion binding on factor IXa activity. *Thromb Haemost* 2006;**95**:936–41.
100. Rezaie AR, He X. Sodium binding site of factor Xa: role of sodium in the prothrombinase complex. *Biochemistry* 2000;**39**:1817–25.
101. Rezaie AR, Kittur FS. The critical role of the 185-189-loop in the factor Xa interaction with Na+ and factor Va in the prothrombinase complex. *J Biol Chem* 2004;**279**:48262–9.
102. Orthner CL, Kosow DP. The effect of metal ions on the amidolytic acitivity of human factor Xa (activated Stuart-Prower factor). *Arch Biochem Biophys* 1978;**185**:400–6.
103. Monnaie D, Arosio D, Griffon N, Rose T, Rezaie AR, Di Cera E. Identification of a binding site for quaternary amines in factor Xa. *Biochemistry* 2000;**39**:5349–54.
104. Underwood MC, Zhong D, Mathur A, Heyduk T, Bajaj SP. Thermodynamic linkage between the S1 site, the Na+ site, and the Ca2+ site in the protease domain of human coagulation factor xa. Studies on catalytic efficiency and inhibitor binding. *J Biol Chem* 2000;**275**:36876–84.
105. Camire RM. Prothrombinase assembly and S1 site occupation restore the catalytic activity of FXa impaired by mutation at the sodium-binding site. *J Biol Chem* 2002;**277**:37863–70.
106. Levigne S, Thiec F, Cherel S, Irving JA, Fribourg C, Christophe OD. Role of the alpha-helix 163-170 in factor Xa catalytic activity. *J Biol Chem* 2007;**282**:31569–79.
107. He X, Rezaie AR. Identification and characterization of the sodium-binding site of activated protein C. *J Biol Chem* 1999;**274**:4970–6.
108. Steiner SA, Amphlett GW, Castellino FJ. Stimulation of the amidase and esterase activity of activated bovine plasma protein C by monovalent cations. *Biochem Biophys Res Commun* 1980;**94**:340–7.
109. Steiner SA, Castellino FJ. Kinetic studies of the role of monovalent cations in the amidolytic activity of activated bovine plasma protein C. *Biochemistry* 1982;**21**:4609–14.
110. Steiner SA, Castellino FJ. Effect of monovalent cations on the pre-steady-state kinetic parameters of the plasma protease bovine activated protein C. *Biochemistry* 1985;**24**:1136–41.
111. Steiner SA, Castellino FJ. Kinetic mechanism for stimulation by monovalent cations of the amidase activity of the plasma protease bovine activated protein C. *Biochemistry* 1985;**24**:609–17.
112. Grinnell BW. Tipping the balance of blood coagulation. *Nat Biotechnol* 1997;**15**:124–5.
113. Page MJ, Di Cera E. Is Na+ a coagulation factor? *Thromb Haemost* 2006;**95**:920–1.
114. Huntington JA. Molecular recognition mechanisms of thrombin. *J Thromb Haemost* 2005;**3**:1861–72.

115. Page MJ, MacGillivray RTA, Di Cera E. Determinants of specificity in coagulation proteases. *J Thromb Haemost* 2005;**3**:2401–8.
116. Krishnaswamy S. Exosite-driven substrate specificity and function in coagulation. *J Thromb Haemost* 2005;**3**:54–67.
117. Krishnaswamy S, Betz A. Exosites determine macromolecular substrate recognition by prothrombinase. *Biochemistry* 1997;**36**:12080–6.
118. Perona JJ, Craik CS. Structural basis of substrate specificity in the serine proteases. *Protein Sci* 1995;**4**:337–60.
119. Perona JJ, Craik CS. Evolutionary divergence of substrate specificity within the chymotrypsin-like serine protease fold. *J Biol Chem* 1997;**272**:29987–90.
120. Hedstrom L. Serine protease mechanism and specificity. *Chem Rev* 2002;**102**:4501–24.
121. Stubbs MT, Oschkinat H, Mayr I, Huber R, Angliker H, Stone SR, et al. The interaction of thrombin with fibrinogen. A structural basis for its specificity. *Eur J Biochem* 1992;**206**:187–95.
122. Bah A, Chen Z, Bush-Pelc LA, Mathews FS, Di Cera E. Crystal structures of murine thrombin in complex with the extracellular fragments of murine protease-activated receptors PAR3 and PAR4. *Proc Natl Acad Sci USA* 2007;**104**:11603–8.
123. Sadasivan C, Yee VC. Interaction of the factor XIII activation peptide with alpha-thrombin. Crystal structure of its enzyme-substrate analog complex. *J Biol Chem* 2000;**275**:36942–8.
124. Rydel TJ, Tulinsky A, Bode W, Huber R. Refined structure of the hirudin-thrombin complex. *J Mol Biol* 1991;**221**:583–601.
125. Vijayalakshmi J, Padmanabhan KP, Mann KG, Tulinsky A. The isomorphous structures of prethrombin2, hirugen-, and PPACK-thrombin: changes accompanying activation and exosite binding to thrombin. *Protein Sci* 1994;**3**:2254–71.
126. Rose T, Di Cera E. Three-dimensional modeling of thrombin-fibrinogen interaction. *J Biol Chem* 2002;**277**:18875–80.
127. Mathews II, Padmanabhan KP, Ganesh V, Tulinsky A, Ishii M, Chen J, et al. Crystallographic structures of thrombin complexed with thrombin receptor peptides: existence of expected and novel binding modes. *Biochemistry* 1994;**33**:3266–79.
128. Baglin TP, Carrell RW, Church FC, Esmon CT, Huntington JA. Crystal structures of native and thrombin-complexed heparin cofactor II reveal a multistep allosteric mechanism. *Proc Natl Acad Sci USA* 2002;**99**:11079–84.
129. Di Cera E, Page MJ, Bah A, Bush-Pelc LA, Garvey LC. Thrombin allostery. *Phys Chem Chem Phys* 2007;**9**:1292–306.
130. Wells CM, Di Cera E. Thrombin is a Na(+)-activated enzyme. *Biochemistry* 1992;**31**:11721–30.
131. Griffon N, Di Stasio E. Thermodynamics of Na+ binding to coagulation serine proteases. *Biophys Chem* 2001;**90**:89–96.
132. Guinto ER, Di Cera E. Large heat capacity change in a protein-monovalent cation interaction. *Biochemistry* 1996;**35**:8800–4.
133. Bah A, Garvey LC, Ge J, Di Cera E. Rapid kinetics of Na+ binding to thrombin. *J Biol Chem* 2006;**281**:40049–56.
134. Botts J, Morales M. Analytical description of the effects of modifiers and of multivalency upon the steady state catalyzed reaction rate. *Trans Faraday Soc* 1953;**49**:696–707.
135. Gianni S, Ivarsson Y, Bah A, Bush-Pelc LA, Di Cera E. Mechanism of Na+ binding to thrombin resolved by ultra-rapid kinetics. *Biophys Chem* 2007;**131**:111–4.
136. Lai MT, Di Cera E, Shafer JA. Kinetic pathway for the slow to fast transition of thrombin. Evidence of linked ligand binding at structurally distinct domains. *J Biol Chem* 1997;**272**:30275–82.

137. Niu W, Chen Z, Bush-Pelc LA, Bah A, Gandhi PS, Di Cera E. Mutant N143P reveals how Na+ activates thrombin. *J Biol Chem* 2009;**284**:36175–85.

138. Page MJ, Di Cera E. Role of Na+ and K+ in enzyme function. *Physiol Rev* 2006;**86**:1049–92.

139. Pineda AO, Savvides SN, Waksman G, Di Cera E. Crystal structure of the anticoagulant slow form of thrombin. *J Biol Chem* 2002;**277**:40177–80.

140. Pineda AO, Zhang E, Guinto ER, Savvides SN, Tulinsky A, Di Cera E. Crystal structure of the thrombin mutant D221A/D222K: the Asp222:Arg187 ion-pair stabilizes the fast form. *Biophys Chem* 2004;**112**:253–6.

141. Marino F, Chen ZW, Ergenekan C, Bush-Pelc LA, Mathews FS, Di Cera E. Structural basis of Na+ activation mimicry in murine thrombin. *J Biol Chem* 2007;**282**:16355–61.

142. Prasad S, Cantwell AM, Bush LA, Shih P, Xu H, Di Cera E. Residue Asp-189 controls both substrate binding and the monovalent cation specificity of thrombin. *J Biol Chem* 2004;**279**:10103–8.

143. Berg DT, Wiley MR, Grinnell BW. Enhanced protein C activation and inhibition of fibrinogen cleavage by a thrombin modulator. *Science* 1996;**273**:1389–91.

144. Rezaie AR, Olson ST. Contribution of lysine 60f to S1' specificity of thrombin. *Biochemistry* 1997;**36**:1026–33.

145. Fuhrmann CN, Daugherty MD, Agard DA. Subangstrom crystallography reveals that short ionic hydrogen bonds, and not a His-Asp low-barrier hydrogen bond, stabilize the transition state in serine protease catalysis. *J Am Chem Soc* 2006;**128**:9086–102.

146. Johnson DJ, Adams TE, Li W, Huntington JA. Crystal structure of wild-type human thrombin in the Na +-free state. *Biochem J* 2005;**392**:21–8.

147. Gandhi PS, Chen Z, Mathews FS, Di Cera E. Structural identification of the pathway of long-range communication in an allosteric enzyme. *Proc Natl Acad Sci USA* 2008;**105**:1832–7.

148. Pineda AO, Chen ZW, Bah A, Garvey LC, Mathews FS, Di Cera E. Crystal structure of thrombin in a self-inhibited conformation. *J Biol Chem* 2006;**281**:32922–8.

149. Fehlhammer H, Bode W, Huber R. Crystal structure of bovine trypsinogen at 1-8 A resolution. II. Crystallographic refinement, refined crystal structure and comparison with bovine trypsin. *J Mol Biol* 1977;**111**:415–38.

150. Wang D, Bode W, Huber R. Bovine chymotrypsinogen A X-ray crystal structure analysis and refinement of a new crystal form at 1.8 A resolution. *J Mol Biol* 1985;**185**:595–624.

151. Reiling KK, Krucinski J, Miercke LJ, Raymond WW, Caughey GH, Stroud RM. Structure of human pro-chymase: a model for the activating transition of granule-associated proteases. *Biochemistry* 2003;**42**:2616–24.

152. Friedrich R, Panizzi P, Fuentes-Prior P, Richter K, Verhamme I, Anderson PJ, et al. Staphylocoagulase is a prototype for the mechanism of cofactor-induced zymogen activation. *Nature* 2003;**425**:535–9.

153. Fersht AR. Conformational equilibria in -and -chymotrypsin. The energetics and importance of the salt bridge. *J Mol Biol* 1972;**64**:497–509.

154. Fersht AR, Requena Y. Equilibrium and rate constants for the interconversion of two conformations of -chymotrypsin. The existence of a catalytically inactive conformation at neutral pH. *J Mol Biol* 1971;**60**:279–90.

155. Rohr KB, Selwood T, Marquardt U, Huber R, Schechter NM, Bode W, et al. X-ray structures of free and leupeptin-complexed human alphaI-tryptase mutants: indication for an alpha-> beta-tryptase transition. *J Mol Biol* 2006;**357**:195–209.

156. Krojer T, Garrido-Franco M, Huber R, Ehrmann M, Clausen T. Crystal structure of DegP (HtrA) reveals a new protease-chaperone machine. *Nature* 2002;**416**:455–9.

157. Jing H, Babu YS, Moore D, Kilpatrick JM, Liu XY, Volanakis JE, et al. Structures of native and complexed complement factor D: implications of the atypical His57 conformation and

self-inhibitory loop in the regulation of specific serine protease activity. *J Mol Biol* 1998;**282**:1061–81.

158. Hink-Schauer C, Estebanez-Perpina E, Wilharm E, Fuentes-Prior P, Klinkert W, Bode W, et al. The 2.2-A crystal structure of human pro-granzyme K reveals a rigid zymogen with unusual features. *J Biol Chem* 2002;**277**:50923–33.
159. Shia S, Stamos J, Kirchhofer D, Fan B, Wu J, Corpuz RT, et al. Conformational lability in serine protease active sites: structures of hepatocyte growth factor activator (HGFA) alone and with the inhibitory domain from HGFA inhibitor-1B. *J Mol Biol* 2005;**346**:1335–49.
160. Carvalho AL, Sanz L, Barettino D, Romero A, Calvete JJ, Romao MJ. Crystal structure of a prostate kallikrein isolated from stallion seminal plasma: a homologue of human PSA. *J Mol Biol* 2002;**322**:325–37.
161. Rickert KW, Kelley P, Byrne NJ, Diehl RE, Hall DL, Montalvo AM, et al. Structure of human prostasin, a target for the regulation of hypertension. *J Biol Chem* 2008;**283**:34864–72.
162. Spraggon G, Hornsby M, Shipway A, Tully DC, Bursulaya B, Danahay H, et al. Active site conformational changes of prostasin provide a new mechanism of protease regulation by divalent cations. *Protein Sci* 2009;**18**:1081–94.
163. Ponnuraj K, Xu Y, Macon K, Moore D, Volanakis JE, Narayana SV. Structural analysis of engineered Bb fragment of complement factor B: insights into the activation mechanism of the alternative pathway C3-convertase. *Mol Cell* 2004;**14**:17–28.
164. Barrette-Ng IH, Ng KK, Mark BL, Van Aken D, Cherney MM, Garen C, et al. Structure of arterivirus nsp4. The smallest chymotrypsin-like proteinase with an alpha/beta C-terminal extension and alternate conformations of the oxyanion hole. *J Biol Chem* 2002;**277**:39960–6.
165. Cavarelli J, Prevost G, Bourguet W, Moulinier L, Chevrier B, Delagoutte B, et al. The structure of Staphylococcus aureus epidermolytic toxin A, an atypic serine protease, at 1.7 A resolution. *Structure* 1997;**5**:813–24.
166. Vath GM, Earhart CA, Rago JV, Kim MH, Bohach GA, Schlievert PM, et al. The structure of the superantigen exfoliative toxin A suggests a novel regulation as a serine protease. *Biochemistry* 1997;**36**:1559–66.
167. Di Cera E. Serine proteases. *IUBMB Life* 2009;**61**:510–5.
168. Wu QY, Sheehan JP, Tsiang M, Lentz SR, Birktoft JJ, Sadler JE. Single amino acid substitutions dissociate fibrinogen-clotting and thrombomodulin-binding activities of human thrombin. *Proc Natl Acad Sci USA* 1991;**88**:6775–9.
169. Marino F, Pelc LA, Vogt A, Gandhi PS, Di Cera E. Engineering thrombin for selective specificity toward protein C and PAR1. *J Biol Chem* 2010;**285**:19145–52.
170. Arosio D, Ayala YM, Di Cera E. Mutation of W215 compromises thrombin cleavage of fibrinogen, but not of PAR-1 or protein C. *Biochemistry* 2000;**39**:8095–101.
171. Feistritzer C, Schuepbach RA, Mosnier LO, Bush LA, Di Cera E, Griffin JH, et al. Protective signaling by activated protein C is mechanistically linked to protein C activation on endothelial cells. *J Biol Chem* 2006;**281**:20077–84.
172. Gruber A, Griffin JH, Harker LA, Hanson SR. Inhibition of platelet-dependent thrombus formation by human activated protein C in a primate model. *Blood* 1989;**73**:639–42.
173. Gruber A, Hanson SR, Kelly AB, Yan BS, Bang N, Griffin JH, et al. Inhibition of thrombus formation by activated recombinant protein C in a primate model of arterial thrombosis. *Circulation* 1990;**82**:578–85.
174. Taylor FB, Chang A, Esmon CT, D'Angelo A, Vigano-D'Angelo S, Blick KE. Protein C prevents the coagulopathic and lethal effects of Escherichia coli infusion in the baboon. *J Clin Invest* 1987;**79**:918–25.
175. Bernard GR, Vincent JL, Laterre PF, LaRosa SP, Dhainaut JF, Lopez-Rodriguez A, et al. Recombinant human protein C Worldwide Evaluation in Severe Sepsis (PROWESS) study

group. Efficacy and safety of recombinant human activated protein C for severe sepsis. *N Engl J Med* 2001;**344**:699–709.
176. Taylor Jr. FB, Peer GT, Lockhart MS, Ferrell G, Esmon CT. Endothelial cell protein C receptor plays an important role in protein C activation in vivo. *Blood* 2001;**97**:1685–8.
177. Esmon CT, Xu J, Gu JM, Qu D, Laszik Z, Ferrell G, et al. Endothelial protein C receptor. *Thromb Haemost* 1999;**82**:251–8.
178. Bae JS, Yang L, Manithody C, Rezaie AR. The ligand occupancy of endothelial protein C receptor switches the protease-activated receptor 1-dependent signaling specificity of thrombin from a permeability-enhancing to a barrier-protective response in endothelial cells. *Blood* 2007;**110**:3909–16.
179. Bae JS, Yang L, Rezaie AR. Receptors of the protein C activation and activated protein C signaling pathways are colocalized in lipid rafts of endothelial cells. *Proc Natl Acad Sci USA* 2007;**104**:2867–72.
180. Hanson SR, Griffin JH, Harker LA, Kelly AB, Esmon CT, Gruber A. Antithrombotic effects of thrombin-induced activation of endogenous protein C in primates. *J Clin Invest* 1993; **92**:2003–12.
181. Griffin JH. Blood coagulation. The thrombin paradox. *Nature* 1995;**378**:337–8.
182. Carter WJ, Myles T, Gibbs CS, Leung LL, Huntington JA. Crystal structure of anticoagulant thrombin variant E217K provides insights into thrombin allostery. *J Biol Chem* 2004;**279**:26387–94.
183. Pineda AO, Chen ZW, Caccia S, Cantwell AM, Savvides SN, Waksman G, et al. The anticoagulant thrombin mutant W215A/E217A has a collapsed primary specificity pocket. *J Biol Chem* 2004;**279**:39824–8.
184. DiBella EE, Scheraga HA. The role of the insertion loop around tryptophan 148 in tthe activity of thrombin. *Biochemistry* 1996;**35**:4427–33.
185. Le Bonniec BF, Guinto ER, Esmon CT. Interaction of thrombin des-ETW with antithrombin III, the Kunitz inhibitors, thrombomodulin and protein C. Structural link between the autolysis loop and the Tyr-Pro-Pro-Trp insertion of thrombin. *J Biol Chem* 1992;**267**:19341–8.
186. Vindigni A, White CE, Komives EA, Di Cera E. Energetics of thrombin-thrombomodulin interaction. *Biochemistry* 1997;**36**:6674–81.
187. Gandhi PS, Page MJ, Chen Z, Bush-Pelc LA, Di Cera E. Mechanism of the anticoagulant activity of the thrombin mutant W215A/E217A. *J Biol Chem* 2009;**284**:24098–105.
188. Page MJ, Di Cera E. Serine peptidases: classification, structure and function. *Cell Mol Life Sci* 2008;**65**:1220–36.
189. Isetti G, Maurer MC. Employing mutants to study thrombin residues responsible for factor XIII activation peptide recognition: a kinetic study. *Biochemistry* 2007;**46**:2444–52.

Regulation of Proteases by Protein Inhibitors of the Serpin Superfamily

STEVEN T. OLSON[*,†] AND PETER G.W. GETTINS[†,‡]

[*]Center for Molecular Biology of Oral Diseases, University of Illinois at Chicago, Chicago, Illinois, USA

[†]The Center for Structural Biology, University of Illinois at Chicago, Chicago, Illinois, USA

[‡]Department of Biochemistry and Molecular Genetics, University of Illinois at Chicago, Chicago, Illinois, USA

I. Overview of the Serpin Superfamily 186
 A. Discovery of Serpins 186
 B. Serpin Classification 187
 C. Biology of Human Serpins 191
 D. Structural Features of Serpins 192
 E. Serpins and Conformational Disease 195
II. Mechanism of Action 197
 A. The Serpin Suicide Substrate Inhibition Mechanism 198
 B. The Michaelis Complex 199
 C. Covalent Complex 203
 D. Cleaved Serpin 208
 E. Stability of the Covalent Complex 208
 F. Energetics of Complex Formation 209
 G. Clearance of Complexes 210
III. Rates of Reaction, Specificity, and Regulation 211
 A. Simple RCL Interactions 211
 B. Exosite Interactions in Binary Complexes 212
 C. Exosites in Ternary Complexes with Cofactors 213
 D. Conformational Change in the Serpin 217
 E. Conformational Change in the Protease 218
 F. Serpins Versus Canonical Inhibitors 219
IV. Engineering Serpins as Therapeutics 221
 A. Recombinant Serpins for Replacement Therapy 221
 B. Engineered Serpin Variants 222
 C. Ovalbumin Variants 228
 D. crmA Variants 228
V. Concluding Remarks 230
 References 231

Progress in Molecular Biology
and Translational Science, Vol. 99
DOI: 10.1016/S1877-1173(11)99005-X

1877-1173/11 $35.00

The serpins comprise an ancient superfamily of proteins, found abundantly in eukaryotes and even in some bacteria and archea, that have evolved to regulate proteases of both serine and cysteine mechanistic classes. Unlike the thermodynamically determined lock-and-key type inhibitors, such as those of the Kunitz and Kazal families, serpins use conformational change and consequent kinetic trapping of an enzyme intermediate to effect inhibition. By combining interactions of both an exposed reactive center loop and exosites outside this loop with the active site and complementary exosites on the target protease, serpins can achieve remarkable specificity. Together with the frequent use of regulatory cofactors, this permits a sophisticated time- and location-dependent mode of protease regulation. An understanding of the structure and function of serpins has suggested that they may provide novel scaffolds for engineering protease inhibitors of desired specificity for therapeutic use.

I. Overview of the Serpin Superfamily

The serpin superfamily of protein protease inhibitors is an ancient family of proteins that are widely distributed in all the three kingdoms of life, the archea, the bacteria, and the eukaryotes, as well as in some viruses.[1,2] They represent a particularly notable advance in the evolution of proteins designed to control the activity of the proteolytic enzymes that regulate a multitude of biologic processes. Indeed, the serpins differ fundamentally from other families of protein protease inhibitors that evolved to regulate proteases with respect to their novel conformational change-based mechanism of irreversible protease inhibition and their ability to fine-tune protease activity in a time- and location-dependent manner that frequently depends on other cofactor molecules.[3] Most other families of protein protease inhibitors, such as the Kunitz and Kazal families, evolved as simple pseudosubstrates of their target proteases capable of forming reversible tight lock-and-key interactions.[4] The marked expansion of serpins in the multicellular eukaryotes has suggested that the serpins are particularly well-suited to controlling the protease cascades that regulate many complex physiologic processes of higher organisms. Their importance is abundantly clear from the severe tissue damage and consequent disease that can result from the dysregulation of protease function in serpin deficiency states.

A. Discovery of Serpins

The serpin superfamily was discovered in the 1980s by noting the homologies between two vertebrate blood proteins, antithrombin and antitrypsin, and the egg white protein, ovalbumin.[5] The family name was later derived as an acronym for *ser*ine *p*rotease *in*hibitor, based on the findings at the time that known family members were inhibitors of serine proteases.[6] It was subsequently found that serpins also function as physiologic inhibitors of cysteine proteases,[7] but the name was retained because its use had become so well established. There are now more than 3000 members of the superfamily that have been identified in various organisms, thus establishing the serpins as the largest known superfamily of protein protease inhibitors.[2] In humans, there are 36 serpins encoded by the genome, many of whose functions have been determined (Table I). Interestingly, as has become clear with several of the human serpins and those of other organisms, a number of serpins have evolved functions that do not involve protease inhibition.

B. Serpin Classification

A systematic classification of the serpins based on their evolutionary relatedness has resulted in the assignment of most serpins to 16 clades, designated by capital letters, A–P, with some residual orphans (Table I).[8,9] Serpins within a clade are given sequential numbers following the clade letter. Of the human serpins, most are members of the A and B clades, these clades accounting for 13 human serpins each. The remaining 10 human serpins are members of seven additional clades, clades C–I. Many of the human serpins have common names that were given before the systematic nomenclature was developed. The genes for these serpins are found on 10 different chromosomes, with 22 of the serpin genes localized in three clusters at 6p25, 14q21.1, and 18q21.3. Within a cluster, all serpins are members of the same clade and may be involved in carrying out a common function. Most of the A clade serpins are in the chromosome 14 cluster and these serpins appear to regulate proteases that mediate the inflammatory responses to injury and pathogen invasion. The B clade serpins are represented by the chromosome 6 and 18 clusters. These clusters appear to have arisen from a single B clade locus that is found in the chicken and that split prior to the divergence of primates and rodents.[10] The B clade serpins, also known as the ov-serpins because of their close relationship to chicken ovalbumin, include serpins that all lack signal peptides and have an intracellular function. Most other human serpin clades have an extracellular localization and are secreted by the specific tissues that express them. Based on a conserved exon–intron organization, vertebrate serpins have been proposed to have evolved from six ancestral serpin genes. These genes arose from primordial serpin genes by gene duplication and subsequent insertion of introns in distinct patterns before the divergence of the vertebrates.[11]

TABLE I

Human Serpins and Their Properties

Serpin	Common name	Cellular localization	Chromsomal localization	P1 residue/target proteases	Cofactors	Diseases associated with deficiency
SERPINA1	Antitrypsin/α_1-proteinase inhibitor	Extracellular	14q32.1	P1Met, neutrophil elastase		Emphysema
SERPINA2	Serpin peptidase inhibitor	Not characterized	14q32.1	Not characterized		
SERPINA3	α_1-Antichymotrypsin	Extracellular	14q32.1	P1Leu, cathepsin G		Lung diseases
SERPINA4	Kallistatin	Extracellular	14q32.1	P1Phe, tissue kallikreins	Heparin	
SERPINA5	Protein C Inhibitor	Extracellular	14q32.1	P1Arg, activated protein C, thrombin, acrosin	Heparin, thrombomodulin	
SERPINA6	Corticosteroid-binding globulin	Extracellular	14q32.1	Noninhibitory		Chronic fatigue
SERPINA7	Thyroxine-binding globulin	Extracellular	Xq22.2	Noninhibitory		Hypothyroidism
SERPINA8	Angiotensinogen	Extracellular	1q42-q43	Noninhibitory		Hypertension
SERPINA9	Centerin	Extracellular	14q32.1			
SERPINA10	Protein Z-dependent protease inhibitor	Extracellular	14q32.1	P1Tyr, factor Xa, factor XIa	Protein Z, lipid, Ca^{2+}	Thrombosis
SERPINA11	Serpin peptidase inhibitor	Not characterized	14q32.1	Not characterized		
SERPINA12	Vaspin	Extracellular	14q32.1	Not characterized		
SERPINA13	Serpin peptidase inhibitor		14q32.1	Pseudogene		
SERPINB1	Monocyte neutrophil elastase inhibitor	Intracellular	6p25	P1Cys, neutrophil elastase, proteinase-3, P1Phe, cathepsin G		

SERPINB2	Plasminogen activator inhibitor-2	Intracellular and extracellular	18q21.3	P1Arg, urokinase plasminogen activator		
SERPINB3	Squamous cell carcinoma antigen-1	Intracellular	18q21.3	P1Ser, cathepsins K, L, S		
SERPINB4	Squamous cell carcinoma antigen-2	Intracellular	18q21.3	P1Leu, cathepsin G, chymase		
SERPINB5	Maspin	Intracellular	18q21.3	Noninhibitory		Tumor progression, complete deficiency is lethal in mice
SERPINB6	Proteinase inhibitor-6	Intracellular	6p25	P1Met, cathepsin G		
SERPINB7	Megsin	Intracellular	18q21.3	P1Lys		
SERPINB8	Proteinase inhibitor-8	Intracellular	18q21.3	P1Arg, furin		
SERPINB9	Proteinase inhibitor-9	Intracellular	6p25	P1Glu, granzyme B		
SERPINB10	Bomapin	Intracellular	18q21.3	P1Arg, unknown		
SERPINB11	Epipin	Intracellular	18q21.3	P1Lys, unknown		
SERPINB12	Yukopin	Intracellular	18q21.3	P1Arg, unknown		
SERPINB13	Hurpin	Intracellular	18q21.3	P1Thr, cathepsin L		
SERPINC1	Antithrombin	Extracellular	1q23-q25	P1Arg, thrombin, factor Xa, factor IXa	Heparin, heparan sulfate	Thrombosis, complete deficiency is lethal in mice
SERPIND1	Heparin cofactor II	Extracellular	22q11	P1Leu, thrombin	Heparin, dermatan sulfate	Thrombosis
SERPINE1	Plasminogen activator inhibitor-1	Extracellular	7q21.3-q22	P1Arg, tissue and urokinase-type plasminogen activators, thrombin	Vitronectin, heparin	Bleeding
SERPINE2	Protease nexin-1	Extracellular	2q33-q35	P1Arg, thrombin	Heparin	
SERPINF1	Pigment epithelium-derived factor	Extracellular	17p13.3	Noninhibitory	Collagen	

(*Continues*)

TABLE I (*Continued*)

Serpin	Common name	Cellular localization	Chromsomal localization	P1 residue/target proteases	Cofactors	Diseases associated with deficiency
SERPINF2	α_2-antiplasmin	Extracellular	17pter-p12	P1Arg, plasmin	Fibrin	Bleeding
SERPING1	C1-inhibitor	Extracellular	11q11-q13.1	P1Arg, C1r, C1s, plasma kallikrein, factor XIIa	Selectins	Angioedema
SERPINH1	Hsp47	Intracellular	11p15	Noninhibitory		Complete deficiency is lethal in mice
SERPINI1	Neuroserpin	Extracellular	3q26	P1Arg, tissue plasminogen activator		
SERPINI2	Pancpin	Extracellular	3q26	P1Ile		Pancreatic insufficiency

C. Biology of Human Serpins

Extracellular human serpins have well-established roles in regulating proteases involved in blood clotting, inflammation, tissue regeneration, and immunity (Table I).[2] The serpins, antithrombin, heparin cofactor II (HCII), protein Z-dependent protease inhibitor (ZPI), protein C inhibitor (PCI), plasminogen activator inhibitor-1 (PAI-1), and α_2-antiplasmin, regulate proteases involved in the formation and dissolution of blood clots. The activity of the neutrophil proteases, neutrophil elastase, and cathepsin G that mediate the inflammatory response are controlled by the serpins, antitrypsin (also known as α_1-proteinase inhibitor) and α_1-antichymotrypsin, respectively. The contact system proteases that control vascular permeability and complement proteases that function in the immune defense system are all inhibited by C1-Inhibitor. The functions of several of these serpins have been defined based on the clinical manifestations of deficiency states (Table I). For example, deficiencies of antithrombin are associated with abnormal blood clotting,[12] deficiencies of PAI-1 correlate with a bleeding disorder,[13] deficiencies of antitrypsin cause progressive destruction of lung tissue and emphysema,[14] and deficiencies of C1-inhibitor cause angioedema that involves compromised vascular barrier function.[15] As the role of these serpins has become better defined in animal models of disease, it has become clear that many serpins produce their physiologic effects not only through protease inhibition but also through interactions with other molecules that reinforce a particular physiologic response. For example, C1-inhibitor interacts with leukocyte and endothelial cell selectins to block these proinflammatory receptors and reinforce the anti-inflammatory effects of contact system protease inhibition.[15]

The functions of the intracellular serpins of the B clade are the least well characterized. Many appear to have cytoprotective functions that protect the cell from proteases in lysosomes or that are stored in secretory compartments.[16] Serpins B1 and B6 protect the neutrophil from the proinflammatory proteases it releases upon activation,[17,18] and serpin B9 protects cytotoxic immune cells from stored granzyme B that is used to induce apoptosis in target cells.[19] Other intracellular serpins, such as serpins B2, B3, B4, B10, and B13, may guard against leakage of lysosomal proteases or interfere with the proapoptotic machinery of the cell through mechanisms yet to be defined. New molecular features have evolved in some of these serpins, such as the CD loop between helices C and D that appears to mediate novel functions like nuclear targeting and redox regulation.[16]

Valuable insights into human serpin function have come from gene knockouts in mice of orthologs of the human genes.[2] The identification of mouse orthologs has been complicated in many cases by the expansion of gene loci corresponding to a particular human clade member in the mouse and the consequent uncertainty in defining a clear ortholog. Of the 23 more definitive mouse orthologs of human serpins that have been targeted, knockouts of only three are embryonic lethal and thus essential for survival. These essential serpins include serpin B5 or maspin, which has a demonstrated tumor suppressor function in humans[20]; serpin C1 or antithrombin, which regulates blood clotting proteases[21]; and serpin H1 or Hsp47, which functions as a chaperone for collagen.[22] Of these, only antithrombin functions as a protease inhibitor, while the other two have evolved noninhibitory functions.

D. Structural Features of Serpins

Serpins are relatively large proteins consisting of an ~350 amino acid core domain that is common to all family members. They may have N- and C-terminal extensions that increase their size up to ~500 amino acids and may contain one or more sites of *N*- and/or *O*-glycosylation.[1] Numerous X-ray structures of serpins have been determined that have established a conserved tertiary structure fold for the superfamily that includes three β-sheets and eight to nine α-helices as well as a prominent exposed reactive center loop (RCL) that provides the bait for target proteases (Fig. 1). About 50 of the ~350 residues constituting the serpin fold are highly conserved across all members of the superfamily and are mostly localized within the hydrophobic core of the protein (Fig. 2).[9] The smallest serpins capable of inhibiting protease are found in viruses and their structures possibly reveal the essential structural elements of serpin function. In the case of the poxvirus serpin, crmA, helix D is absent and helix A is considerably shortened.[23,24] Structures of serpins from some archaea lack helix G.[25] These structural features are therefore presumably dispensable to the operation of the serpin protease inhibitory machinery.

A key feature of the serpin fold and its ability to function as a protease inhibitor is that it is metastable. This metastability first became clear when a more stable fold of the protein, the latent conformation, was shown to be spontaneously adopted by some serpins.[26] The transformation to the latent form was accompanied by a loss of protease inhibitor function, demonstrating that the native metastable fold is critical for serpins to function as protease inhibitors. The X-ray structure of the latent form of the serpin, PAI-1, revealed that the exposed RCL of the native serpin, which is anchored through its C-terminal end as strand 1 of the C β-sheet, had been stripped away from the C sheet and become inserted into the center of a larger β-sheet, the A sheet, that constitutes one face of the core domain (Fig. 3).[27] This transformation resulted in an ~20 °C increase in the melting temperature from the 50–60 °C characteristic of native serpins,[28] conclusively demonstrating that the native serpin fold is metastable and is capable of adopting a more stable fold. Cleavage in the RCL induces a similar conformational change in the serpin to a more stable state in which the N-terminal end of the cleaved RCL has inserted into sheet A (Fig. 3).[29] In this case, however, the increase in stability is much greater, the melting temperature increasing to > 100 °C and the enthalpy

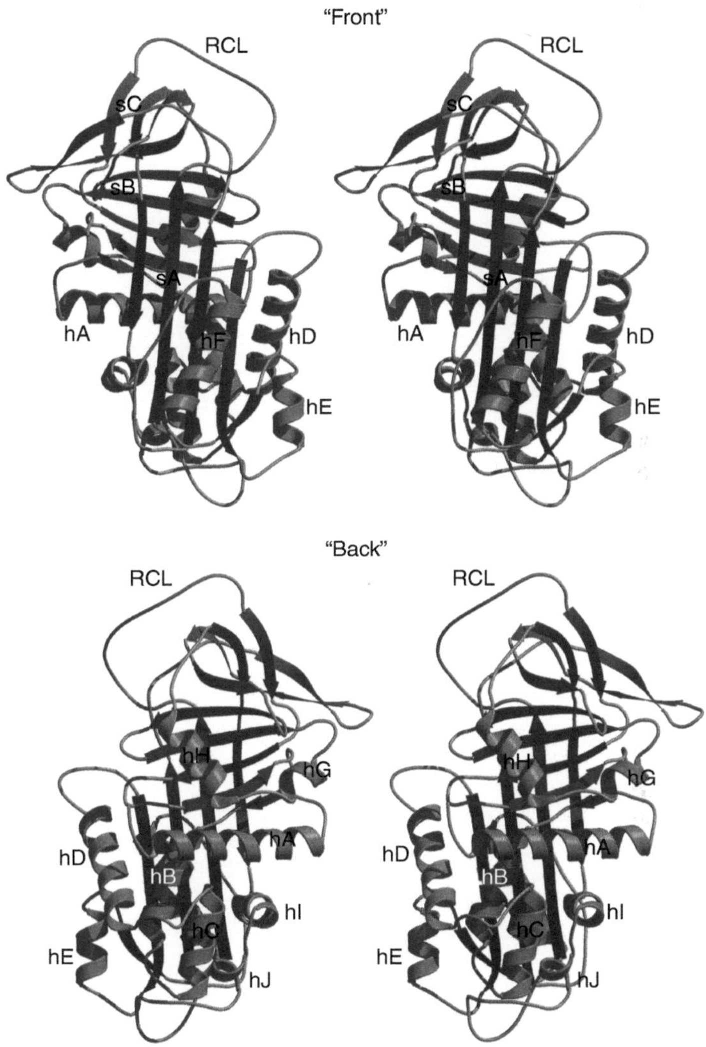

FIG. 1. Native serpin structure. Front and back stereo views of the structure of the serpin, α_1PI in its native state (pdb 1QLP), in ribbon representation. The ~350 amino acid core domain consists of three beta sheets, nine alpha helices (A–J), and a prominent reactive center loop (RCL) that protrudes from the top end of the molecule. Beta sheets are red (sheet A), blue (sheet B), and green (sheet C) and helices are yellow. Reprinted with permission from Ref. 1. Copyright 2002 American Chemical Society. (See Color Insert.)

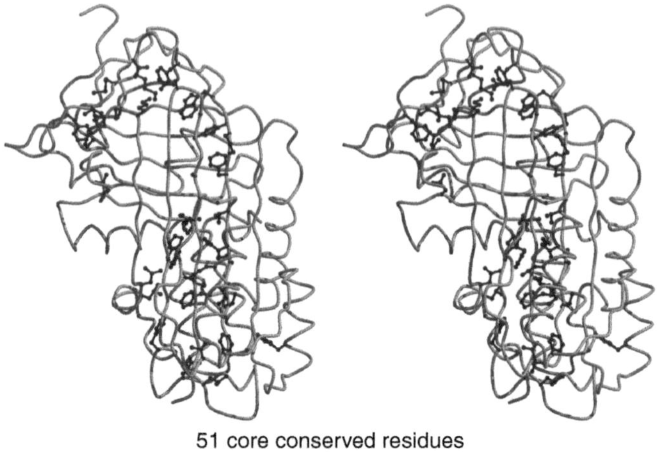

FIG. 2. Conserved residues in serpins. Shown is a stereo view of the RCL-cleaved α_1PI structure in backbone representation (gray, pdb 7API) depicting the location of the 51 highly conserved residues in > 70% of serpins in ball-and-stick representation (blue) that were identified by Irving *et al.*[9] Most of the conserved residues are part of the hydrophobic core. Reprinted with permission from Ref. 1. Copyright 2002 American Chemical Society. (See Color Insert.)

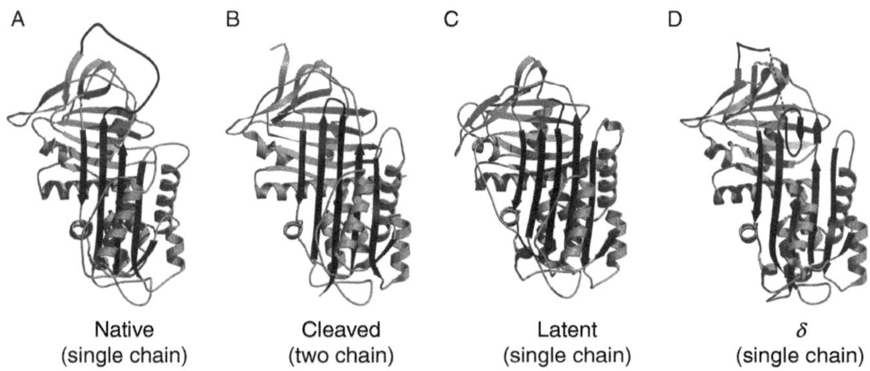

FIG. 3. Conformational states of serpins. Comparison of X-ray structures of the metastable native serpin conformation (A; α_1PI, pdb 1QLP) and the more stable RCL-cleaved (B; α_1PI, pdb 7API), and latent (C; PAI-1, pdb 1C5G) conformations. Also shown is the unusual δ-conformation of an α_1-antichymotrypsin L55P variant with a mutation in the shutter region (D; pdb 1QMN). All structures are shown in ribbon representation with the RCL colored blue, the A β-sheet red, and the remainder of the structure in gray. Reprinted with permission from Ref. 1. Copyright 2002 American Chemical Society. (See Color Insert.)

change indicating a highly exothermic transition.[30,31] The greater stability of the cleaved over the latent state reflects the fact that the C-terminal end of the RCL remains a part of sheet C and is not affected by cleavage.

For serpins to preferentially fold to and be trapped in a metastable, rather than the most stable folded state, the folding pathway must be kinetically, rather than thermodynamically, controlled. This kinetic control of folding appears to result from the C sheet forming before the A sheet so as to create a large kinetic barrier to RCL incorporation into the A sheet and adoption of the more stable latent fold.[1] The transformation to the latent state does not occur under physiologic conditions for most serpins in order to maintain protease inhibitory function. However, it appears to be a normal physiologic process for PAI-1,[32] and possibly also for antithrombin, as a means of regulating their function.[33] The lifetime of the metastable fold of PAI-1 can be prolonged by binding to the extracellular matrix protein, vitronectin.[34] Remarkably, serpins from thermophilic organisms have been shown to contain N- or C-terminal extensions that function to stabilize the native metastable fold.[35]

E. Serpins and Conformational Disease

The design of serpins as metastable proteins to enable them to function as protease inhibitors has come at the cost of an increased susceptibility to mutations that cause conformational instability and loss of function. Such mutations result in a class of conformational diseases known as the serpinopathies in which the mutant serpin polymerizes and, in some cases, forms cytotoxic inclusion bodies.[36] These mutations are localized in a "shutter" region at the end of sheet A most distal from the RCL that controls the opening of the sheet.[37] An unusual inactive conformation adopted by such a shutter region mutant, termed the δ-conformation, was revealed by an X-ray structure of the mutant serpin and has been the basis for proposed mechanisms of polymerization.[38] In the δ-state, the RCL has partially inserted into the proximal end of sheet A with a concomitant loss of the RCL–sheet C interaction, while the F helix, which overlays sheet A, has unraveled by several strands to insert into the distal end of sheet A (Fig. 3). Based on this structure, shutter mutations are thought to make the A sheet more prone to opening and adopting the δ-state. Given the observation that a polymerogenic intermediate is involved in the heat-induced polymerization of normal and mutant serpins,[39] the polymerization of shutter mutants *in vivo* has been suggested to involve the stabilization of the δ-conformation, which seeds and propagates polymer formation through intermolecular RCL–A sheet and RCL–C sheet linkages (Fig. 4A).[36]

More recently, the structure of a domain swap dimer of antithrombin has suggested an alternative polymerization mechanism for shutter mutants.[41] In this case, polymerization occurs through an intermediate on the folding pathway in which most secondary structures have formed, except for strand 5 of

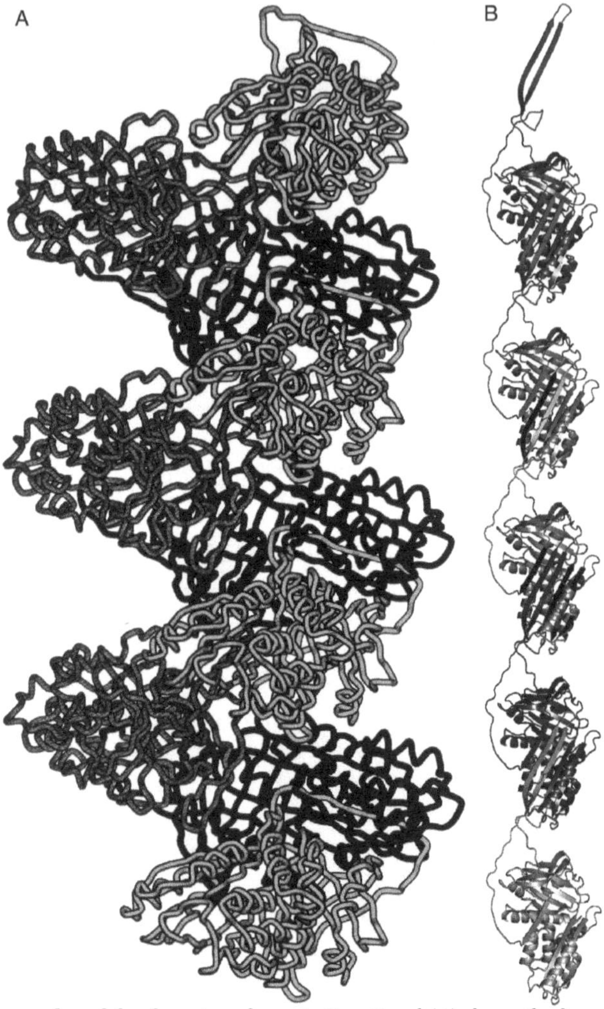

FIG. 4. Proposed models of serpin polymerization. Panel (A) shows the loop-sheet model of serpin polymerization proposed by Elliott *et al.*[40] in which the RCL of one native serpin molecule is inserted into β-sheet A of a second molecule. Each consecutive molecule in the polymer is represented by a different color. Reprinted by permission from Macmillan Publishers Ltd[40], copyright 1996. Panel (B) shows the more recent model proposed by Yamasaki *et al.*[41] in which strand 5 of sheet A and the RCL, the last elements of the serpin to fold, insert as paired antiparallel β-strands from one partially folded serpin molecule into a second partially folded molecule. In the normal folding pathway, the RCL completes sheet C before strand 5 can pair with the RCL and insert into sheet A so that strand 5 alone inserts into sheet A, leaving the RCL to adopt its native exposed state anchored to sheet C. Reprinted by permission from Macmillan Publishers Ltd[41], copyright 2008. (For interpretation of the references to color in this figure legend, the reader is referred to the Web version of this chapter.)

sheet A and the RCL, including the C-terminal end that makes up strand 1 of sheet C. The native structure is normally reached in this model by the RCL completing sheet C and strand 5 completing the formation of sheet A, leaving the RCL unable to insert because of its anchoring to sheet C. The shutter mutations are proposed to increase the lifetime of the intermediate and favor an intermolecular insertion of strand 5 and the RCL into the A sheet of another serpin molecule. This occurs in a manner that pairs strand 5 and the RCL as antiparallel β-strands before the RCL can complete sheet C and allows the paired strands to insert into sheet A as in the latent serpin structure (Fig. 4B). This mechanism is supported by earlier findings that the more common natural antitrypsin shutter mutant forms polymers through a long-lived folding intermediate rather than through an unstable native serpin fold, since mutant monomers are as stable as wild-type monomers.[42]

In the case of antitrypsin, whose plasma concentrations are normally the highest of all serpins circulating in the blood and whose plasma concentrations greatly increase when inflammation is triggered, shutter mutations result in cytotoxic polymers forming in liver cells and consequent cirrhosis of the liver.[43] The decreased secretion of such mutants by the liver also results in decreased plasma levels of antitrypsin and progressive lung disease. Similar mutations in antithrombin, C1-inhibitor, and antichymotrypsin result in serpin deficiency but no liver disease since these serpins are expressed by the liver at much lower levels and thus produce lower levels of polymer that do not accumulate. However, like antitrypsin, shutter region mutations in neuroserpin, which is expressed in neurons of the brain, are associated with the formation of cytotoxic polymers that produce dementia.[36]

II. Mechanism of Action

Unlike most other protein protease inhibitors, serpins are capable of inhibiting proteases belonging to different mechanistic classes, as well as to different clades within a given class.[1] While the most numerously documented examples are of serpins inhibiting serine proteases of the chymotrypsin family, there are examples of serpins inhibiting furin-like serine proteases[44] and of serpins inhibiting cysteine proteases of both the cathepsin[45] and caspase families.[46] Furthermore, some individual serpins are able to efficiently inhibit proteases of more than one class.[47] This behavior is readily understandable in the light of the unique mechanism of inhibition by serpins, in which the reaction between the protease and the serpin, as substrate, is drastically slowed at the stage of formation of the acyl-intermediate that is common to both serine and cysteine proteases.

A. The Serpin Suicide Substrate Inhibition Mechanism

Serpins inhibit proteases of both serine and cysteine classes by a branched pathway suicide substrate inhibition mechanism, outlined in Fig. 5. Protease and serpin form an initial noncovalent encounter complex (Michaelis complex), involving specific recognition between the protease and elements in the serpin, most usually within the exposed RCL, but additionally involving exosites and/or cofactors. The protease then proceeds to cleave the RCL. Although the four families of protease mentioned above use different configurations of catalytic

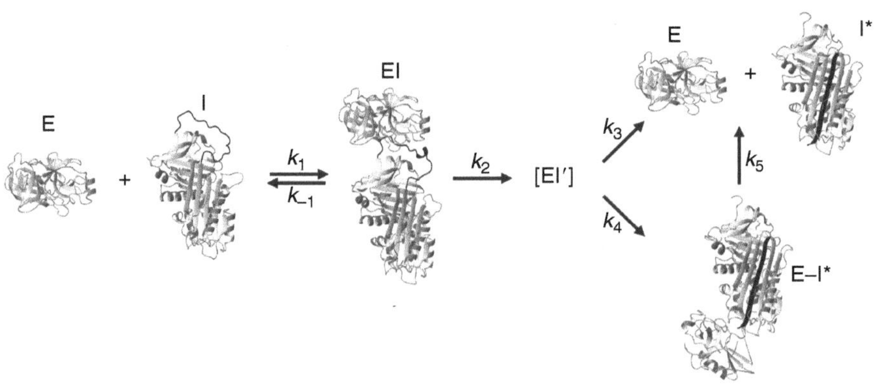

FIG. 5. Basic serpin suicide substrate, branched pathway, mechanism. The protease (E) recognizes residues within the reactive center loop (RCL) of the serpin (I) and associates, with second-order rate constant of k_1, to form an initial noncovalent Michaelis-like complex (EI) by simply docking, with no need for any major conformational change in either protease or serpin RCL. This complex can dissociate with a rate constant of k_{-1}, but is more likely to react further in a substrate-like manner (i.e., $k_2 \gg k_{-1}$) to give cleavage of the scissile peptide bond and formation of an acyl-enzyme intermediate (EI′) in which the serine in the enzyme active site is covalently linked to the cleaved peptide carbonyl. Cleavage of the RCL allows the RCL to insert into β-sheet A, with concomitant dragging of the protease from the top to the bottom of β-sheet A. Two outcomes are possible and are in competition with each other. One is the complete insertion of the RCL and consequent compromise of the activity of the protease, resulting in the acyl-enzyme intermediate being kinetically trapped (E–I*). This trapping occurs with a rate constant, k_4. If, however, the protease can complete the hydrolysis reaction before its activity is fully compromised, products of cleaved serpin (I*) and free protease are formed, with a rate constant of k_3. The relative amounts of cleaved serpin and covalent, inhibited complex are determined by the ratio of these rate constants (k_3/k_4). This ratio leads to a "stoichiometry of inhibition" (SI), representing the number of mols of serpin required to form 1 mol of covalent complex that is given as $1 + k_3/k_4$, which deviates from the value of 1 in direct relation to the rate of the hydrolysis reaction relative to loop insertion. Finally, since the covalent complex is only a kinetically trapped intermediate, it can still form the same products of I* and E by noncatalyzed hydrolysis, with a rate constant of k_5. Typically, k_5 is many orders of magnitude slower than k_3 and results in the complex being stable for days to weeks. This research was originally published in The Journal of Biological Chemistry[48]. Copyright the American Society of Biochemistry and Molecular Biology.

residues, each employs an analogous cleavage mechanism that is illustrated for serine proteases of the chymotrypsin family (Fig. 6). Importantly for the serpin mechanism, each protease, whether cysteine or serine, proceeds via a covalent acyl-enzyme intermediate that is formed upon cleavage of the RCL. Cleavage of the RCL permits the serpin to undergo a remarkable conformational change in which the covalently attached protease is translocated to the distal end of the serpin as the cleaved RCL to which the protease is covalently linked inserts into the principal β-sheet of the serpin.[50] As a result of the translocation, the protease is squeezed against the bottom of the serpin, thereby distorting the active site of the protease and kinetically trapping the acyl-enzyme intermediate, which thus represents the inhibited serpin–protease complex.[51] The inhibition is kinetic in that the acyl-enzyme intermediate can still be hydrolyzed to give free, cleaved serpin, and release of active protease, though the rate of release is usually many orders of magnitude less than for the normal substrate–protease reaction.[52]

Because the inhibition mechanism requires completion of a conformational change subsequent to RCL cleavage, there is a time window between the first formation of the acyl-intermediate and its final kinetic trapping during which the protease retains some or all of its catalytic efficacy and can complete the normal substrate hydrolysis of the RCL. As a consequence, the serpin pathway is a branched one, with one branch leading to kinetically trapped acyl-enzyme complex and the other directly to cleaved serpin. The relative overall rates of the two branches determine the efficiency of the inhibition. This efficiency is defined as the stoichiometry of inhibition (SI), which is the number of molecules of serpin required to inhibit one molecule of protease and represents the inverse of the fraction of the first-formed acyl-intermediate that ends up as trapped complex. In terms of the overall rate constants for the two branches of the pathway (k_3 and k_4), the SI is given as Eq. (1)[53]:

$$\mathrm{SI} = (k_3 + k_4)/k_4 \tag{1}$$

For a reaction that leads overwhelmingly to complex, the value of SI is close to 1, while for one that leads predominantly to cleaved serpin, that is, mostly a substrate reaction, the SI is very much greater than 1. (In the terminology generally used to characterize the efficiency of suicide substrate inactivators of enzymes, the SI represents the number of turnovers an enzyme undergoes before it is inactivated.)

B. The Michaelis Complex

X-ray structures exist for Michaelis complexes of five different serpins (antithrombin, α_1-protease inhibitor, Manduca sexta 1B, PCI, and HCII), with one or more of four different proteases (trypsin, thrombin, and factors IXa Xa; Figs. 7 and 9; Table II).[54,55,58–63] In addition, several of the complexes have heparin bound as an allosteric activator. A common feature of all of the

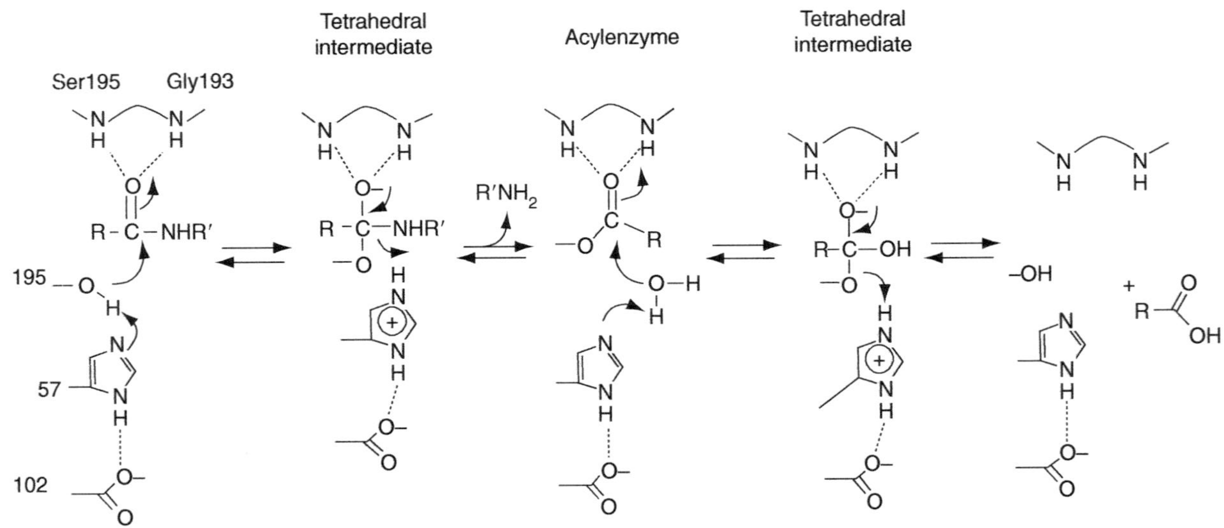

Fig. 6. Two-stage mechanism by which serine and cysteine family proteases catalyze peptide bond cleavage. Shown is the mechanism of proteolysis of peptide bonds by the catalytic triad, Ser195, His57, and Asp102, of chymotrypsin family proteases. The mechanism is similar for cysteine proteases, except that a Cys replaces the catalytic Ser. Proteolysis begins by the binding of the peptide substrate to the protease active site to form the Michaelis complex. In the first stage, the scissile peptide bond is attacked by the hydroxyl group of Ser195, which is activated as a nucleophile by hydrogen bonding to the His57-Asp102 dyad. The attack produces the tetrahedral intermediate in which the Ser195 oxygen is covalently linked to the substrate carbonyl carbon, the hydroxyl hydrogen has been transferred to His57, and the anionic carbonyl oxygen is stabilized by hydrogen bonds with the backbone NH groups of Gly193 and Ser195, known as the oxyanion hole. The tetrahedral intermediate then collapses to the acyl-intermediate in which the substrate peptide bond has been cleaved, the His57 hydrogen transferred to the leaving amine group, and the Ser195 hydroxyl has formed an ester bond with the substrate carbonyl. In the second stage of proteolysis, His57 activates an incoming water molecule for nucleophilic attack on the carbonyl carbon of the acyl-intermediate with protonation of His57 and formation of a second tetrahedral intermediate whose oxyanion is again stabilized by the oxyanion hole residues. The tetrahedral intermediate then collapses, resulting in the cleavage of the ester bond, transfer of the His57 hydrogen to regenerate the Ser195 hydroxyl, and conversion of the substrate carbonyl to a carboxyl group that then dissociates from the active site. Reprinted with permission from Ref. 49. Copyright 2002 American Chemical Society.

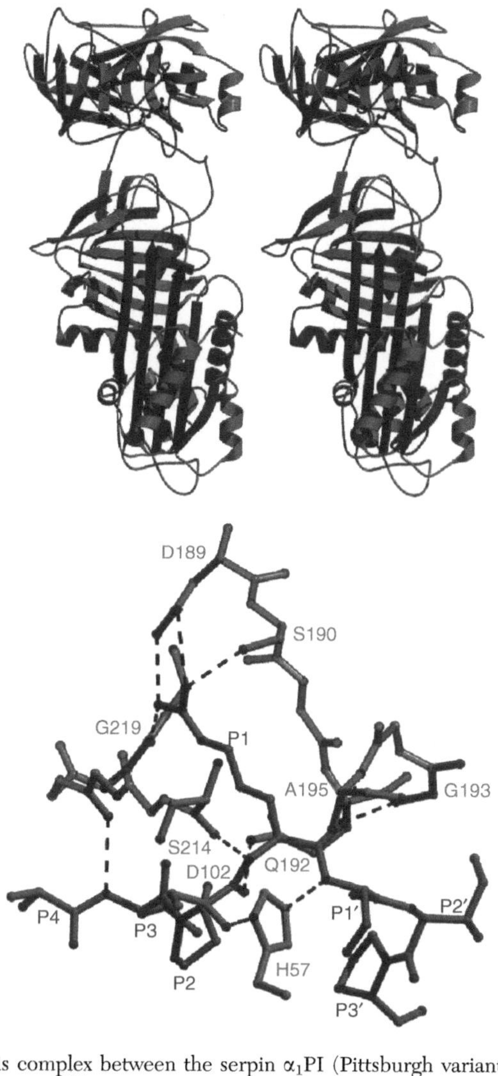

FIG. 7. Michaelis complex between the serpin α_1PI (Pittsburgh variant with a P1 Arg) and bovine trypsin (pdb 1OPH).[54] Top: stereo ribbon diagram; bottom: stick figure of the contacts between the active site of the protease and the RCL of the serpin. Besides the P1 arginine side chain fitting into the S1 pocket of the protease and engaging the aspartate (D189) that helps determine trypsin's specificity, most of the contacts involve H-bonds between the extended canonical conformation of the serpin RCL backbone with the protease. No change in conformation is necessary for the serpin RCL to interact with the protease in this manner. This research was originally published in The Journal of Biological Chemistry.[54] Copyright the American Society of Biochemistry and Molecular Biology.

TABLE II
X-RAY STRUCTURES OF SERPIN–PROTEASE COMPLEXES

Serpin	Protease	Complex	pdb code	References
α_1PI	Pancreatic elastase	Covalent	2D26	56
α_1PI	Bovine trypsin	Covalent	1EZX	57
α_1PI Pittsburgh	S195A bovine trypsin	Noncovalent	1OPH	54
Manduca sexta serpin 1K	S195A rat trypsin	Noncovalent	1K9O	58
Heparin cofactor II–heparin	S195A thrombin	Noncovalent	1JMO	59
Antithrombin–heparin	Anhydrothrombin	Noncovalent	1SR5	55
Antithrombin–heparin	S195A thrombin	Noncovalent	1TB6	60
Antithrombin–heparin	S195A factor Xa	Noncovalent	2GD4	61
Antithrombin–heparin	S195A factor IXa	Noncovalent	3KCG	62
PCI-heparin	S195A thrombin	Noncovalent	3B9F	63

complexes is that the protease is docked with the RCL at the expected site of cleavage of the scissile bond, with the P1* side chain in the S1 pocket of the protease. Moreover, the RCL backbone on either side of this peptide is in the canonical conformation of an optimal peptide substrate and forms hydrogen bonds with the protease, in the manner found for other families of protein protease inhibitor that employ a reversible noncovalent complex mechanism. Where the complexes differ is in the precise orientation of the protease relative to the body of the serpin. This difference is seen for the same protease with different serpins, for the same serpin with different proteases, and even for the same serpin:protease pair. This variability most likely reflects the flexibility of the linkers that attach the portion of the RCL that docks with the protease to the serpin body rather than being of functional significance.

The simplest complex to understand is that between S195A trypsin and a P1 Arg variant of α_1PI (Fig. 7).[54] Superpositioning of the complexed serpin with free serpin and of the complexed protease with free protease shows that complex formation is a simple rigid body docking, with no conformational change necessary in either component. The serpin RCL has an extended canonical conformation. Contacts between serpin and protease are limited to residues P2–P2′ in the serpin RCL. Because the RCL extends well out from the body of the serpin and because the loops that surround the active site of trypsin are short compared to those of the blood coagulation serine proteases, there are no additional interactions between the protease and the serpin.

*Nomenclature of Schecter and Berger[64] in which peptide substrate residues are designated P1, P2, P3, etc. on the N-terminal side of the scissile peptide bond of the substrate and P1′, P2′, P3′, etc. on the C-terminal side of the bond. Corresponding protease binding pockets for the substrate residues in the active site are designated S1, S2, S3 and S1′, S2′, S3′, etc.

A closely related Michaelis complex, and indeed the first one whose structure was determined by X-ray crystallography, is that between Manduca sexta 1B (P1 mutant to Lys) and S195A trypsin.[58] As with the trypsin:α_1PI complex, the protease engages the serpin RCL in canonical conformation, though the contacts are more extensive, extending from P4 to P4′. In addition, the orientation of the trypsin differs slightly from that in the trypsin:α_1PI complex so that the 140s loop of trypsin contacts more distal parts of the RCL (P9 and P10), though these additional contacts most likely result from the relatively unconstrained positioning of the RCL relative to the body of the serpin and the dictates of crystal packing.

Evidence for the mobility of the RCL relative to the serpin body is directly provided by linewidths of NMR resonances from the RCL of α_1PI, both alone and in complex with S195A trypsin[65] and with anhydroelastase,[48] which report higher mobility than the serpin body and relatively minor constraint imposed by complex formation with either protease (Fig. 8). In addition, two complexes of the same ternary complex of S195A (or anhydro) thrombin with antithrombin and heparin have somewhat different orientations of the protease relative to the serpin, even though the principal contacts between protease and serpin RCL are the same (Fig. 9).[55,60]

C. Covalent Complex

Even before the first detailed structural studies on the final serpin:protease complex, it was known for inhibition of serine proteases of the chymotrypsin family that the complex represents a covalent acyl-enzyme intermediate that deacylates very slowly and in which the P1 side chain is no longer in the same location in the S1 pocket as in the Michaelis complex.[68] First FRET[69] and then NMR[66] and X-ray crystallography[57] showed that the covalent complex is composed of a serpin moiety that has undergone a major conformational change, with movement of the protease by over 70 Å from the top of the serpin to the bottom, and with the expected covalent linkage between the active site serine of the protease and the carbonyl of the P1 residue of the cleaved peptide bond of the serpin RCL (Fig. 10). X-ray structures of covalent complexes of α_1PI with trypsin[57] and later with elastase[56] showed that the active site has been distorted such that the acyl-intermediate is no longer appropriately positioned relative to the catalytic histidine for efficient catalysis and that the P1 side chain has been extracted from the S1 pocket of the protease. In contrast to the noncovalent Michaelis complex, in which there appears to be considerable freedom of movement of the protease and docked RCL relative to the serpin body, the covalent complex behaves as a single rigid species. This difference is clearly evident in the large increase in linewidth of resonances from residues within the serpin body upon complex formation, indicating an increase in rotational correlation time, arising presumably from the effective increase in

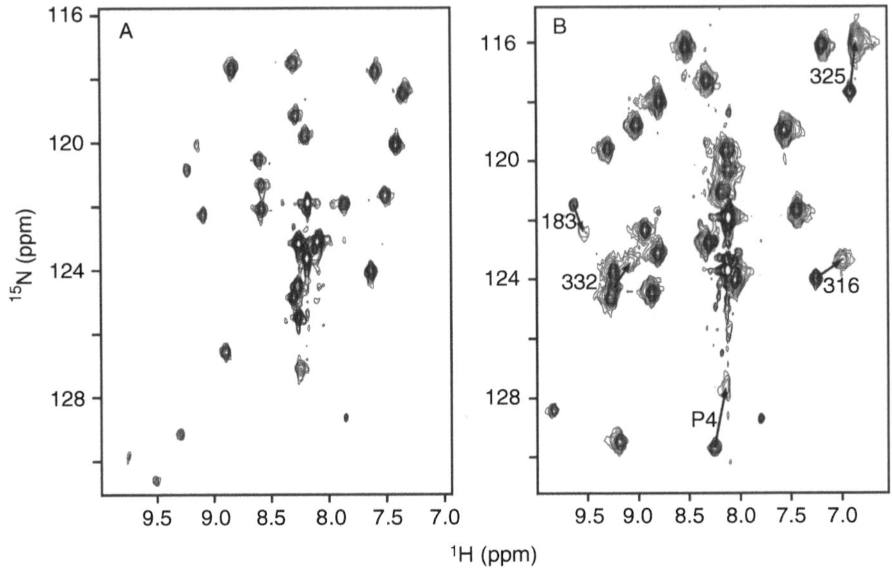

FIG. 8. Demonstration of the conformation of the serpin moiety in the noncovalent and covalent complexes with trypsin by comparison of 2D HSQC spectra of ^{15}N-alanine-labeled native or cleaved serpin with spectra of the appropriate complex (no NMR-visible label in the protease). Left panel: spectrum of native serpin (red) superimposes almost perfectly with that of the noncovalent complex. Right panel: spectrum of cleaved serpin (red) superimposes well with that of the covalent complex. The few differences in chemical shift result from a direct contact between the protease and the serpin (Ala residues 316 and 325 for the covalent complex and Ala 284 for the noncovalent complex) or from small perturbations at the bottom of β-sheet A in the covalent complex [Ala residues 183, 332, and 355 (P4)]. Note the minimal line broadening for the noncovalent complex, but much greater broadening for the covalent complex.[65,66] Reprinted from Ref. 67. Copyright 2004, with permission from Elsevier. (For interpretation of the references to color in this figure legend, the reader is referred to the Web version of this chapter.)

mass caused by the presence of the protease (Fig. 8).[66] The placement of the protease in the complex is at the bottom of β-sheet A, but to one side, where it is held closely against the body of the serpin. The same location is seen both by NMR and X-ray crystallography for the trypsin:α_1PI complex, as well as for the elastase:α_1PI complex, though the protease is rotated about 30° in the latter. Although providing a much lower resolution picture, limited FRET measurements on several PAI-1:protease complexes also suggest a similar location for each protease relative to the serpin as for α_1PI.[70]

The serpin moiety in the complex is almost indistinguishable in structure from that of the RCL-cleaved serpin, with the cleaved RCL inserted into β-sheet A as the fourth of a six-stranded, perfectly antiparallel β-sheet (Figs. 8 and 10). Small chemical shift differences of NMR resonances from

FIG. 9. Role of heparin (green and black) in bridging between thrombin (red) and antithrombin (yellow) to promote formation of the Michaelis complex. The structure (pdb 1SR5)[55] shows specific binding of the high affinity pentasaccharide motif (black) to the heparin-binding site of antithrombin and of the more distal end of the heparin chain to exosite II of thrombin. Reprinted by permission from Macmillan Publishers Ltd,[55] copyright 2004. (See Color Insert.)

alanines at the bottom of the sheet in complex compared with simple cleaved serpin may reflect subtle perturbations of hydrogen bonds within the sheet that are not resolved in the relatively low resolution of the two X-ray structures of complexes.

The structure and properties of the serine protease in the covalent complex is still somewhat uncertain. In the first X-ray structure of a complex, that of the complex of α_1PI with trypsin, only about 60% of the protease was well defined in the electron density maps, with one whole lobe of the structure missing.[57] It was suggested at the time that this reflected a more molten globule-like structure for the missing domain. However, it appears from SDS-PAGE analysis of this complex that the trypsin may have one or more cleavages within it, so that an alternative explanation for the "missing" electron density is that there are multiple conformations present for that lobe, resulting in no single well-defined structure. In keeping with such an interpretation, the second X-ray structure of a covalent complex, that of α_1PI with pancreatic elastase, gives a

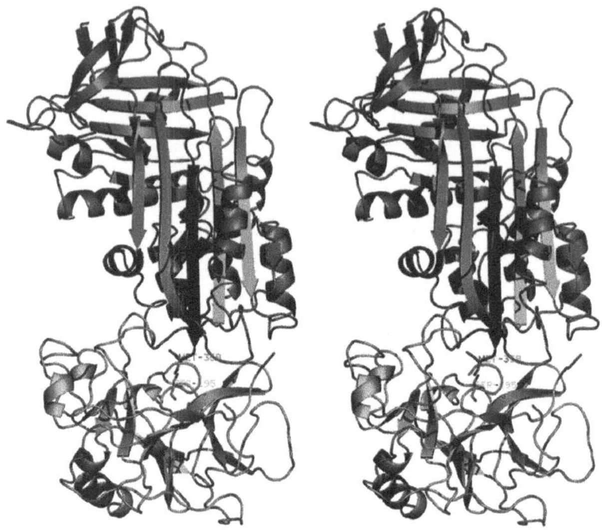

FIG. 10. Stereo view of the covalent, kinetically trapped acyl-enzyme intermediate complex of α_1PI (pink) with pancreatic elastase (cyan) (pdb 2D26).[56] Sheet A of the serpin is highlighted in yellow and the cleaved RCL that has inserted into sheet A is colored red. Within the protease active site, the acyl ester linkage between the catalytic serine and the P1 residue has been displaced relative to the remainder of the catalytic apparatus, rendering the enzyme inactive and trapping the complex kinetically. Other differences in structure between the complex and the constituent free protease and cleaved serpin moieties are minor and limited to the contact interface loops. This research was originally published in The Journal of Biological Chemistry.[56] Copyright the American Society of Biochemistry and Molecular Biology. (See Color Insert.)

completely defined conformation for the protease, with only a few residues in loops at the contact interface with conformations that differ from those in the free protease (Fig. 10).[56] Nevertheless, there is clear evidence from NMR,[66] proteolytic susceptibility studies,[71] and studies of altered exosite function[72] for conformational disorder induced in some regions of the protease in the complex.

Much less is known about the structures of complexes of serpins with cysteine proteases, and that is limited to complexes with caspases. In a FRET study on complexes formed between the viral serpin, crmA, and either the serine protease, granzyme B, or the cysteine protease, caspase 1, it was shown that both involve similar translocation of the protease from the top to the bottom of the serpin, as seen in the X-ray structures of α_1PI complexes with trypsin and elastase.[73] Since granzyme B is structurally related to trypsin and elastase, it serves as a positive control for the experiments with caspase 1 and implies that the same mechanism of kinetic trapping is used for caspases as for serine proteases. In another study, complexes formed between crmA and caspases 1, 6, and 8 were examined by a variety of analytical approaches.[74] Mass spectrometry provided evidence for the presence of a covalent link, that is, a thiol ester, between the serpin and the protease active site cysteine, as well as for the dissociation of the caspase tetramer induced by complex formation, such that only the p20 (20 kDa) active site-containing subunit remained associated with the serpin in the case of caspases 1 and 8. For caspase 6, both the p10 and p20 subunits remained associated with each other. Analytical ultracentrifugation also supported the dissociation of the p10 subunit and hence disruption of the protease tetramer for complexes with caspases 1 and 8 (Fig. 11).

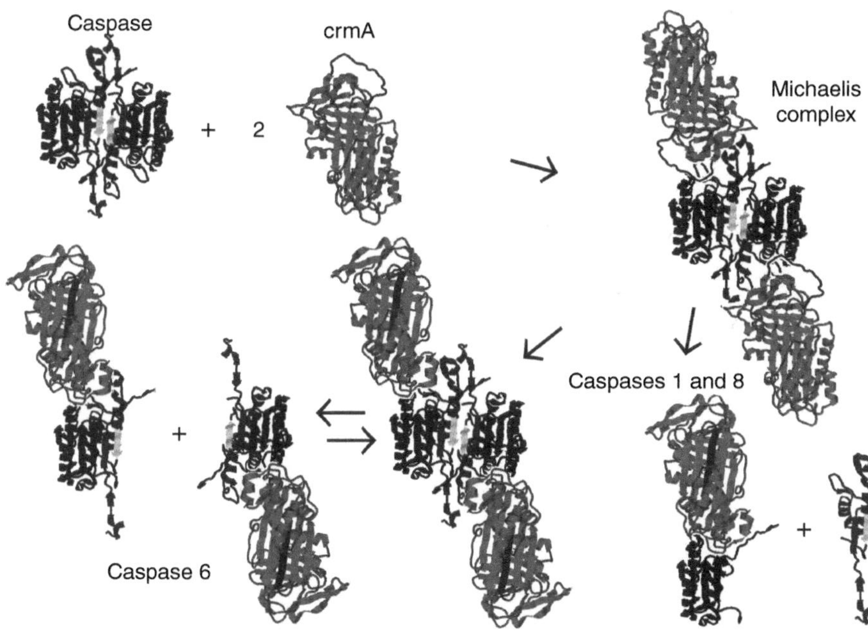

FIG. 11. Inferred organization of the Michaelis and covalent complexes formed between serpins (here crmA) and caspases.[74] Active caspases are bivalent heterotetramers composed of two copies each of p10 (red) and p20 (blue) subunits that can interact with 2 mol of serpin. As a result of progression to the covalent thioester intermediate, conformational changes within the caspase are manifested at the two subunit interfaces, such that dissociation either of the p10 subunit from the serpin-p20 unit (caspases 1 and 8) or of one p10 subunit from the other (caspase 6) occurs. This research was originally published in The Journal of Biological Chemistry.[74] Copyright the American Society of Biochemistry and Molecular Biology. (For interpretation of the references to color in this figure legend, the reader is referred to the Web version of this chapter.)

D. Cleaved Serpin

The cleaved form of inhibitory serpins can be generated in two ways from reaction of the serpin with protease. The first is directly from that fraction of the first-formed acyl-intermediate that travels down the substrate branch of the pathway (Fig. 5), and the second is from slow deacylation of the covalent, kinetically trapped complex. Both routes lead to the same species, which differs from the native state in having the cleaved RCL inserted into the center of β-sheet A as a sixth strand. The mechanistic significance of this structure, which, as mentioned above, is equivalent to the serpin moiety in the covalent complex, is that it is thermodynamically very much more stable than the native form. Thus, whereas native serpins typically denature at around 55–60 °C, their cleaved forms only do so well above 100 °C.[30,75,76] The source of the extra stability is thought to be from the additional hydrogen bonds formed by insertion of the previously exposed RCL to form a perfect antiparallel β-sheet, as well as from better packing of core residues.[77] Although the exact coupling mechanism is still to be worked out (see below), this extra stability is believed to be the driving force for favorable formation of the covalent complex in which the protease active site is distorted.

E. Stability of the Covalent Complex

Because the covalent serpin–protease complex, whether involving a serine or cysteine protease, represents a kinetically trapped intermediate on the normal substrate cleavage pathway, there is a finite rate at which the intermediate will be hydrolyzed to complete the cleavage reaction and liberate functional protease and cleaved serpin. The rate at which this will occur at a given pH depends on the residual catalytic activity of the protease as well as any contribution from uncatalyzed acid or base hydrolysis. The residual catalytic activity, in turn, will depend on the extent of distortion of the active site as a result of compression of the protease against the body of the serpin. A study on the rates and pH dependence of deacylation of complexes of both antithrombin and α_1PI complexes showed that the effect of complex formation was to eliminate any contribution from the catalytic histidine (His 57) to the deacylation rate, so that deacylation was determined, in that conformational state, exclusively by base-catalyzed hydrolysis.[52] This base-mediated hydrolysis corresponded, at pH 7.4, to rates of deacylation of antithrombin:thrombin and factor Xa complexes of less than 10^{-6} s^{-1}. That the elimination of catalytic activity depends on the degree of compression of the protease against the serpin was elegantly shown by a study in which the position of the scissile bond within the RCL of both a P1 Arg α_1PI variant and PAI-1 was shifted forward or backward by up to three residues, thereby shortening or lengthening the stretch of RCL available for insertion into β-sheet A upon complex formation.[78] The results bore out the expectations, such that shortening the available peptide increased distortion (up to two residues) until the RCL was so short that the protease could no longer reach the bottom of the serpin and be compressed. In this way, the half-life of the α_1PI:factor Xa complex of several weeks was modestly increased by shortening by one or two residues, respectively. Conversely, by increasing the length of the inserted RCL, the degree of compression was greatly reduced so that the half-life for the same complex decreased to 6 min or to 53 s by addition of one or two residues, respectively. In a more limited study on crmA, it was similarly shown that there is a precise relationship between RCL length from the cleavage point and the ability to form stable complex.[79]

This critical relationship between the length of polypeptide available for insertion into β-sheet A upon complex formation and the stabilization of the covalent complex through distortion of the protease active site explains why the scissile bond is always either 16 or 17 residues C-terminal from the "fulcrum point" for insertion into the sheet, if stable complex is to be achieved. For proteases that attack elsewhere in the RCL, no inhibition is ever seen.

While it might seem that the most stable complex would always be most desirable, this appears not to be the case in some instances. Thus, the complexes between both single- and double-chain forms of tissue plasminogen activator (tPA) and neuroserpin, which are believed to be physiologically relevant pairs, decay in minutes.[80] In contrast, the complexes between the same proteases and the serpin PAI-1 are orders of magnitude more stable. This difference in complex stability suggests that transient inhibition may actually be desirable in some instances and provides another example of the advantages of the serpin mechanism over canonical inhibitors. A less dramatic example of the relatively rapid rate of deacylation is provided by the complex formed between ZPI and factor Xa, which dissociates over a time period of 2–6 h at physiological pH[81] compared to the many days required for complexes of the same protease with antithrombin to dissociate.[52]

F. Energetics of Complex Formation

Serpins are most unusual in folding into a metastable conformation from which a more stable latent conformer can later form through partial disruption of β-sheet C and expansion of β-sheet A by insertion of the RCL as an additional strand in the same way as in the formation of cleaved serpin (Fig. 3).[27] The trade-off between additional hydrogen bonding in sheet A, but loss of interactions in sheet C, results in a relatively modest increase in stability.[28] In forming cleaved serpin, there is no such trade-off and the resulting stabilization is very much greater.[30,31] This favorable energy difference of ~ 60 kcal mol^{-1} is more than sufficient to bring about the distortion in the active site needed to kinetically trap the acyl-intermediate and bring about inhibition. The only question is how the energy derived from insertion of the RCL as the sixth strand of β-sheet A is coupled to protease distortion, since the latter most likely must occur after the former has gone to completion. One proposal is that helix F, which lies across the path that the protease must traverse in moving between the two poles of the serpin during inhibition, is temporarily moved away during protease transit in an energy-requiring manner. Return of the helix to its original location across β-sheet A following passage of the protease might then return the stored energy at the point where it is needed.[82] While this proposal envisioned a rigid body movement of helix F, all that is needed is an energy-requiring displacement that could also involve an unfolding of the helix. In keeping with that idea, an X-ray structure of α_1-antichymotrypsin, albeit of a nonfunctional variant, shows a partial unfolding of helix F and insertion into β-sheet A (the δ-conformation of Fig. 3).[38] It is possible that such partial insertion represents part of the energy storage, while the remainder results from the necessary unfolding of the helix. Without doubt, however, helix F is an important component of the inhibition mechanism, since a variant of PAI-1 lacking the helix folded properly, but reacted with protease exclusively as a substrate.[83]

G. Clearance of Complexes

Serpin–protease complexes are cleared from circulation with a half-life of minutes, whereas native and cleaved forms of the serpin require much longer.[84] This differential clearance was thought to result from serpins possessing a cryptic epitope for binding to clearance receptors that became exposed only upon complex formation.[85–87] It now appears that this is not the case; instead, both native and cleaved forms of serpins bind, but additional contacts between receptor and protease, when serpin forms covalent complexes, increase the affinity to a value that ensures preferential binding of complexes at the concentrations encountered physiologically.[88] The receptors involved are members of the LDL receptor family, with tight binding and clearance reported to LRP1, LRP2, and VLDLR.[89–91] LDL receptor family members are constitutively active and endocytose ligands such as serpin:protease complexes via clathrin-coated pits. Whereas ligands such as $\alpha_2$macroglobulin:protease complexes have been shown to signal via LRP1,[92] it has not been clear yet whether serpin:protease complexes do the same.

Binding of the serpin moiety involves the engagement of two or possibly three lysine residues by an equivalent number of small calcium-dependent cysteine-rich CR domains within the receptor. Since much of the binding energy comes from engagement of lysine on the ligand, and the requirement for suitably separated lysines is relatively easily met, the receptor(s) is able to bind a wide range of serpin:protease complexes as well as other unrelated ligands.[93]

III. Rates of Reaction, Specificity, and Regulation

Canonical inhibitors form reversible noncovalent complexes with target proteases. The extent of inhibition is determined by the affinity of the complex and the concentration of each species. In contrast, the Michaelis complex between a serpin and protease, once formed, usually leads irreversibly to the kinetically trapped intermediate. The effectiveness of inhibition by a serpin is thus determined by the rates at which the Michaelis complex is formed and undergoes acylation, modified by the SI of the interaction, which reflects the relative rates of the subsequent two branches of the suicide inhibitor pathway. Since the rate of Michaelis complex formation may depend on the presence of cofactors, specificity and regulation by serpins is a complex process that must take these into account.

A. Simple RCL Interactions

The simplest serpin reactions are those that involve interactions solely between the protease and residues within the RCL, with no conformational change necessary in either component. Here, there is abundant evidence for the primary importance of the P1 residue, with flanking residues contributing to a lesser extent, depending on the normal substrate preferences of the protease. Thus, α_1PI, with P1 methionine, inhibits its target neutrophil elastase with a second-order rate constant of $6.5 \times 10^7\ M^{-1}\ s^{-1}$, but inhibits human trypsin, with its P1 preference for lysine or arginine, three orders of magnitude slower.[94] Conversely, the P1 Arg variant of α_1PI inhibits trypsin with a rate constant of $1.8 \times 10^6\ M^{-1}\ s^{-1}$, but neutrophil elastase three orders of magnitude slower.[95] For many other serpins, there are similar readily understandable relationships between the P1 residue and the rate and preference for reaction with different proteases. A good example is the set of 12 serpins produced by gene-1 in *Manduca sexta*, where the alternate exon usage for the RCL-encoding exon produces 12 serpins with identical sequences in the body and differing only in the RCL sequence. The protease preference then principally reflects the nature of the P1 residue, with serpin 1A (P1 Arg) reacting rapidly with trypsin, and plasmin and serpin 1H, with P1 tyrosine, reacting rapidly with chymotrypsin. The clear implication of this observation is that mutagenesis of the P1 residue can drastically alter the specificity of a serpin.

Structurally, this crucial specificity-determining role of the P1 residue can be understood from the simple docking of the serpin RCL with the protease active site. The best example for which an X-ray structure of the Michaelis complex is available is that of S195A trypsin with the P1 Arg variant of α_1PI (Fig. 7).[54] Neither serpin nor protease undergoes a conformational change upon complex formation, and the RCL conformation is already the preferred canonical one, ready to hydrogen-bond with the protease. The P1 residue has been mutated to the preferred arginine of trypsin and occupies the S1 specificity pocket of trypsin. The second-order rate constant for the reaction is less than one order of magnitude slower than the cleavage reaction between trypsin and substrate peptides.[96]

B. Exosite Interactions in Binary Complexes

While all serpins can be expected, from the fundamental requirements of the inhibitory mechanism, to engage their target proteases within the RCL, the interaction is often subject to modification, both in rate and location, by additional interactions outside of the RCL. These are collectively termed exosite interactions and can be either favorable or unfavorable, and involve determinants solely within the serpin:protease pair, or involve one or more cofactors. We first consider exosite involvement in binary complexes.

Although the reaction between α_1PI and its target neutrophil elastase, described above, clearly takes advantage of the suitable P1 methionine to promote rapid reaction, the second-order rate constant is actually ~1000-fold faster than for reaction of the protease with a peptide substrate with the same P4–P4′ sequence.[97] By contrast, pancreatic elastase, which has similar substrate specificity as neutrophil elastase, reacts ~100-fold slower than the neutrophil enzyme and is therefore more like the peptide substrate. This difference suggests that there is an additional, favorable exosite interaction between neutrophil elastase and α_1PI. In keeping with this conclusion, an attempt to convert α_1-antichymotrypsin into a potent neutrophil elastase inhibitor by swapping residues P3–P3′ from α_1PI into antichymotrypsin gave a rate constant of only 10^5 M^{-1} s^{-1} with neutrophil elastase,[98] suggesting that the RCL sequence alone is insufficient to account for the extremely fast rate between α_1PI and neutrophil elastase. The nature and location of the exosite are, however, unknown.

Other examples of involvement of exosites in binary complexes are the reaction of PAI-1 with tPA and of kallistatin with kallikrein. In the case of the reaction of PAI-1 with tPA, the 37-loop of tPA, which flanks the active site, interacts with residues in the RCL of PAI-1, but outside of the primary interaction site. This exosite interaction involves residues C-terminal to the scissile bond, with specific involvement of Glu residues at positions P4′ and P5′. Replacement of these glutamates with alanine causes a 13-fold reduction in the rate of Michaelis complex formation.[99] Remarkably, a snake has exploited this need for residues in the 37-loop for effective inhibition of tPA by PAI-1, by evolving its tPA such that it *lacks* the loop, thereby partly evading inhibition by the prey's PAI-1.[100] In the case of kallistatin, an exosite is needed to ensure specificity for kallikreins over other Arg-specific proteases. Thus, a P1 Phe is used rather than P1 Arg, thereby contributing to overall low reactivity with all Arg-specific proteases, unless there is a compensating factor. This factor is provided, but only in the case of kallikreins, by a basic patch between helix H and strand 2C of the serpin, which is able to interact favorably with kallikreins, but not with other Arg-specific proteases.[101]

C. Exosites in Ternary Complexes with Cofactors

There are now several very well-established examples of cofactors being used to achieve either protease- or location-specific inhibition by serpins.

A good recent example of location-specific use of a cofactor is the inhibition of membrane-associated factor Xa by the serpin, ZPI, in complex with membrane-associated cofactor, protein Z. Although antithrombin, when bound to heparin, is a rapid inhibitor of freely circulating factor Xa, there is also a need for the specific inhibition of factor Xa at the site of its generation on the platelet membrane surface. Even though factor Xa is an Arg-specific protease, ZPI, like kallistatin, has an aromatic residue at P1 (here Tyr) to help achieve its specificity. The presence of the poorly recognized P1 Tyr ensures that, in the absence of compensating factors, reaction between ZPI and any Arg-specific protease, including factor Xa, will be slow. To enhance reactivity with membrane-associated factor Xa, ZPI binds tightly to protein Z, which has a membrane-binding Gla domain like factor Xa itself, and so becomes membrane-associated, where a specific interaction between the two Gla domains promotes formation of the ternary complex and presents the active site of factor Xa to the RCL of ZPI.[102] This interaction is strengthened by a specific interaction between a critical Arg in the 140 loop of factor Xa (Arg 143)[103] with a conserved Glu (Glu 313) of ZPI. Together, these features ensure that only when factor Xa is membrane-associated will it interact rapidly with ZPI, while also ensuring that most other Arg-specific proteases will not[104] (Fig. 12). An exception is factor XIa, which reacts with the ZPI:PZ complex without needing to be membrane-associated, though here, the additional interaction with Glu 313 appears to be necessary.[104,105]

The theme of adopting a seemingly inappropriate P1 residue to achieve specificity is also present in HCII, which is a specific inhibitor of thrombin over other Arg-specific proteases. HCII has a P1 Leu and is as fast an inhibitor of chymotrypsin as of thrombin in the absence of a glycosaminoglycan (GAG) cofactor, though both these rates are sufficiently low ($\sim 4 \times 10^3\ M^{-1}\ s^{-1}$) as to be physiologically unimportant. While replacement of the P1 residue with Arg gives the expected increase in reactivity with thrombin, with an $\sim$100-fold increase in rate constant, it also increases rates of inhibition against other Arg-specific proteases, resulting in an effective but nonspecific inhibitor even in the absence of GAGs.[106] Instead, HCII achieves its specificity for thrombin by exploiting the presence of two basic exosites on thrombin, but not other proteases, on opposite sides of the protease and close to the active site. Heparin and dermatan sulfate bind to exosite II of thrombin more tightly than they do to HCII and form a thrombin:GAG complex. The extended GAG moiety can then

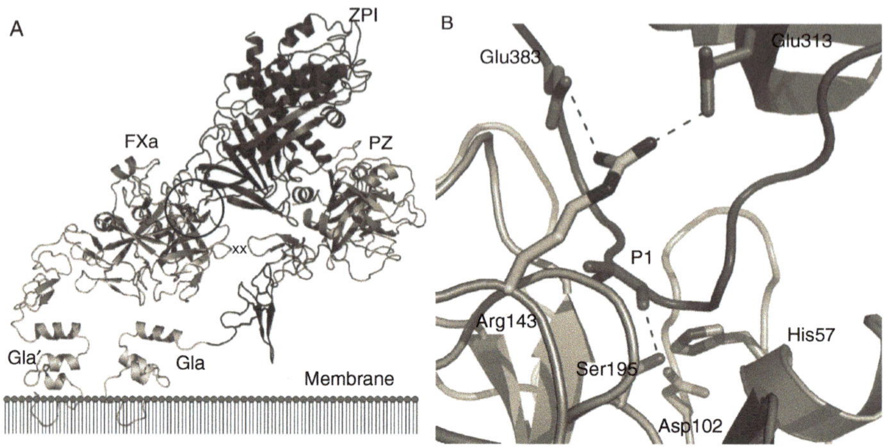

FIG. 12. Model of the ternary complex between membrane-associated PZ:ZPI complex with membrane-associated factor Xa.[104] Panel (A) shows the entire complex, with interaction between Gla domains of factor Xa and PZ. Panel (B) shows a close-up of interactions between the factor Xa active site and the serpin RCL, highlighting the stabilizing interactions between Arg 143 of the protease and Glu residues 313 and 383 from the serpin. This research was originally published in The Journal of Biological Chemistry.[104] Copyright the American Society of Biochemistry and Molecular Biology.

bind to HCII at a basic site to which the highly acidic N-terminal tail of HCII is thought to be bound. The displaced acidic tail is then available to engage exosite I of thrombin and further promote association of serpin and protease by forming a second bridge between HCII and thrombin[107–109] (Fig. 13). This two-armed grip of thrombin is only available to HCII and is sufficiently favorable that it more than offsets the poor interaction between thrombin and the P1 Leu, to give a second-order rate constant that has been enhanced ~4000-fold to 10^7 M^{-1} s^{-1}.

A somewhat different strategy is adopted by antithrombin to achieve selectivity as an inhibitor of procoagulant over anticoagulant Arg-specific proteases, and then only when heparin is also present. Unlike HCII, antithrombin has a P1 Arg and thus might be expected to be a quite fast inhibitor of Arg-specific proteases. However, this is not the case, with second-order rate constants for inhibition of the principal targets, factors IXa, Xa, and thrombin, being 6×10^1, 6×10^3, and 9×10^3 M^{-1} s^{-1}, respectively, in the absence of heparin, while the rate with the anticoagulant protease-activated protein C (APC) is a remarkably low 0.08 M^{-1} s^{-1}.[110] For inhibition of each of these classes of protease, pro- and anticoagulant, these rate constants appear to represent different types of repression that can be reversed for the former by binding to heparin, but only slightly for the latter.

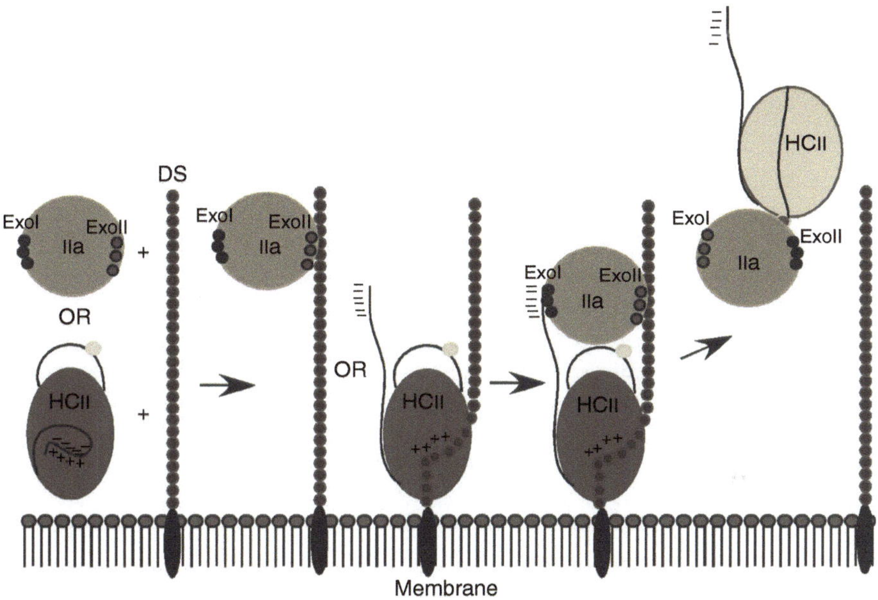

FIG. 13. Schematic representation of the role of dermatan sulfate or heparin and of the N-terminal tail of HCII in ensuring specific anti-IIa-directed rate acceleration of the reaction between HCII and thrombin (IIa). Both IIa and HCII bind only weakly to DS or heparin to form binary complexes. These more rapidly associate with the remaining species (HCII or IIa) to give a ternary Michaelis complex in which bidentate bridging between IIa and HCII occurs via DS and the N-terminal HCII tail. Following covalent complex formation, the affinity for DS is reduced and the complex dissociates. This research was originally published in The Journal of Biological Chemistry.[3] Copyright the American Society of Biochemistry and Molecular Biology.

In the case of APC, it seems that cooperation between residues within the RCL of antithrombin ensures that, despite a P1 Arg, the resulting rate constant for inhibition is exceedingly low. Thus, on the one hand, several serpins with P1 Arg, but otherwise quite different RCL sequences, can inhibit APC 5×10^3 to 5×10^5 fold faster than antithrombin.[111] On the other hand, the antithrombin RCL sequence from P7–P3′, whether present in antithrombin itself or in α_1-antichymotrypsin[112] or in α_1PI,[113] gives a very low rate of reaction with APC. However, while there is clearly something special about the antithrombin RCL, many of the residues appear to have little effect, with the exception that residues P2 (Gly) and P3′ (Asn) are both necessary to produce the very low rate of reaction with APC.[113] The reason for this requirement is not known, though the X-ray structure of APC does show a unique pocket for the P3′

residue that is not present in other chymotrypsin family members, together with an S2 pocket that can accommodate large residues and which would therefore not interact favorably with a P2 Gly.[114]

For the inhibition of the procoagulant proteases, factors IXa, Xa, and thrombin, antithrombin uses both positive and negative exosites on the serpin, a heparin-induced conformational change within the serpin, and template bridging between the serpin and the protease to enhance the rate constants by up to 100,000-fold. Antithrombin is unique amongst serpins in having an RCL whose N-terminal end is partially inserted into β-sheet A in the resting state.[115,116] In this state, the exposed portion of the RCL is effectively shorter and thereby held closer to the body of the serpin. Binding of a specific heparin pentasaccharide at a site distant from the RCL causes major conformational changes in antithrombin, which collectively result in a several 100-fold increase in the rate of reaction of antithrombin with factors IXa and Xa, though only a ~ twofold change for reaction with thrombin.[117,118] Part of the conformational changes is an expulsion of the buried hinge of the RCL, allowing the RCL to move away from the serpin body. However, it is now thought that this change alone is insufficient to account for the large rate increase for factors IXa and Xa and that, in addition, part of the conformational change results in a reduction in a repulsive interaction between the protease and the body of the serpin that may reside in strand 3C.[119] Additionally, there is a favorable exosite interaction between Arg 150 in both factors Xa[120] and IXa[121] and an exosite on antithrombin involving Tyr 253,[122] though this may contribute to binding in both native and heparin-activated states.[119] Thus, conformational activation by heparin results from a reduced repulsive interaction between the protease and the surface of antithrombin that is further mitigated by outward movement of the antithrombin RCL through expulsion of the buried N-terminal hinge. In this view of activation, the P1 Arg and the exosite at Tyr 253 contribute to binding in both native and activated states. In the reaction with thrombin, the conformational changes induced in antithrombin by heparin contribute little to the rate of reaction. Instead, long-chain heparin that can bridge between the high affinity heparin-binding site of antithrombin and the heparin-binding exosite II of thrombin serves to enhance the rate of reaction ~ 1000-fold by a template mechanism analogous to that for the HCII:thrombin reaction.[123] This mechanism is clearly evident from the crystal structure of the ternary anhydrothrombin:antithrombin:heparin complex[55] (Fig. 9). Since there are also homologous heparin-binding sites on both factors Xa and IXa, bridging heparin also contributes to the enhancement of these reactions, though to a lesser extent than for the reaction with thrombin.[118,124]

While exosites clearly provide important determinants of serpin–protease recognition in the Michaelis encounter complex, their stabilizing effect on the Michaelis complex also favors its progression to the acyl-intermediate complex.

In the case of heparin–antithrombin–protease and PAI-1–protease reactions, exosites were shown to stabilize the Michaelis complex to an extent that greatly favored acylation and caused the reaction to be limited by the diffusion-controlled encounter of serpin and protease.[125,126]

D. Conformational Change in the Serpin

Two different examples have already been given of functional conformational changes induced in a serpin to promote specific inhibition. One was the displacement of the N-terminal tail of HCII by GAG binding to make it available to attach to exosite I of thrombin and thereby promote Michaelis complex formation. The second was specific long-range conformational changes in antithrombin that reduce unfavorable interactions between the serpin and the target proteases by a combination of alteration of the repulsive surface and extension of the RCL to permit the protease to move further away from the site of repulsion while still being able to engage the favorable P1 Arg and Tyr 253 exosite. While each of these changes is specific to a single serpin, a change that is common to all serpins that act as protease inhibitors results from the insertion of the cleaved RCL into β-sheet A that occurs as a necessary part of the inhibition mechanism. This represents a major conformational alteration that affects not only β-sheet A, which is transformed from a 5-stranded partly parallel, partly antiparallel sheet into a perfectly antiparallel six-stranded sheet, but also the helices that pack against the sheet on both sides. In three instances, it is known that this conformational change is used to greatly weaken the interaction with cofactor, so that the serpin–protease complex is likely to dissociate from it under physiological conditions, making the cofactor a catalyst rather than a one-use species. One instance is antithrombin, where the affinity for the specific heparin pentasaccharide falls by three orders of magnitude between the native and cleaved forms of the serpin. This large decrease in affinity is expected to make it easier for the ternary Michaelis complex to progress to the translocated acyl-enzyme intermediate, both by reducing the energetic penalty of heparin displacement as the RCL inserts and by allowing the protease to move away from the top of the serpin.[127] Indeed, when the affinity of heparin for both serpin and protease is increased, for example, by reducing the ionic strength, the rate of loop insertion is reduced so that more of the initially formed acyl-intermediate proceeds to cleaved serpin, which is reflected in an increased SI.[128] A second example is the estimated three orders of magnitude reduction in affinity of the cofactor protein Z for the serpin, ZPI, once the serpin has undergone conformational change to the loop-inserted form.[81] This change in affinity is expected to result in the dissociation of protein Z from the inhibited ZPI:factorXa complex *in vivo* and make it available for further use. Such dissociation is particularly important, given the limited quantity of protein Z available, all of which is normally bound to ZPI in plasma.[129] A final example is the several 100-fold reduction in affinity of the cofactor, vitronectin, for PAI-1 when it forms complexes with its target proteases. In this case, the cofactor acts to localize and stabilize active PAI-1 at extracellular matrix sites where its target proteases act, so release of PAI-1–protease complexes from the limiting cofactor sites in the matrix is critical to maintain functional PAI-1 at these sites.[130]

A quite different consequence of the major conformational changes that occur in the serpin following reaction with the protease, whether the resulting species is a covalent complex or a cleaved serpin, is the generation of neoepitopes. While it is no longer thought that the conformational changes are used to bring about efficient clearance by LRP1 or other LDL receptor family members,[88] they may be utilized for signaling purposes. As examples: (i) α_1-antichymotrypsin:chymotrypsin complex has been shown to inhibit activation of NADPH oxidase, and therefore is involved in regulation of superoxide production by neutrophils following their stimulation by fMet-Leu-Phe,[131] (ii) α_1PI: neutrophil elastase complexes possess neutrophil chemotactic properties,[132] and (iii) neoepitopes in cleaved or latent antithrombin possess antiangiogenic activity.[133] The receptor involved in such processes is not, however, established.

E. Conformational Change in the Protease

It was pointed out in Section II that there are functional consequences to both serine and cysteine proteases of forming kinetically trapped complexes with serpins. In both cases, the changes in protease function appear to arise from the compression between serpin and protease that is needed to effect active site distortion and kinetic trapping of the acyl-intermediate. In chymotrypsin family serpins, this active site distortion has been shown to result in greatly enhanced susceptibility of surface regions to undergo proteolysis, even though there are only minimal structural differences visible in the crystal structure of the elastase complex.[71,134] Although it is not clear whether this enhanced proteolytic susceptibility has a physiological purpose, it may be that it serves to irreversibly modify the protease component, such that any protease that dissociates from the final acyl-enzyme complex has been rendered inactive. The conformational distortion of the protease has also been shown to inactivate regulatory exosites that mediate protease biologic activities. For example, exosite 1 of thrombin, which mediates binding to thrombomodulin, fibrin, and PAR receptors, is inactivated in serpin–thrombin complexes.[72,135] This inactivation is likely to be important in allowing only functional thrombin to bind such receptors and prevent receptor blockade by inactive serpin–thrombin complexes. For members of the heterotetrameric intracellular caspases, inhibition by serpin (crmA) results in long-range conformational perturbations that disturb the interface between p10 and p20 subunits and, for some, between p10 and p10 subunits, such that dissociation of the tetramer to either dimers or monomers occurs

(Fig. 11).[74] For caspases where the p10 subunits dissociate from the p20 subunits, it seems that the former aggregate *in vitro*. If such aggregation occurred *in vivo*, it would provide a means of ensuring that enzymatic activity is not recovered from any acyl-enzyme complex that is deacylated.

F. Serpins Versus Canonical Inhibitors

Canonical inhibitors from multiple families have many virtues. They are typically small proteins or, if large, contain multiple inhibitory domains, and represent a modest investment in their biosynthesis. Specificity is often determined by residues within an exposed RCL, which is held in a rigid canonical conformation by the presence of disulfide bonds close to the ends. And since the conformation is usually optimal for docking with the target protease, the entropic penalty of forming a complex is minimized (Fig. 14). Complex

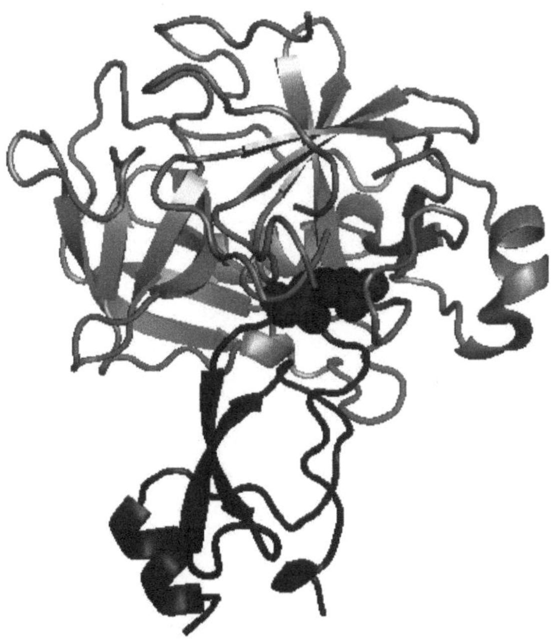

FIG. 14. Structure of a canonical inhibitor–protease complex. X-ray structure of the Kunitz family canonical inhibitor, APPI, in complex with bovine chymotrypsin in ribbon representation (pdb 1CA0). The inhibitor is in red and the protease in yellow. The P1 Arg of the inhibitor is shown in space-filling representation and protrudes into the S1 pocket of the protease. The orientation is the same as in Fig. 7 to illustrate the similar mode of interaction of serpin and canonical inhibitors in the Michaelis complex but a large difference in size of the two inhibitor families. (See Color Insert.)

formation is noncovalent and the extent is determined by the affinity of the inhibitor for the protease and the concentration of species. Affinities can easily be in the subnanomolar range. While many canonical inhibitors are found in solution, there are others that are membrane-anchored, perhaps as one domain of a much larger protein, such as the Kunitz domain in amyloid precursor protein.[136] In some cases, noncovalent inhibitors may interact reversibly with proteases in a manner that blocks the active site but not through a canonical substrate-like mode of interaction. Hirudin, a potent inhibitor of thrombin found in the leech, is a 65-residue polypeptide with a folded N-terminal core and an extended acidic tail. It binds with femtomolar affinity to thrombin using the N-terminal three residues to bind to P2 and P3 subsites within the active site and with the acidic C-terminal tail bound separately to exosite I.[137]

Although noncovalent inhibitors have advantages in terms of their economy of synthesis and functional versatility, serpins are very widely used by most organisms, including examples from bacteria and archaea. Despite their larger size and disadvantage of being prone to polymerization, the widespread prevalence of serpins argues for their distinct advantages, at least in certain circumstances. Perhaps the most useful is the conversion of the initially formed Michaelis complex into an essentially irreversible covalent complex as a result of the greatly increased thermodynamic stability of the loop-inserted state for the serpin. This conversion occurs irrespective of whether the initial complex forms slowly or rapidly; so complex formation, *per se*, is liberated to a large extent from the affinity within the Michaelis complex. A good example is the inhibition of the complement proteases C1r and C1s by the serpin, C1 inhibitor. The second-order rate constants for these processes are low, being in the range of 10^3–10^4 M^{-1} s^{-1},[138] and so must reflect relatively weak, though specific, interactions between the serpin and the protease. Nevertheless, C1 inhibitor is the principal inhibitor of these proteases. A canonical inhibitor with similar specificity would probably be too weak to form significant amounts of complex, unless its concentration was extremely high. Another possible advantage of serpins is that the lifetime of the covalent complex is dependent on the degree of distortion of the active site, and, in turn, on the relationship between RCL length and loops at the interface between the serpin and the protease. Documented stabilities range from half-lives of weeks[52] to minutes,[80] yet these can be manipulated.[78] Thus, there may be benefits to a short-lived intermediate to give a time-dependent inhibition, especially if the inhibited protease is itself involved in signaling. A third advantage is the use of cofactors such as heparin or protein Z. While these are not precluded from use by canonical inhibitors, the larger size of the serpin and the possibilities of modifying its conformation, which would be expected to be much less for the smaller and more rigid canonical domains, probably make their use by serpins much more versatile. A number of examples have been given above for how cofactors may tailor serpin specificity to a given protease in a given location. Finally, there are possibilities afforded by the generation of neoepitopes in either the cleaved serpin or in its protease complex that are more extensive than for canonical inhibitors. Since the latter are rigid body interactions, neoepitopes are largely restricted to the interface region of the complex. In contrast, while this region is also a potential source of neoepitopes in serpins, much of the remainder of the serpin surface also represents alterations in conformation from the native state.

IV. Engineering Serpins as Therapeutics

A. Recombinant Serpins for Replacement Therapy

Normal and variant serpins have been engineered as recombinant proteins for the purposes of treating diseases whose pathogenesis is associated with dysfunctional protease regulation. As discussed earlier, serpin deficiencies can result in tissue damage and disease due to an overactivity of the proteases they regulate. One treatment for such diseases is replacement of the deficient serpin, usually with a recombinant protein to restore the natural protease/inhibitor balance.

Lung diseases that develop in antitrypsin-deficient individuals due to excess neutrophil elastase activity can be treated by replacement of antitrypsin.[139] Similarly, recombinant serpin B1, also known as monocyte neutrophil elastase inhibitor, has proved effective in reducing the excess neutrophil protease activity and lung damage associated with inflammatory diseases.[140] Antithrombin replacement therapy has been approved to reduce the risk of thrombotic diseases in individuals with congenital antithrombin deficiency and to treat thrombotic complications in patients with acquired antithrombin deficiency due to the consumptive coagulopathy accompanying sepsis or cardiac surgery.[141] C1-inhibitor has been shown to be an effective treatment for the protease destruction accompanying inflammatory diseases such as sepsis and ischemia–reperfusion injury.[15] In this case, the effectiveness of C1-inhibitor arises not only from neutralization of proteases of the complement and kallikrein–kinin system, but also from activities that do not involve protease inhibition, such as the blocking of leukocyte–endothelial cell interactions and neutralization of bacterial lipopolysaccharide. PAI-1 has been shown to be an effective therapy for preventing hypoxic-ischemic brain injury in newborns through its inhibition of tPA, a protease thought to be responsible for this pathologic condition due to its activation of matrix metalloprotease 9 (MMP9) in the brain.[142]

Nonhuman serpins have also proved useful in treating human disease. Viral serpins protect viruses from the proinflammatory proteases of their host organisms that provide defense against pathogen invasion. SERP1 from the myxoma poxvirus inhibits human coagulation and fibrinolytic proteases and has been shown to have potent anti-inflammatory actions and efficacy in the treatment of human inflammatory diseases induced by vascular injury.[143] The specific inhibition of the proapoptotic caspases and granzyme B by the poxvirus serpin, crmA, to block an infected cell from undergoing apoptosis has made this serpin a valuable reagent in apoptosis research.[144]

B. Engineered Serpin Variants

An understanding of how serpins achieve specificity for their target proteases and are able to adjust the level of protease activity at the time and place needed through interactions with other molecules, as discussed in the previous sections, has led to efforts to engineer novel types of serpins for tailor-made purposes (Table III). We discuss these efforts in the context of the particular serpins that have been used as scaffolds for engineering. In some cases, the choice of the scaffold is based on the desire to exploit interaction sites outside the RCL that are unique for a particular serpin.

1. α_1PI Variants

The ability to express α_1PI in large amounts in bacteria has made this serpin a favorite scaffold to employ for the purposes of engineering novel functions (Table III). In one of the earliest examples, an improved α_1PI was engineered for replacement therapy in patients with antitrypsin deficiency. The P1 Met residue was targeted for modification because of its susceptibility to oxidation by endogenous oxidants released by neutrophils as well as exogenous ones introduced by cigarette smoking.[145] Oxidation of the P1 residue reduces the antielastase activity of the serpin by 2000-fold, which contributes to the pathogenesis of emphysema in cigarette smokers and to the accelerated onset of emphysema in smokers who have an inherited deficiency in antitrypsin.[94] An engineered P1 Met to Val variant was found to retain the antielastase activity of the wild-type serpin, to be resistant to oxidation, and to be a superior therapeutic to the wild-type serpin in an animal model of inflammatory lung disease.[145]

An early indication of the critical importance of the P1 residue for serpin specificity was the discovery of the natural P1 Arg-Pittsburgh variant of antitrypsin.[146] This variant lost its elastase inhibitory activity and acquired a potent thrombin inhibitory activity, accounting for the early death of the affected individual due to bleeding. The Pittsburgh variant subsequently prompted

TABLE III

ENGINEERED SERPIN VARIANTS WITH THERAPEUTIC POTENTIAL

Serpin	Mutations	Altered function	Potential therapy
α_1PI	RCL P1M → V	Nonoxidizable P1 residue	α_1PI deficiency
α_1PI	RCL P7–P3′ FLEAIPMSIP → AVVIAGRSLN	High thrombin/activated protein C selectivity	Thrombosis
α_1PI	s3C K222 → Y L224 → E, RCL P4–P2 AIP → IAG	Factor Xa selective inhibitor	Thrombosis
α_1PI	RCL P4A → R, P1M → R	Furin inhibitor	Infectious disease, squamous cell carcinomas
α_1PI	RCL P4–P2′ AIPMSI → RRRRSA	PACE4/PC6 selective inhibitor	Metastatic cancer
α_1PI	RCL P10–P3′ GAMFLEAIPMSIP → AATAVVGFGSSPA	Cysteine cathepsin targets	
PAI-1	RCL P3–P1′ SARM → YSKW, SARM → YSYM	tPA selective inhibitor	Thrombosis
PAI-1	RCL ARM → RKA	uPA selective inhibitor	Cancer
PAI-1	RCL P1R → V, P4V → A	Matrix-bound neutrophil elastase target	Inflammatory lung disease
PAI-1	RCL P1R → F	Matrix-bound cathepsin G target	Inflammatory lung disease
PAI-1	N150 → H, K154 → T, Q319 → L, M354 → I	Stable plasminogen activator inhibitor	PAI-1 deficiency
PAI-1	RCL P1R → A	Noninhibitory, retains vitronectin binding only in native form	
PAI-1	Q123 → K	Plasminogen activator inhibitor, does not bind vitronectin	
PAI-1	RCL P12A → R P14T → R	Noninhibitory, retains vitronectin binding in both native and RCL-cleaved forms	Diabetic nephropathy, blocks intimal hyperplasia after ischemia–reperfusion injury of blood vessels

attempts to engineer a novel antithrombotic serpin as a potential therapy for thrombotic diseases.[147] A challenge in this effort was the finding that the Pittsburgh variant effectively inhibited both procoagulant coagulation proteases and the principal anticoagulant protease, APC, as discussed in the previous section. This lack of discrimination contrasted with the natural inhibitor of procoagulant proteases, antithrombin, which poorly inhibited APC even in the presence of the activator, heparin. To produce an engineered antitrypsin that approached the selectivity of antithrombin for inhibiting thrombin over APC, it was found necessary to replace the P7–P3′ residues of antitrypsin with those of antithrombin.[113]

The identification of exosites in antithrombin that are responsible for enhancing the reactivity of the heparin-activated serpin with factors Xa and IXa prompted efforts to exploit these exosite interactions to engineer antitrypsin to be a specific factor Xa inhibitor.[148] Addition of two key antithrombin exosite residues in strand 3 of sheet C in the serpin body to the P1 Arg-Pittsburgh variant of antitrypsin was found to greatly enhance reactivity with factors Xa and IXa. This reactivity enhancement was comparable to that resulting from allosteric activation of antithrombin by a heparin pentasaccharide and was dependent on Arg150 of factor Xa, the complementary exosite in the protease. Since the antithrombin exosites alone produced an $\sim$100-fold greater reactivity of the engineered antitrypsin with factor over Xa over IXa, but only a modestly greater reactivity with factor Xa over thrombin, further changes in the P2–P4 residues to those in antithrombin were made. Although increasing factor Xa reactivity a modest twofold, this change greatly reduced thrombin reactivity to result in an $\sim$1000-fold better inhibitor of factor over Xa than thrombin. Although not tested, the selectivity of this engineered serpin for factor Xa over aPC would be expected to require the replacement of additional RCL residues with those of antithrombin, based on the findings discussed above.

Because serpins are metastable, attempts have been made to engineer a more stable variant that retains its protease inhibitory functions. Random mutagenesis with a selection of mutants that retained protease inhibitor function when subjected to heat denaturation resulted in the identification of seven stabilizing mutations which, when combined, produced a more stable α_1PI variant that retained normal protease inhibitory function.[149] These mutations all involved hydrophobic core residues, which were suggested to be responsible for strain in the metastable fold but to have this strain relieved when the serpin adopts the more stable latent and cleaved folds. Notably, other mutations that stabilized the metastable fold were found to decrease inhibitory activity, and these were associated with those elements of the fold that were critical for A sheet opening and insertion of the cleaved RCL in the protease inhibitory mechanism.[77]

α_1PI has been used as a scaffold to create novel inhibitors of the seven mammalian proprotein convertases, serine proteases of the subtilisin family, which function both intracellularly and extracellularly to process numerous precursor forms of proteins to their biologically active state.[150] These enzymes are involved in prohormone, growth factor, and proprotease processing but are also hijacked by viruses and bacteria to process viral proteins and bacterial toxins and, hence, are targets for antipathogen therapeutics. Furin is the prototype protease of the family, all of which are characterized by cleaving an RXXR sequence. Engineering the RIPR recognition sequence into the P1–P4 residues of α_1PI resulted in an efficient inhibitor of furin ($k_{ass} > 10^6\ M^{-1}\ s^{-1}$) and several other proprotein convertases.[151] This variant, known as α_1PDX, has been shown to have antitumor properties that correlate with the inhibition of furin processing of MMP and growth factor substrates by squamous cell carcinoma cell lines.[152] Additional substitutions of RCL residues in the P4–P2′ sites showed that specificity and selectivity for inhibiting two other secreted proprotein convertases, PACE4 and PC6, could be obtained with the sequence, RRRRSA.[153] The P2′ Ala substitution was shown to be particularly important for the selectivity of the serpin for these proprotein convertases over furin. Such α_1PI variants could have therapeutic value as cancer drugs, since PACE4 processing of extracellular matrix proteoglycans has been shown to drive the metastatic progression of certain types of tumors.

One final example of the use of antitrypsin for engineering novel target protease specificities is a study that replaced the RCL with that of serpin B3, an established inhibitor of lysosomal cathepsin-type cysteine proteases.[154] Variants incorporating homologous RCL residues together with zero to three of the RCL insertions of serpin B3 were shown to acquire the ability to efficiently inhibit cathepsins, thereby validating that the RCL sequence is a principal determinant of the ability of serpin B3 to inhibit cathepsins and that serpins can be engineered to target cysteine proteases as well as serine proteases.

2. PAI-1 Variants

PAI-1 has been another serpin in which efforts have been made to engineer novel types of variants for understanding its *in vivo* function and for use as potential therapeutics (Table III). Again, ease of bacterial expression of this serpin has enabled high throughput screening of many types of variants. Saturation mutagenesis of the P1- and P1′-reactive bond residues in the RCL revealed the critical importance of a P1 Arg or Lys for inhibition of urokinase-type plasminogen activator (uPA), but tolerance for many types of P1′ substitutions.[155] By contrast, the other major PAI-1 target protease, tPA, was an effective inhibitor of P1 variants with a wide range of substitutions, with P1

Tyr and P1 His being most active.[156] By incorporating additional changes in P2 and P3 residues, efficient and selective PAI inhibitors of tPA and uPA were obtained. The P3–P1′ sequences, YSKW and YSYM, were thus rapid inhibitors of tPA ($2–3 \times 10^6$ M^{-1} s^{-1}) but inactive with uPA, whereas an RKA P2–P1′ sequence resulted in an ~100-fold more selective inhibitor of uPA over tPA.

A P1/P4 Val/Ala PAI-1 variant and a P1 Phe variant were shown to change PAI-1 protease specificity to neutrophil elastase and cathepsin G, respectively.[157] Because the natural serpin inhibitors of these proteases, α_1PI and α_1-antichymotrypsin, lack high affinity binding sites for vitronectin and heparin, ligands that localize PAI-1 to the matrix, studies were performed to determine the efficacy of the PAI-1 variants in neutralizing the neutrophil proteases in inflammatory lung disease. The PAI-1 variants were found to be superior to the natural serpin inhibitors in reducing neutrophil protease activity in an animal model of the disease, supporting the idea that the engineered inhibitor could more effectively access neutrophil proteases that bound to matrix sites.

PAI-1 is unique among inhibitory serpins in having a poorly stable native conformation, presumably due to a need to control the inhibitory activity of this secreted serpin. Random mutagenesis with screening for heat stable variants resulted in the identification of a 70-fold more stable variant with four mutations in the hydrophobic core.[32] This variant retained protease inhibitor function, although with a modestly reduced inhibition rate and enhanced substrate reactivity. The mutations were found to involve hydrophobic core residues that were changed from nonconsensus to consensus serpin residues, suggesting that modifications in core packing evolved in this serpin to reduce the kinetic barrier for the latency transition and allow regulation of its function. This mutant has been of value in biophysical studies of native PAI-1 as well as in understanding the role of PAI-1 instability in carrying out its *in vivo* functions.[158] A subsequent random mutagenesis study identified 10 mutations that, when combined, resulted in a variant with > 200-fold enhanced stability but, in this case, the protease inhibitory function was indistinguishable from wild type.[159] These mutations occurred throughout the molecule but involved subdomains, which had been shown to control the latency transition.

Since elevated PAI-1 levels are associated with an increased risk of atherosclerosis and heart attack as well as with obesity and diabetes, inhibition of PAI-1 activity has been the desired therapy, and thus, variants that interfere with native PAI-1 function have been sought.[158] PAI-1 mutants designed to assess the contributions of protease inhibition and vitronectin cofactor binding to PAI-1 *in vivo* functions have resulted in novel variants with potential therapeutic value (Fig. 15). Vitronectin binding stabilizes the active conformation of PAI-1 and localizes PAI-1 activity to the extracellular matrix, but the binding masks an RGD sequence in the cofactor protein. Conversion of PAI-1 to the more stable latent, cleaved, or protease-complexed forms releases vitronectin

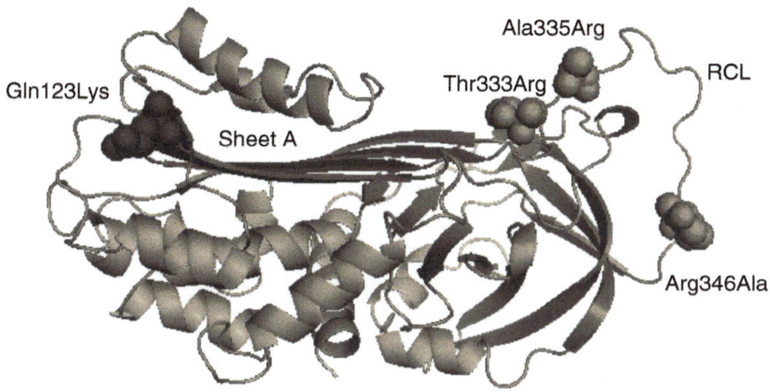

FIG. 15. Engineered PAI-1 variants. X-ray structure of the stable variant of native PAI-1 engineered by Berkenpas *et al.*[32] (pdb 1B3K) in ribbon representation with beta sheets in magenta, helices in cyan, and loops in pink. The residues mutated in PAI-1 variants of potential therapeutic value listed in Table III are shown in space-filling representation. The PAI-1 A variant has the RCL P1 Arg346 mutated to Ala and is inactive as a protease inhibitor, but retains vitronectin-binding function in native PAI-1, but not in RCL-cleaved PAI-1. The PAI-1 K variant has a mutation of Gln123 to Lys in the vitronectin binding site and is active as a protease inhibitor, but has greatly reduced affinity for vitronectin. The PAI-1 R variant has the RCL P12 Ala335 and P14 Thr333 residues both mutated to Arg and has lost PAI-1 inhibitory function, but retains vitronectin-binding function even in RCL-cleaved PAI-1, since the cleaved RCL cannot insert into sheet A. (For interpretation of the references to color in this figure legend, the reader is referred to the Web version of this chapter.)

and exposes the RGD sequence to induce cell migration.[160] A P1 Ala variant was designed to inactivate the protease inhibitor function but retain the vitronectin binding function of the native state and loss of this function in the cleaved state. Another Q123K variant in the vitronectin binding site, known as PAI-1K, abrogated vitronectin binding without affecting protease inhibitory function. To provide a variant that inactivated protease inhibitor function but retained high affinity vitronectin binding even with RCL cleavage, a variant known as PAI-1R was designed with Arg residues at both the P12 and P14 RCL positions. These substitutions blocked the serpin RCL–sheet A interaction and thereby inactivated PAI-1 inhibitory function without affecting vitronectin binding. These variants were used to demonstrate an important role for vitronectin binding independent of protease inhibition in mediating PAI inhibition of angiogenesis,[161] PAI-1 inhibition of vascular intima hyperplasia,[162] and PAI-1 inhibition of extracellular matrix degradation in the kidney.[163] In animal models of diabetic nephropathy, the PAI-1R variant was shown to have a dominant-negative effect in blocking the fibrotic damage caused by excess PAI-1 bound to matrix vitronectin. The testing of PAI-1R in animal models of vascular injury has also suggested its therapeutic value in preventing restenosis following ischemia–reperfusion injury by blocking vascular intima hyperplasia with the added benefit of not increasing the risk of thrombosis.

C. Ovalbumin Variants

As a noninhibitory serpin, ovalbumin has presented a challenge to understanding the minimal requirements of the serpin protease inhibitory machinery and, as a result, has stimulated efforts to engineer changes in the serpin that convert it back to a protease inhibitor. Based on the lack of an appropriate P1 residue and a nonconsensus P14 Arg RCL hinge residue expected to block RCL insertion into sheet A, initial protein engineering efforts changed the P1 Ala to Arg to allow recognition by trypsin and changed P14, P12, and P10 RCL hinge residues to consensus residues of other inhibitory serpins.[164] Trypsin was able to efficiently cleave the variants and to promote partial, but not full, RCL insertion. Surprisingly, a subsequent study, which determined the structure of an RCL-cleaved P14 Thr ovalbumin variant, found the RCL to be fully inserted into sheet A.[165] However, a P1Arg/P14Thr variant was unable to inhibit trypsin but was cleaved by the protease as a substrate. The cleaved serpin variant could be refolded to the RCL-inserted form with a long-lived intermediate metastable state detectable on the folding pathway, implying that RCL insertion into sheet A was too slow for a protease to be trapped.[166] Evidence was obtained that a charged acidic residue, a natural disulfide, and a strong F helix–sheet A interaction interfered with the opening of sheet A and all contributed to the slow RCL insertion. These findings suggested that the principal impediment to RCL insertion into sheet A was a rigidification of the A sheet and overlying F helix structures that are critical for facile A sheet expansion. The only successful attempt to convert ovalbumin to a protease inhibitor was to create a chimera in which more than half of the structure was substituted with the inhibitory serpin sequence of the related B clade serpin, PAI-2.[167] Although correcting the protease inhibition defect, this chimera seemed to reveal ovalbumin structures that were not important for the RCL conformational change rather than showing those that were.

D. crmA Variants

Because the viral serpin crmA represents one of the smallest serpins capable of protease inhibition and can be expressed in high yields in bacteria, it would seem an ideal scaffold for engineering novel types of specificities for desired protease targets. However, unlike antitrypsin and PAI-1, whose scaffolds can be engineered to yield new target protease specificities that retain efficient rates of inhibition and minimal substrate reactivity, use of crmA as a scaffold may be problematic. In one study, the engineering of a P1 Asp to Arg crmA variant showed the expected loss in the ability to inhibit granzyme B and caspases but gain in the ability to inhibit thrombin and factor Xa.[79] However, the latter trypsin-like proteases were inhibited, with slow rate constants of 10^2–10^4 M^{-1} s^{-1} and with elevated substrate reactivity. Moreover, trypsin was not inhibited but efficiently cleaved the P1Arg crmA as a substrate. Surprisingly, elastase was inhibited by wild type or P1Arg crmAs through the P1′–P2′ Cys-Ala bond, but with a high inhibition stoichiometry, consistent with two adjacent RCL sites being capable of protease inhibition. Shifting the P1Arg residue toward the distal end of the RCL resulted in a substrate reactivity with all proteases. The difficulty in utilizing crmA as a scaffold for engineering novel protease inhibitors may relate to its RCL scissile bond being one residue closer to the A sheet hinge than other serpins, a consequence of the smaller size of this serpin and the need for the cleaved RCL to be shorter to conformationally deform a translocated protease. Comparison of cleaved antitrypsin and crmA structures suggests that the shorter RCL in crmA does not insert as far into the A sheet as other serpins and, hence, would be expected to conformationally distort tethered proteases only with dimensions that allow the protease to reach the distal edge of sheet A and be wedged under the sheet (Fig. 16).[23]

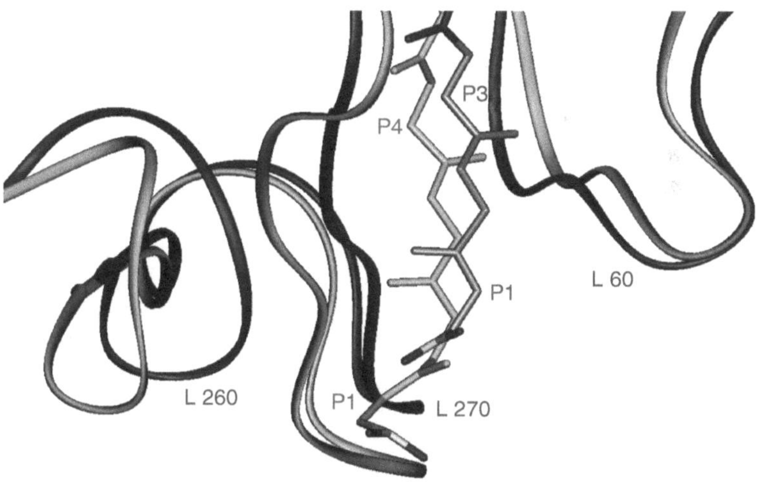

FIG. 16. Superposition of RCL-cleaved α_1PI and crmA structures. The backbone of the distal end of sheet A is shown for cleaved α_1PI (red, pdb 7API) and cleaved crmA (green, pdb 1F0C) with the inserted RCL backbone shown in stick representation and the P1 terminal carboxyl group atoms colored red for oxygen and white for carbon. The P3–P1 residues of crmA were modeled, since the RCL is cleaved at P4 in the structure. Three loop structures are labeled. The P1 residue of α_1PI can be seen to extend further into the sheet than the P1 residue of crmA. Reprinted with permission from Ref. 23. Copyright 2000 Wiley. (For interpretation of the references to color in this figure legend, the reader is referred to the Web version of this chapter.)

V. Concluding Remarks

The serpin superfamily of protein protease inhibitors has evolved a unique conformational trapping mechanism of protease inhibition that is fundamentally different from that of the canonical lock-and-key inhibitors and that allows for novel modes of protease regulation. A common theme is the use of both RCL and exosite determinants on the serpin or on associated cofactors to lure a protease target to form an initial Michaelis encounter complex at an appropriate time and place. For serpins such as α_1PI that do not utilize cofactors, exosites on the serpin ensure a specific recognition and rapid inhibition of neutrophil elastase when this protease is encountered at inflammatory sites of neutrophil activation. Serpins that do make use of cofactors employ a more complex mode of protease regulation. In this case, where the cofactor is made available determines the location a serpin will encounter its protease target. Heparan sulfate and dermatan sulfate on the luminal and subluminal sides of blood vessels serve to localize antithrombin and HCII reactions with blood-clotting proteases to sites of vessel injury. Vitronectin localizes PAI-1 reactions with plasminogen activators to extracellular matrix sites of tissues. An activated platelet membrane localizes the reaction of ZPI with factor Xa at an injury site when ZPI is complexed with an additional cofactor, protein Z, to enable the serpin to bind the membrane. Not only do cofactors ensure that serpins encounter their target proteases at the right location but also the cofactor itself can provide exosites for binding the protease to markedly enhance the rate of the serpin–protease encounter through template-bridging effects. The cofactor may also provide large enhancements of the rate of protease inhibition by acting as an allosteric activator of the serpin. In such cases, cofactor activation makes the serpin exosites competent to engage a target protease, as in the cases of heparin activation of antithrombin or HCII. This cofactor requirement for exosite function has the added advantage of greatly downregulating serpin reactivity in the absence of the cofactor, especially in cases where the serpin has an unfavorable P1 residue, such as HCII or ZPI.

Another distinguishing feature of serpins is their action as irreversible suicide substrate inhibitors of their target proteases. In this mechanism of action, they exploit the two-stage acyl-intermediate mechanism of serine and cysteine proteases by linking proteolysis of the serpin RCL bait to a major serpin conformational change that traps the protease at the acyl-intermediate stage of proteolysis. The trapping is kinetic in that it results in the conformational deformation of the protease and a consequent dramatic slowing of the deacylation stage of proteolysis. The irreversibility of the trapping results from the energetically favorable cleavage of the serpin by protease to form the acyl-intermediate and the energetically downhill transformation of the cleaved metastable serpin to a more stable folded state.

While exosite interactions function to promote and stabilize the initial serpin–protease encounter complex and favor progression to the acyl-intermediate,[125] these interactions must be broken once the acyl-intermediate has formed to allow the protease to be translocated and conformationally trapped.[126] The need to disrupt exosite interactions in the acyl-intermediate may limit the efficiency of its conformational trapping, since deacylation of the acyl-intermediate to generate cleaved serpin competes with trapping. When exosites are limited to the serpin, the trapping conformational change may be fast enough to compete with deacylation and allow efficient inhibition. When cofactors provide additional exosites, disruption of the exosite interactions may be facilitated by coupling the conformational change to large reductions in the affinity for cofactors to allow efficient conformational trapping.[81,128,130]

The kinetically trapped serpin–protease complexes are usually eliminated by receptors of the LRP family but, in some cases, may dissociate on a physiologic time-scale to provide transient inhibition as a means of regulating the level of protease activity needed. New insights into the biology of serpins have suggested that serpins such as PAI-1 and C1 inhibitor have evolved sites for interaction with receptors and other types of molecules that not only localize serpin antiprotease function, but also reinforce this function by acting as receptor antagonists or through signaling. Engineered serpin variants that have been designed based on an understanding of the complex mechanisms by which serpins and their cofactors regulate protease activity in time and space have proved to be effective therapeutics in animal models of various human diseases. Thus, serpin engineering shows considerable promise for creating novel therapeutics that can normalize the dysfunctional protease regulation, which is the root cause of many diseases.

References

1. Gettins P. Serpin structure, mechanism, and function. *Chem Rev* 2002;**102**:4751–803.
2. Silverman GA, Whisstock JC, Bottomley SP, Huntington JA, Kaiserman D, Luke CJ, et al. Serpins flex their muscle: putting the clamps on proteolysis in diverse biological systems. *J Biol Chem* 2010;**285**:24299–305.
3. Gettins PGW, Olson ST. Exosite determinants of serpin specificity. *J Biol Chem* 2009;**284**:20441–5.
4. Bode W, Huber R. Natural protein proteinase inhibitors and their interaction with proteinases. *Eur J Biochem* 1992;**204**:433–51.
5. Hunt LT, Dayhoff MO. A Surprising new protein superfamily containing ovalbumin, antithrombin-III, and alpha$_1$-proteinase inhibitor. *Biochem Biophys Res Commun* 1980;**95**:864–71.
6. Carrell RW, Travis J. α1-Antitrypsin and the serpins: variation and countervariation. *Trends Biochem Sci* 1985;**10**:20–4.

7. Ray CA, Black RA, Kronheim SR, Greenstreet TA, Sleath PR, Salvesen GS, et al. Viral inhibition of inflammation: cowpox virus encodes an inhibitor of the interleukin-1beta converting enzyme. *Cell* 1992;**69**:597–604.
8. Silverman GA, Bird PI, Carrell RW, Church FC, Coughlin PB, Gettins PGW, et al. The serpins are an expanding superfamily of structurally similar but functionally diverse proteins. *J Biol Chem* 2001;**276**:33293–6.
9. Irving JA, Pike RN, Lesk AM, Whisstock JC. Phylogeny of the serpin superfamily: implications of patterns of amino acid conservation for structure and function. *Genome Res* 2000;**10**:1845–64.
10. Benarafa C, Remold-O'Donnell E. The ovalbumin serpins revisited: perspective from the chicken genome of clade B serpin evolution in vertebrates. *Proc Natl Acad Sci USA* 2005;**102**:11367–72.
11. Ragg H, Lokot T, Kamp P-B, Atchley WR, Dress A. Vertebrate serpins: construction of a conflict-free phylogeny by combining exon-intron and diagnostic site analyses. *Mol Biol Evol* 2001;**18**:577–84.
12. Van Boven HH, Lane DA. Antithrombin and its inherited deficiency states. *Semin Hematol* 1997;**34**:188–204.
13. Fay WP, Shapiro AD, Shih JL, Schleef RR, Ginsburg D. Complete deficiency of plasminogen-activator inhibitor type 1 due to a frameshift mutation. *N Engl J Med* 1992;**327**:1729–33.
14. Carrell RW, Jeppsson JO, Laurell CB, Brennan SO, Owen MC, Vaughan L, et al. Structure and variation of human α_1-antitrypsin. *Nature* 1982;**298**:329–34.
15. Davis III AE, Mejia P, Lu F. Biological activities of C1 inhibitor. *Mol Immunol* 2008;**45**:4057–63.
16. Silverman GA, Whisstock JC, Askew DJ, Pak SC, Luke CJ, Cataltepe S, et al. Human clade B serpins (ov-serpins) belong to a cohort of evolutionarily dispersed intracellular proteinase inhibitor clades that protect cells from promiscuous proteolysis. *Cell Mol Life Sci* 2004;**61**:301–25.
17. Cooley J, Takayama TK, Shapiro SD, Schechter NM, Remold-O'Donnell E. The serpin MNEI inhibits elastase-like and chymotrypsin-like serine proteases through efficient reactions at two active sites. *Biochemistry* 2001;**40**:15762–70.
18. Scott FL, Hirst CE, Sun J, Bird CH, Bottomley SP, Bird PI. The intracellular serpin proteinase inhibitor 6 is expressed in monocytes and granulocytes and is a potent inhibitor of the azurophilic granulae protease, cathepsin G. *Blood* 1999;**93**:2089–97.
19. Bird CH, Sutton VR, Sun J, Hirst CE, Novak A, Kumar S, et al. Selective regulation of apoptosis: the cytotoxic lymphocyte serpin proteinase inhibitor 9 protects against granzyme B-mediated apoptosis without perturbing the Fas cell death pathway. *Mol Cell Biol* 1998;**18**:6387–98.
20. Gao F, Shi HY, Daughty C, Cella N, Zhang M. Maspin plays an essential role in early embryonic development. *Development* 2004;**131**:1479–89.
21. Ishiguro K, Kojima T, Kadomatsu K, Nakayama Y, Takagi A, Suzuki M, et al. Complete antithrombin deficiency in mice results in embryonic lethality. *J Clin Invest* 2000;**106**:873–8.
22. Nagai N, Hosokawa M, Itohara S, Adachi E, Matsushita T, Hosokawa N, et al. Embryonic lethality of molecular chaperone Hsp47 knockout mice is associated with defects in collagen biosynthesis. *J Cell Biol* 2000;**150**:1499–505.
23. Simonovic M, Gettins PGW, Volz K. Crystal structure of viral serpin crmA provides insights into its mechanism of cysteine proteinase inhibition. *Protein Sci* 2000;**9**:1423–7.
24. Renatus M, Zhou Q, Stennicke HR, Snipas SJ, Turk D, Bankston LA, et al. Crystal structure of the apoptotic suppressor CrmA in its cleaved form. *Structure* 2000;**8**:789–97.

25. Irving JA, Cabrita LD, Rossjohn J, Pike RN, Bottomley SP, Whisstock JC. The 1.5 A crystal structure of a prokaryote serpin: controlling conformational change in a heated environment. *Structure* 2003;**11**:387–97.
26. Hekman CM, Loskutoff DJ. Endothelial cells produce a latent inhibitor of plasminogen activators that can be activated by denaturants. *J Biol Chem* 1985;**260**:11581–7.
27. Mottonen J, Strand A, Symersky J, Sweet RM, Danley DE, Geoghegan KF, et al. Structural basis of latency in plasminogen activator inhibitor-1. *Nature* 1992;**355**:270–3.
28. Lawrence DA, Olson ST, Palaniappan S, Ginsburg D. Engineering plasminogen activator inhibitor 1 mutants with increased functional stability. *Biochemistry* 1994;**33**:3643–8.
29. Loebermann H, Tokuoka R, Deisenhofer J, Huber R. Human α_1-Proteinase inhibitor crystal structure analysis of two crystal modifications, molecular model and preliminary analysis of the implications for function. *J Mol Biol* 1984;**177**:531–56.
30. Kaslik G, Kardos J, Szabo L, Zavodszky P, Westler WM, Markely JL, et al. Effects of serpin binding on the target proteinase: global stabilization, localized increased structural flexibility, and conserved hydrogen bonding at the active site. *Biochemistry* 1997;**36**:5455–74.
31. Boudier C, Bieth JG. The reaction of serpins with proteinases involves important enthalpy changes. *Biochemistry* 2001;**40**:9962–7.
32. Berkenpas MB, Lawrence DA, Ginsberg D. Molecular evolution of plasminogen activator inhibitor-1 functional stability. *EMBO J* 1995;**14**:2969–77.
33. Zhou A, Huntington JA, Carrell RW. Formation of the antithrombin heterodimer in vivo and the onset of thrombosis. *Blood* 1999;**94**:3388–96.
34. Declerck PJ, De Mol M, Alessi MC, Baudner S, Paques EP, Preissner KT, et al. Purification and characterization of a plasminogen activator inhibitor 1 binding protein from human plasma identification as a multimeric form of S protein (vitronectin). *J Biol Chem* 1988;**263**:15454–61.
35. Whisstock JC, Silverman GA, Bird PI, Bottomley SP, Kaiserman D, Luke CJ, et al. Serpins flex their muscle: structural insights into target peptidase recognition, polymerization and transport functions. *J Biol Chem* 2010;**285**:24307–12.
36. Gooptu B, Lomas DA. Conformational pathology of the serpins: themes, variations, and therapeutic strategies. *Annu Rev Biochem* 2009;**78**:147–76.
37. Zhou A, Stein P, Huntington J, Carrell R. Serpin polymerization is prevented by a hydrogen bond network that is centered on His-334 and stabilized by glycerol. *J Biol Chem* 2003;**278**:15116–22.
38. Gooptu B, Hazes B, Chang WW, Dafforn TR, Carrell RW, Read RJ, et al. Inactive conformation of the serpin alpha1-antchymotrypsin indicates two-stage insertion of the reactive loop: implications for inhibitory function and conformational disease. *Proc Natl Acad Sci USA* 2000;**97**:67–72.
39. Dafforn TR, Mahadeva R, Elliott PR, Sivasothy P, Lomas DA. A kinetic mechanism for the polymerization of alpha1-antitrypsin. *J Biol Chem* 1999;**274**:9548–55.
40. Elliott P, Lomas D, Carrell R, Abrahams J. Inhibitory conformation of the reactive loop of antitrypsin. *Nat Struct Biol* 1996;**3**:676–81.
41. Yamasaki M, Li W, Johnson DJD, Huntington JA. Crystal structure of a stable dimer reveals the molecular basis of serpin polymerization. *Nature* 2008;**455**:1255–8.
42. Yu MH, Lee KN, Kim J. The Z type variation of human alpha 1-antitrypsin causes a protein folding defect. *Nat Struct Biol* 1995;**2**:363–7.
43. Lomas DA, Evans DL, Finch JT, Carrell RW. The mechanism of Z α_1-antitrypsin accumulation in the liver. *Nature* 1992;**357**:605–7.
44. Dahlen JR, Jean F, Thomas G, Foster DC, Kisiel W. Inhibition of soluble recombinant furin by human proteinase inhibitor 8. *J Biol Chem* 1998;**273**:1851–4.

45. Schick C, Pemberton PA, Shi GP, Kamachi Y, Cataltepe S, Bartuski AJ, et al. Cross-class inhibition of the cysteine proteinases cathepsins K, L, and S by the serpin squamous cell carcinoma antigen 1: a kinetic analysis. *Biochemistry* 1998;**37**:5258–66.
46. Zhou Q, Snipas S, Orth K, Muzio M, Dixit VM, Salvesen GS. Target protease specificity of the viral serpin CrmA. Analysis of five caspases. *J Biol Chem* 1997;**272**:7797–800.
47. Quan LT, Caputo A, Bleackley RC, Pickup DJ, Salvesen GS. Granzyme B is inhibited by the cowpox virus serpin cytokine response modifier A. *J Biol Chem* 1995;**270**:10377–9.
48. Dobo J, Gettins PGW. α1-Proteinase inhibitor forms initial non-covalent and final covalent complexws with elastase analogously to other serpin-proteinase pairs, suggesting a common mechanism of inhibition. *J Biol Chem* 2004;**279**:9264–9.
49. Hedstrom L. Serine protease mechanism and specificity. *Chem Rev* 2002;**102**:4501–23.
50. Stratikos E, Gettins PGW. Formation of the covalent serpin-proteinase complex involves translocation of the proteinase by more than 70 A and full insertion of the reactive center loop into β-sheet A. *Proc Natl Acad Sci USA* 1999;**96**:4808–13.
51. Lawrence DA, Ginsburg D, Day DE, Berkenpas MB, Verhamme IM, Kvassman J, et al. Serpin-protease complexes are trapped as stable acyl-enzyme intermediates. *J Biol Chem* 1995;**270**:25309–12.
52. Calugaru SV, Swanson R, Olson ST. The pH dependence of serpin-proteinase complex dissociation reveals a mechanism of complex stabilization involving inactive and active conformational states of the proteinase which are perturbable by calcium. *J Biol Chem* 2001;**276**:32446–55.
53. Patston PA, Gettins P, Beechem J, Schapira M. Mechanism of serpin action: evidence that C1 inhibitor functions as a suicide substrate. *Biochemistry* 1991;**30**:8876–82.
54. Dementiev A, Simonovic M, Volz K, Gettins PGW. Canonical inhibitor-like interactions explain reactivity of α_1-proteinase inhibitor Pittsburgh and antithrombin with proteinases. *J Biol Chem* 2003;**278**:37881–7.
55. Dementiev A, Petitou M, Herbert J-M, Gettins PGW. The ternary complex of antithrombin-anhydrothrombin-heparin reveals the basis of inhibitor specificity. *Nat Struct Mol Biol* 2004;**11**:863–7.
56. Dementiev A, Dobo J, Gettins PGW. Active-site distortion is sufficient for proteinase inhibition by serpins. *J Biol Chem* 2006;**281**:3452–7.
57. Huntington JA, Read RJ, Carrell RW. Structure of a serpin-protease complex shows inhibition by deformation. *Nature* 2000;**407**:923–6.
58. Ye S, Cech AL, Belmares R, Bergstrom RC, Tong Y, Corey DR, et al. The structure of a Michaelis serpin-protease complex. *Nat Struct Biol* 2001;**8**:979–83.
59. Baglin T, Carrell R, Church F, Esmon C, Huntington J. Crystal structures of native and thrombin-complexed heparin cofactor II reveal a multistep allosteric mechanism. *Proc Natl Acad Sci USA* 2002;**99**:11079–84.
60. Li W, Johnson DJ, Esmon CT, Huntington JA. Structure of the antithrombin-thrombin-heparin ternary complex reveals the antithrombotic mechanism of heparin. *Nat Struct Mol Biol* 2004;**11**:857–62.
61. Johnson DJ, Li W, Adams TE, Huntington JA. Antithrombin-S195A factor Xa-heparin structure reveals the mechanism of antithrombin activation. *EMBO J* 2006;**25**:2029–37.
62. Johnson DJD, Langdown J, Huntington JA. Molecular basis of factor IXa recognition by heparin-activated antithrombin revealed by a 1.7 A structure of the ternary complex. *Proc Natl Acad Sci USA* 2010;**107**:645–50.
63. Li W, Adams TE, Nangalia J, Esmon CT, Huntington JA. Molecular basis of thrombin recognition by protein C inhibitor revealed by the 1.6-A structure of the heparin-bridged complex. *Proc Natl Acad Sci USA* 2008;**105**:4661–6.

64. Schechter I, Berger A. On the size of the active site in proteases. I. Papain. *Biochem Biophys Res Commun* 1967;**27**:157–62.

65. Peterson FC, Gordon NC, Gettins PGW. Formation of a noncovalent serpin-proteinase complex involves no conformational change in the serpin. Use of ^{1}H-^{15}N HSQC NMR as a sensitive nonperturbing monitor of conformation. *Biochemistry* 2000;**39**:11884–92.

66. Peterson FC, Gettins PGW. Insight into the mechanism of serpin-proteinase inhibition from 2D (^{1}H-^{15}N) NMR studies of the 69 kDa α_1-proteinase inhibitor pittsburgh-trypsin covalent complex. *Biochemistry* 2001;**40**:6284–892.

67. Gettins PGW, Backovic M, Peterson FC. Use of NMR to study serpin function. *Methods* 2004;**32**:120–9.

68. Futamura A, Stratikos E, Olson ST, Gettins PGW. Change in environment of the P1 side chain upon progression from the Michaelis complex to the covalent serpin-proteinase complex. *Biochemistry* 1998;**37**:13110–9.

69. Stratikos E, Gettins P. Major proteinase movement upon stable serpin-proteinase complex formation. *Proc Natl Acad Sci USA* 1997;**94**:453–8.

70. Backovic M, Stratikos E, Lawrence DA, Gettins PGW. Structural similarity of the covalent complexes formed between the serpin plasminogen activator inhibitor-1 and the arginine-specific proteinases trypsin, LMW urokinase, HMW urokinase and tPA: use of site-specific fluorescent probes of local environment. *Protein Sci* 2002;**11**:1182–91.

71. Stavridi ES, O'Malley K, Lukacs CM, Moore WT, Lambris JD, Christianson DW, et al. Structural change in α-chymotrypsin induced by complexation with α_1-antichymotrypsin as seen by enhanced sensitivity to proteolysis. *Biochemistry* 1996;**35**:10608–15.

72. Bock PE, Olson ST, Björk I. Inactivation of thrombin by antithrombin is accompanied by inactivation of regulatory exosite 1. *J Biol Chem* 1997;**272**:19837–45.

73. Swanson R, Raghavendra MP, Zhang W, Froelich C, Gettins PGW, Olson ST. Serine and cysteine proteases are translocated to similar extents upon formation of covalent complexes with serpins. Fluorescence perturbation and fluorescence resonance energy transfer mapping of the protease binding site in CrmA complexes with granzyme B and caspase-1. *J Biol Chem* 2007;**282**:2305–13.

74. Dobo J, Swanson R, Salvesen GS, Olson ST, Gettins PGW. Cytokine response modifier a inhibition of initiator caspases results in covalent complex formation and dissociation of the caspase tetramer. *J Biol Chem* 2006;**281**:38781–90.

75. Gettins P, Harten B. Properties of thrombin- and elastase-modified human antithrombin III. *Biochemistry* 1988;**27**:3634–9.

76. Bruch M, Weiss V, Engel J. Plasma serine proteinase inhibitors (serpins) exhibit major conformational changes and a large increase in conformational stability upon cleavage at their reactive sites. *J Biol Chem* 1988;**263**:16626–30.

77. Seo EJ, Im H, Maeng J-S, Kim KE, Yu M-H. Distribution of the native strain in human α_1-antitrypsin and its association with protease inhibitor function. *J Biol Chem* 2000;**275**:16904–9.

78. Zhou A, Carrell RW, Huntington JA. The serpin inhibitory mechanism is critically dependent on the length of the reactive center loop. *J Biol Chem* 2001;**276**:27541–7.

79. Tesch LD, Raghavendra MP, Bedsted-Faarvang T, Gettins PGW, Olson ST. Specificity and reactive loop length requirements for CrmA inhibition of serine proteases. *Protein Sci* 2005;**14**:533–42.

80. Barker-Carlson K, Lawrence D, Schwartz B. Acyl-enzyme complexes between tissue-type plasminogen activator and neuroserpin are short-lived in vitro. *J Biol Chem* 2002;**277**:26852–32768.

81. Huang X, Swanson R, Broze Jr. GJ, Olson ST. Kinetic characterization of the protein Z-dependent protease inhibitor reaction with blood coagulation factor Xa. *J Biol Chem* 2008;**283**:29770–83.

82. Gettins PGW. The F-helix of serpins plays an essential, active role in the proteinase inhibition mechanism. *FEBS Lett* 2002;**523**:2–6.

83. Vleugels N, Gils A, Bijnens A-P, Knockaert I, Declerck P. The importance of helix F in plasminogen activator inhibitor-1. *Biochim Biophys Acta* 2000;**1476**:20–6.

84. Mast AE, Enghild JJ, Pizzo SV, Salvesen G. Analysis of the plasma elimination kinetics and conformational stabilities of native, proteinase-complexed, and reactive site cleaved serpins: comparison of α_1-proteinase inhibitor, α_1-antichymotrypsin, antithrombin III, α_2-antiplasmin, angiotensinogen, and ovalbumin. *Biochemistry* 1991;**30**:1723–30.

85. Joslin G, Fallon RJ, Bullock J, Adams SP, Perlmutter DH. The SEC receptor recognizes a pentapeptide neodomain of alpha 1-antitrypsin-protease complexes. *J Biol Chem* 1991;**266**:11282–8.

86. Stefansson S, Muhammad S, Cheng X-F, Battey FD, Strickland DK, Lawrence DA. Plasminogen activator inhibitor-1 contains a cryptic high affinity binding site for the low density lipoprotein receptor-related protein. *J Biol Chem* 1998;**273**:6358–66.

87. Horn IR, Van den Berg BMM, Moestrup SK, Pannekoek H, van Zonneveld AJ. Plasminogen activator inhibitor 1 contains a cryptic high affinity receptor binding site that is exposed upon complex formation with tissue-type plasminogen activator. *Thromb Haemost* 1998;**80**:822–8.

88. Jensen JK, Dolmer K, Gettins PGW. Specificity of binding of the low density lipoprotein receptor-related protein (LRP) to different conformational states of the clade E serpins PAI-1 and PN-1. *J Biol Chem* 2009;**284**:17989–97.

89. Kasza A, Petersen HH, Heegaard CW, Oka K, Christensen A, Dubin A, et al. Specificity of serine proteinase-serpin complex binding to very-low-density lipoprotein receptor and α2-macroglobulin receptor low-density-lipoprotein-receptor-related protein. *Eur J Biochem* 1997;**248**:270–81.

90. Gonias SL, Fuchs HE, Pizzo SV. A unique pathway for the plasma elimination of α_2-antiplasmin-protease complexes in mice. *Thromb Hemost* 1982;**48**:208–10.

91. Kounnas MZ, Church FC, Argraves WS, Strickland DK. Cellular internalization and degradation of antithrombin III-thrombin, heparin cofactor II-thrombin, and α_1-antitrypsin-trypsin complexes is mediated by the low density lipoprotein receptor-related protein. *J Biol Chem* 1996;**271**:6523–9.

92. Lillis AP, Mikhailenko I, Strickland DK. Beyond endocytosis: LRP function in cell migration, proliferation and vascular permeability. *J Thromb Haemost* 2005;**3**:1884–93.

93. Lillis AP, van Duyn LB, Murphy-Ullrich JE, Strickland DK. LDL receptor-related protein 1: unique tissue-specific functions revealed by selective gene knockout studies. *Phys Rev* 2008;**88**:887–918.

94. Beatty K, Bieth J, Travis J. Kinetics of association of serine proteinases with native and oxidized α-1-proteinase inhibitor and α-1-antichymotrypsin. *J Biol Chem* 1980;**255**:3931–4.

95. Travis J, Matheson NR, George PM, Carrell RW. Kinetic studies on the interaction of $á_1$-proteinase inhibitor (Pittsburgh) with trypsin-like serine proteinases. *Biol Chem Hoppe Seyler* 1986;**367**:853–9.

96. Bianchini EP, Louvain VB, Marque P-E, Juliano MA, Juliano L, Le Bonniec BF. Mapping of the catalytic groove preferences of factor Xa reveals an inadequate selectivity for its macromolecule substrates. *J Biol Chem* 2002;**277**:20527–34.

97. McRae B, Nakajima K, Travis J, Powers JC. Studies on reactivity of human leukocyte elastase, cathepsin G, and porcine pancreatic elastase toward peptides including sequences related to the reactive site of α_1-protease inhibitor (α_1-antitrypsin). *Biochemistry* 1980;**19**:3973–8.

98. Plotnick MI, Schechter NM, Wang ZM, Liu X, Rubin H. Role of the P6-P3′ region of the serpin reactive loop in the formation and breakdown of the inhibitory complex. *Biochemistry* 1997;**36**:14601–8.
99. Ibarra CA, Blose GE, Christian TD, Shore JD. The contribution of the exosite residues of plasminogen activator inhibitor-1 to proteinase inhibition. *J Biol Chem* 2004;**279**:3643–50.
100. Braud S, Le Bonniec BF, Bon C, Wisner A. The stratagem utilized by the plasminogen activator from the snake *Trimeresurus stejnegeri* to escape serpins. *Biochemistry* 2002;**41**:8478–84.
101. Chen VC, Chao L, Chao J. A postively charged loop on the surface of kallistatin functions to enhance tissue kallikrein inhibitor by acting as a secondary binding site for kallikrein. *J Biol Chem* 2000;**275**:40371–7.
102. Rezaie AR, Bae J-S, Manithody C, Qureshi SH, Yang L. Protein Z-dependent protease inhibitor binds to the C-terminal domain of protein Z. *J Biol Chem* 2008;**283**:19922–6.
103. Rezaie AR, Manithody C, Yang L. Identification of factor Xa residues critical for interaction with protein Z-dependent protease inhibitor. *J Biol Chem* 2005;**280**:32722–8.
104. Huang X, Dementiev A, Olson ST, Gettins PGW. Basis for the specificity and activation of the serpin protein Z-dependent proteinases inhibitor as an inhibitor of membrane-associated factor Xa. *J Biol Chem* 2010;**285**:20399–409.
105. Rezaie AR, Sun M-F, Gailani D. Contributions of basic amino acids in the autolysis loop of factor XIa to serpin specificity. *Biochemistry* 2006;**45**:9427–33.
106. Derechin VM, Blinder MA, Tollefsen DM. Substitution of arginine for Leu^{444} in the reactive site of heparin cofactor II enhances the rate of thrombin inhibition. *J Biol Chem* 1990;**265**:5623–8.
107. Sheehan JP, Tollefsen DM, Sadler JE. Heparin cofactor II is regulated allosterically and not primarily by template effects. Studies with mutant thrombins and glycosaminoglycans. *J Biol Chem* 1994;**269**:32747–51.
108. Van Deerlin VM, Tollefsen DM. The N-terminal acidic domain of heparin cofactor II mediates the inhibition of alpha-thrombin in the presence of glycosaminoglycans. *J Biol Chem* 1991;**266**:20223–31.
109. Verhamme I, Bock PE, Jackson CM. The preferred pathway of glycosaminoglycan-accelerated inactivation of thrombin by heparin cofactor II. *J Biol Chem* 2004;**279**:9785–95.
110. Olson ST, Swanson R, Raub-Segall E, Bedsted T, Sadri M, Petitou M, et al. Accelerating ability of synthetic oligosaccharides on antithrombin inhibition of proteinases of the clotting and fibrinolytic systems. Comparison with heparin and low-molecular-weight heparin. *Thromb Haemost* 2004;**92**:929–39.
111. Hermans JM, Stone SR. Interaction of activated protein C with serpins. *Biochem J* 1993;**295**:239–45.
112. Djie M, Stone S, Le Bonniec B. Intrinsic specificity of the reactive site loop of antitrypsin, antichymotrypsin, antithrombin III, and protease nexin I. *J Biol Chem* 1997;**272**:16268–73.
113. Hopkins PCR, Pike RN, Stone SR. Evolution of serpin specificity: cooperative interactions in the reactive-site loop sequence of antithrombin specifically restrict the inhibition of activated protein C. *J Mol Evol* 2000;**51**:507–15.
114. Mather T, Oganessyan V, Hof P, Huber R, Foundling S, Esmon CT, et al. The 2.8 Å crystal structure of Gla-domainless activated protein C. *EMBO J* 1996;**15**:6822–31.
115. Huntington JA, Olson ST, Fan B, Gettins PGW. Mechanism of heparin activation of antithrombin. Evidence for reactive center loop preinsertion with expulsion upon heparin binding. *Biochemistry* 1996;**35**:8495–503.
116. Jin L, Abrahams JP, Skinner R, Petitou M, Pike RN, Carrell RW. The anticoagulant activation of antithrombin by heparin. *Proc Natl Acad Sci USA* 1997;**94**:14683–8.

117. Olson ST, Björk I, Sheffer R, Craig PA, Shore JD, Choay J. Role of the antithrombin-binding pentasaccharide in heparin acceleration of antithrombin-proteinase reactions. Resolution of the antithrombin conformational change contribution to heparin rate enhancement. *J Biol Chem* 1992;**267**:12528–38.
118. Bedsted T, Swanson R, Chuang Y-J, Bock PE, Björk I, Olson ST. Heparin and calcium ions dramatically enhance antithrombin reactivity with factor IXa by generating new interaction exosites. *Biochemistry* 2003;**42**:8143–52.
119. Gettins PGW, Olson ST. Activation of antithrombin as a factor IXa and Xa inhibitor involves mitigation of repression rather than positive enhancement. *FEBS Lett* 2009;**583**:3397–400.
120. Manithody C, Yang L, Rezaie AR. Role of basic residues of the autolysis loop in the catalytic function of factor Xa. *Biochemistry* 2002;**41**:6780–8.
121. Yang L, Manithody C, Olson ST, Rezaie AR. Contribution of basic residues of the autolysis loop to the substrate and inhibitor specificity of factor IXa. *J Biol Chem* 2003;**278**:25032–8.
122. Izaguirre G, Olson ST. Residues Tyr253 and Glu255 in strand 3 of β-sheet C of antithrombin are key determinants of an exosite made accessible by heparin activation to promote rapid inhibition of factors Xa and IXa. *J Biol Chem* 2006;**281**:13424–32.
123. Olson ST, Björk I. Predominant contribution of surface approximation to the mechanism of heparin acceleration of the antithrombin-thrombin reaction. Elucidation from salt concentration effects. *J Biol Chem* 1991;**266**:6353–64.
124. Rezaie AR. Calcium enhances heparin catalysis of the antithrombin-factor Xa reaction by a template mechanism: evidence that calcium alleviates Gla domain antagonism of heparin binding to factor Xa. *J Biol Chem* 1998;**273**:16824–7.
125. Izaguirre G, Swanson R, Raja SM, Rezaie AR, Olson ST. Mechanism by which exosites promote the inhibition of blood coagulation proteases by heparin-activated antithrombin. *J Biol Chem* 2007;**282**:33609–22.
126. Olson ST, Swanson R, Verhamme I, Kvassman J, Shore JD. Resolution of Michaelis complex, acylation and conformational change steps in the reactions of the serpin PAI-1 with trypsin and tissue plasminogen activator. *Biochemistry* 2001;**40**:11742–56.
127. Olson ST, Shore JD. Transient kinetics of heparin-catalyzed protease inactivation by antithrombin III. The reaction step limiting heparin turnover in thrombin neutralization. *J Biol Chem* 1986;**261**:13151–9.
128. Olson ST. Heparin and ionic strength-dependent conversion of antithrombin III from an inhibitor to a substrate of alpha-thrombin. *J Biol Chem* 1985;**260**:10153–60.
129. Tabatabai A, Fiehler R, Broze Jr. GJ. Protein Z circulates in plasma in a complex with protein Z-dependent protease inhibitor. *Thromb Haemost* 2001;**85**:655–60.
130. Lawrence DA, Palaniappan S, Stefansson S, Olson ST, Francis-Chmura AM, Shore JD, et al. Characterization of the binding of different conformational forms of plasminogen activator inhibitor-1 to vitronectin. Implications for the regulation of pericellular proteolysis. *J Biol Chem* 1997;**272**:7676–80.
131. Schuster MG, Enriquez PM, Curran P, Cooperman BS, Rubin H. Regulation of neutrophil superoxide by antichymotrypsin-chymotrypsin complexes. *J Biol Chem* 1992;**267**:5056–9.
132. Banda MJ, Rice AG, Griffin GL, Senior RM. α_1-Proteinase inhibitor is a neutrophil chemoattractant after proteolytic inactivation by macrophage elastase. *J Biol Chem* 1988;**263**:4481–4.
133. Richard B, Swanson R, Schedin-Weiss S, Ramirez B, Izaguirre G, Gettins PGW, et al. Characterization of the conformational alterations, reduced anticoagulant activity, and enhanced antiangiogenic activity of prelatent antithrombin. *J Biol Chem* 2008;**283**:14417–29.
134. Egelund R, Petersen TE, Andreasen PA. A serpin-induced extensive proteolytic susceptibility of urokinase-type plasminogen activator implicates distortion of the proteinase substrate binding pocket and oxyanion hole in the serpin inhibitory mechanism. *Eur J Biochem* 2001;**268**:673–85.

135. Fredenburgh JC, Stafford AR, Weitz JI. Conformational changes in thrombin when complexed by serpins. *J Biol Chem* 2001;**276**:32768.
136. Scheidig AJ, Hynes TR, Pelletier LA, Wells JA, Kossiakoff AA. Crystal structures of bovine chymotrypsin and trypsin complexed to the inhibitor domain of Alzheimer's amyloid β-protein precursor (APPI) and basic pancreatic trypsin inhibitor (BPTI): engineering of inhibitors with altered specificities. *Protein Sci* 1997;**6**:1806–24.
137. Grütter JG, Priestle JP, Rahuel J, Grossenbacher H, Bode W, Hofsteenge J, et al. Crystal structure of the thrombin-hirudin complex: a novel mode of serine protease inhibition. *EMBO J* 1990;**9**:2361–5.
138. Sim RB, Arlaud G, Colomb M. Kinetics of reaction of human C1-inhibitor with the human complement system proteases C1r and C1s. *Biochim Biophys Acta* 1980;**612**:433–49.
139. Alpha1-Antitrypsin Deficiency Registry Study Group. A registry of patients with severe deficiency of alpha1-antitrypsin. *Chest* 1994;**106**:1223–32.
140. Benarafa C, Priebe GP, Remold-O'Donnell E. The neutrophil serine protease inhibitor serpinb1 preserves lung defense functions in Pseudomonas aeruginosa infection. *J Exp Med* 2007;**204**:1901–9.
141. Adiguzel C, Iqbal O, Demir M, Fareed J. European community and US-FDA approval of recombinant human antithrombin produced in genetically altered goats. *Clin Appl Thromb Hemost* 2009;**15**:645–51.
142. Yang D, Nemkul N, Shereen A, Jone A, Dunn RS, Lawrence DA, et al. Therapeutic administration of plasminogen activator inhibitor-1 prevents hypoxic-ischemic brain injury in newborns. *J Neurosci* 2009;**29**:8669–74.
143. Dai E, Guan H, Liu L, Little S, McFadden G, Vaziri S, et al. Serp-1, a viral anti-inflammatory serpin, regulates cellular serine proteinase and serpin responses to vascular injury. *J Biol Chem* 2003;**278**:18563–72.
144. Salvesen GS, Dixit VM. Caspases: intracellular signaling by proteolysis. *Cell* 1997;**91**:443–6.
145. George PM, Vissers MCM, Travis J, Winterbourn CC, Carrell RW. A genetically engineered mutant of α1-antitrypsin protects connective tissue from neutrophil damage and may be useful in lung disease. *Lancet* 1984;**2**:1426–8.
146. Owen MC, Brennan SO, Lewis JH, Carrell RW. Mutation of antitrypsin to antithrombin. alpha 1-antitrypsin Pittsburgh (358 Met to Arg), a fatal bleeding disorder. *N Engl J Med* 1983;**309**:694–8.
147. Hopkins PCR, Crowther DC, Carrell RW, Stone SR. Development of a novel recombinant serpin with potential antithrimbotic properties. *J Biol Chem* 1995;**270**:11866–71.
148. Izaguirre G, Rezaie AR, Olson ST. Engineering functional antithrombin exosites in alpha1-proteinase inhibitor that specifically enhnace the inhibiiton of factor Xa and factor IXa. *J Biol Chem* 2009;**284**:1550–8.
149. Lee KN, Im H, Kang SW, Yu M-H. Characterization of a human α1-antitrypsin variant that is as stable as ovalbumin. *J Biol Chem* 1998;**273**:2509–16.
150. Jean F, Stella K, Thomas L, Liu G, Xiang Y, Thomas G. α1-Antitrypsin Protland, a bioengineered serpin highly selective for furin: application as an antipathogenic agent. *Proc Natl Acad Sci USA* 1998;**95**:7293–8.
151. Dufour EK, Denault J-B, Hopkins PCR, Leduc R. Serpin-like properties of α1-antitrypsin Portland towards furin convertase. *FEBS Lett* 1998;**426**:41–6.
152. Bassi DE, DeCicco RL, Mahloogi H, Zucker S, Thomas G, Klein-Szanto AJP. Furin inhibition results in absent or decreased invasiveness and tumorigenicity of human cancer cells. *Proc Natl Acad Sci USA* 2001;**98**:10326–31.
153. Tsuji A, Kanie H, Makise H, Yuasa K, Nagahama M, Matsuda Y. Engineering of α1-antitrypsin variants selective for subtilisin-like proprotein convertases PACE4 and PC6: importance of

the P2′ residue in stable complex formation of the serpin with proprotein convertase. *Protein Eng Des Sel* 2007;**20**:163–70.
154. Irving JA, Pike RN, Dai W, Bromme D, Worrall DM, Silverman GA, et al. Evidence that serpin architecture intrinsically supports papain-like cysteine protease inhibition: engineering α1-antitrypsin to inhibit cathepsin proteases. *Biochemistry* 2002;**41**:4998–5004.
155. Sherman PM, Lawrence DA, Yang AY, Vandenberg ET, Paielli D, Olson ST, et al. Saturation mutagenesis of the plasminogen activator inhibitor-1 reactive center. *J Biol Chem* 1992;**267**:7588–95.
156. Sherman PM, Lawrence DA, Verhamme IM, Paielli D, Shore JD, Ginsburg D. Identification of tissue-type plasminogen activator-specific plasminogen activator inhibitor-1 mutants. *J Biol Chem* 1995;**270**:9301–6.
157. Stefansson S, Yepes M, Gorlatova N, Day DE, Moore EG, Zabaleta A, et al. Mutants of plasminogen activator inhibitor-1 designed to inhibit neutrophil elastase and cathepsin G are more effective in vivo than their endogenous inhibitors. *J Biol Chem* 2004;**279**:29981–7.
158. Cale JM, Lawrence DA. Structure-function relationships of plasminogen activator inhibitor-1 and its potential as a therapeutic agent. *Curr Drug Targets* 2007;**8**:971–81.
159. Stoop AA, Eldering E, Dafforn TR, Read RJ, Pannekoek H. Differential structural requirements for plasminogen activator inhibitor-1 (PAI-1) during latency transition and proteinase inhibition as evidenced by phage-displayed hypermutated PAI-1 libraries. *J Mol Biol* 2001;**305**:773–83.
160. Stefansson S, Lawrence D. The serpin PAI-1 inhibits cell migration by blocking integrin binding to vitronectin. *Nature* 1996;**383**:441–3.
161. Stefansson S, Petitclerc E, Wong MK, McMahon GA, Brooks PC, Lawrence DA. Inhibition of angiogenesis in vivo by plasminogen activator inhibitor-1. *J Biol Chem* 2001;**276**:8135–41.
162. Wu J, Peng L, McMahon GA, Lawrence DA, Fay WP. Recombinant plasminogen activator inhibitor-1 inhibits intimal hyperplasia. *Arterioscler Thromb Vasc Biol* 2009;**29**:1565–70.
163. Huang Y, Border WA, Lawrence DA, Noble NA. Mechanisms underlying the antifibrotic properties of noninhibitory PAI-1 (PAI-1R) in experimental nephritis. *Am J Physiol Renal Physiol* 2009;**297**:F1045–54.
164. Huntington J, Fan B, Karlsson K, Deinum J, Lawrence D, Gettins P. Serpin conformational change in ovalbumin. Enhanced reactive center loop insertion through Hinge Region mutations. *Biochemistry* 1997;**36**:5432–40.
165. Yamasaki M, Arii Y, Mikami B, Hirose M. Loop-inserted and thermostabilized structure of P1-P1′ cleaved ovalbumin mutant R339T. *J Mol Biol* 2002;**315**:113–20.
166. Onda M, Nakatani K, Takehara S, Nishiyama M, Takahashi N, Hirose M. Cleaved serpin refolds into the relaxed state via a stressed conformer. *J Biol Chem* 2008;**283**:17568–78.
167. McCarthy BJ, Worrall DM. Analysis of serpin inhibitory function by mutagenesis of ovalbumin and generation of chimeric ovalbumin/PAI-2 fusion proteins. *J Mol Biol* 1997;**267**:561–9.

Multiple Roles of Metalloproteinases in Neurological Disorders

Yi Yang, Jeff W. Hill, and
Gary A. Rosenberg

Department of Neurology, University of New Mexico Health Sciences Center, Albuquerque, New Mexico, USA

I. Introduction 242
II. Biology of Matrix Metalloproteinases 243
III. MMPs in Neuroinflammation 247
IV. Tissue Inhibitors of Metalloproteinases and Apoptosis 248
V. The Neurovascular Unit, Tight Junction Proteins, and MMPs 250
VI. MMPs and Cell Death 251
VII. Neurological Disorders 253
A. Multiple Sclerosis 253
B. Bacterial Meningitis 254
C. Hypoxia/Ischemia 254
D. Vascular Cognitive Impairment and MMP-Mediated Inflammation 255
VIII. Conclusions 257
References 259

Once thought to mainly act in brain to remodel the extracellular matrix, the family of metalloproteinases is important in many normal and pathological processes in the nervous system. Matrix metalloproteinases (MMPs) and A disintegrin and metalloproteinases (ADAMs) are the two major families of metalloproteinases in the brain. MMPs are comprised of several related enzymes that act on extracellular molecules. Normally, they are important in angiogenesis and neurogenesis in development. In neuroinflammatory illnesses, they disrupt the basal lamina and tight junction proteins to open the blood–brain barrier (BBB). ADAMs are important in neuroinflammation through activation of tumor necrosis factor-α (TNF-α) and their action as secretases that modulate the action of receptors on the cell surface. Four tissue inhibitors of metalloproteinases (TIMPs) are the main inhibitors of the MMPs and ADAMs. Recently, MMPs were found to affect DNA repair processes by an unexpected intranuclear action. MMPs and ADAMs have been implicated in the pathophysiology of neurodegenerative diseases such as Alzheimer's disease and vascular cognitive impairment. Growing literature on the functions of MMPs and ADAMs in the central nervous system is opening up new and exciting areas of research that may lead to novel approaches to treatment of neurological diseases.

Progress in Molecular Biology
and Translational Science, Vol. 99
DOI: 10.1016/S1877-1173(11)99006-1

I. Introduction

Regulation of the extracellular matrix (ECM) by proteases and protease inhibitors is a fundamental biological process for normal growth, development, and repair, but can lead to cell death under pathological conditions. Several classes of enzymes participate in cellular processes in the extracellular space, cytoplasm, and nucleus. These are the neutral and acid proteases that act on a wide variety of cellular components. Neutral proteases, which are active at normal cellular pH, are important in both normal and pathological cellular functions. The break down of proteins in the ECM occurs normally during the process of growing new blood vessels and cells. When an injury signal is released by infection, ischemia, or trauma, disruption of the ECM begins to remove the damaged proteins. Similarly, enzymes are essential in the processes that remove intracellular debris in the cytoplasm and nucleus.

The family of metalloproteinases, including the matrix metalloproteinases (MMPs) and A disintegrin and metalloproteinases (ADAMs), act on all components of the ECM and have recently been shown to have a role in the nucleus. Serine proteases, including the two plasminogen activators, tissue plasminogen activator (tPA) and urokinase plasminogen activator (uPA), act in a wide variety of circumstances; they are involved in clotting mechanisms and activation of other enzymes during angiogenesis and neurogenesis. In addition to the enzymes, there are inhibitors of both classes that include the tissue inhibitors of metalloproteinases (TIMPs) and plasminogen activator inhibitors (PAIs).

Another class of neutral proteases is the cysteine proteases that include the caspases and the cathepsins, which are involved in cell death. Normally, the highly integrated action of the MMPs and PAs remodels components of the matrix and performs essential functions at the cell surface involved in signaling, cell survival, and cell death. Neuroinflammation involves a large number of molecular events from the activation of genes through the production of enzymes that act as the common final pathway in many types of injury. During the inflammatory response induced in infection, autoimmune reactions, and hypoxia/ischemia, MMPs and PAs act in unison to break down the ECM, open the blood–brain barrier (BBB), cause brain edema and hemorrhage, and to prevent normal cell signaling and metabolism, eventually leading to cell death.

This selective review will focus mainly on the role of the MMP gene family and will include the PAs in their contributory role in the neuroinflammatory response. Animal studies suggest that agents that block the action of the MMPs may have therapeutic value in the major illnesses that have been associated with neuroinflammation. Several recent reviews have been published detailing the basic biology and the role of MMPs and TIMPs in the central nervous system.[1–4]

II. Biology of Matrix Metalloproteinases

MMPs are a family of zinc-dependent endopeptidases which cleave components of the ECM, including collagen, gelatins, laminin, casein, and others.[5] The MMP family of proteins shares common structural features: an N-terminal signal sequence; autoinhibitory propeptide; and catalytic, hemopexin-like, and transmembrane domains (Fig. 1). The N-terminal signal sequence is responsible for subcellular targeting and extracellular export of the enzyme. The autoinhibitory propeptide domain contains a highly conserved PRCGVPD sequence which assists in binding of the cysteine thiol to an active site Zn^{2+}, thereby blocking interaction of substrates with the catalytic site.[6] Although some MMPs and ADAMs are constitutively expressed, the proteins remain latent until activated by free radicals or other proteolytic enzymes that disrupt the propeptide cysteine-active site zinc interaction or cleave off the propeptide region and make the active site available for cleavage of substrates. While early studies emphasized the role of the MMPs in the extracellular space where they are secreted, accumulating evidence suggests that they localize and act at sites within cells, including the nucleus.[7,8]

MMPs are divided into major subgroups based on which components of the ECM are degraded: collagenases, gelatinases, stromelysins, matrilysin, metalloelastases, and membrane-type MMPs (MT-MMPs). Collagenases

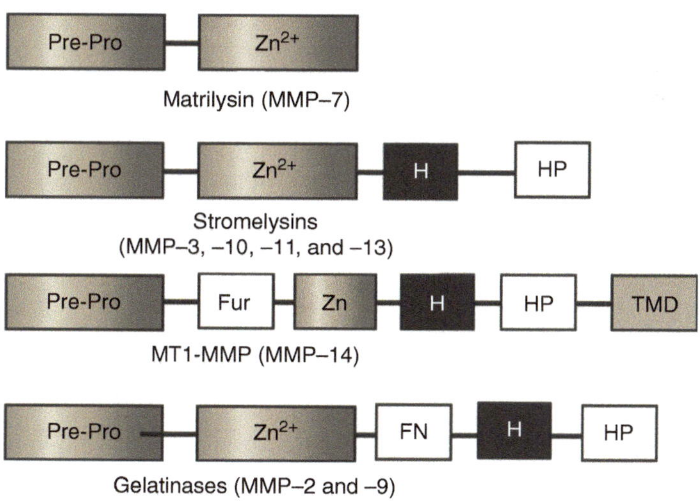

FIG. 1. Matrix metalloproteinases (MMP) with major roles in the central nervous system. Matrilysin (MMP-7) is the smallest MMP with a pre-pro region and a zinc-active site. Stromelysins have an additional hinge region (H) and hemopexin (HP). MT1-MMP adds a transmembrane region and the gelatinases have a furin-binding site (FN).

(MMP-1, -8, and -13) degrade fibrillar collagens type I, II, and III, which are major components of artery walls, skin, bone, and cartilage. Gelatinases, the major enzymes responsible for the degradation of type IV collagen and gelatins, proteolyze molecules in basement membranes, facilitate angiogenesis and neurogenesis, and contribute to cell death. Gelatinase A (MMP-2) and gelatinase B (MMP-9) have been studied extensively due to their elevation in ischemic brain and roles in neuronal remodeling and cell death (Table I).[9] Both MMP-2 and -9 are able to proteolytically activate proMMPs; MMP-2 activates proMMP-1, -2, and -13, while MMP-9 activates proMMP-2, -9, and -13. Stromelysins (MMP-3, -10, and -11) degrade multiple components of basement membranes and stroma and are distinguished from collagenases by their inability to cleave type I collagens. MMP-3 and -10 activate proMMP-1, -7, -8, and -9. Matrilysins (MMP-7 and -26) are distinct among MMPs due to their small size and lack of a hemopexin domain. In addition to common MMP substrates, MMP-7 also degrades proteoglycans and participates in the activation of other MMPs, including proMMP-1, -2, and -9. Metalloelastase (MMP-12) is expressed primarily by macrophages. Macrophages of MMP-12-deficient mice have reduced capacity to degrade ECM components and are unable to penetrate basal membranes *in vivo* and *in vitro*. MMP-12 therefore appears to be essential for macrophage ECM proteolysis and tissue invasion.[10] MT-MMPs

TABLE I
MAJOR METALLOPROTEINASES IN THE CENTRAL NERVOUS SYSTEM AND THEIR ENDOGENOUS INHIBITORS

Name	Function	Inhibitor
Gelatinases (MMP-2/-9)	Disruption of the BBB, angiogenesis, neurogenesis, remodeling of the basal lamina, regeneration of axons, remyelination, apoptosis	All TIMPs
Stromelysins (MMP-3/-10) and matrilysin (MMP-7)	Proteolysis of proteins in the ECM, disruption of the BBB, angiogenesis, synaptic remodeling, glutamate receptor proteolysis, apoptosis	All TIMPs
Metalloelastase (MMP-12)	Proteolyis of the ECM and basal lamina	All TIMPs
Membrane-type MMP (MT1-MMP or MMP-14)	Forms trimolecular complex with TIMP-2 and proMMP-2 for activation of MMP-2 at the cell surface	TIMP-3
ADAM10	α-Secretase in amyloid precursor protein proteolysis, degrades NOTCH protein, acts as sheddase at the cell surface for growth factors, integrins, etc.	TIMP-1 and TIMP-3
ADAM17 (TACE)	α-Secretase, forms 17 kDa TNF-α from 28 kDa form, sheddase for TNFRs at the cell surface	TIMP-3

MMP, matrix metalloproteinases; ADAM, A disintegrin and metalloproteinase; BBB, blood–brain barrier; TIMP, tissue inhibitor of metalloproteinases; ECM, extracellular matrix; TNF, tumor necrosis factor; TACE, TNF-α converting enzyme; TNFRs, TNF receptors.

contain a unique C-terminal transmembrane domain responsible for their cell-surface localization where they act as sheddases and perform important functions, including the activation of other proteases and processing of growth factors and their receptors. Stromelysins and MT-MMPs contain a furin-like cleavage site between the propeptide and catalytic domain, and can be activated intracellularly in the Golgi network by the proconvertase furin and at the cell surface by the serine protease plasmin present in the blood.[11]

Constitutive expression of MMP-2 provides an ongoing, well-controlled remodeling of the ECM. MMP-2 remains in the pro or latent form until activated by a molecular cascade that involves a trimolecular complex made up of MMP-2, TIMP-2, and MT-MMP. This reaction occurs close to the cell surface where it provides local proteolysis without involvement of the surrounding tissues.[12] The early events in the molecular cascade of hypoxia/ischemia probably involve the induction of furin by hypoxia inducible factor-1α (HIF-1α).[13] Furin is an activator of MT-MMP which is required for the activation of MMP-2. Since MMP-2 is constitutively present in large quantities in the normal brain and is found in astrocytes and cerebrospinal fluid (CSF), the rate-limiting step is activation, making MT-MMP critical in the process.

The major inducible MMPs that have been identified in the neuroinflammatory response are MMP-2, -3, and -9. In the brain, a number of cell types express MMP-9, including endothelial cells (ECs), neurons, reactive astrocytes, and microglia. Studies to identify the factors involved in induction of MMPs in inflammation have used lipopolysaccharide (LPS), a potent stimulus for MMP-3 and -9, in cultures of brain astrocytes and microglia.[14] Intracerebral injection of LPS disrupts the BBB through the action of MMP-9.[15] Additionally, intracerebral injection of LPS significantly elevates mRNA levels of MMP-2 and -3, and tumor necrosis factor-α (TNF-α).[16] Neutrophils contain vesicles filled with active MMP-9. When they arrive at a site of inflammation, the MMP-9 can be released to act. Synthesis of protein is not necessary in a neutrophil-mediated process. LPS injection into the brain results in an increase in MMP-9 protein, but not in MMP-9 mRNA, suggesting that the source of MMP-9 is most likely invading neutrophils, which had prepackaged the MMP-9. This possibility is supported by experimental data indicating that brain vascular ECs and infiltrating neutrophils are major sources of MMP-9 after focal cerebral ischemia.[17] C-reactive protein, a marker of inflammation and predictor of vascular events, upregulates MMP-2 mRNA expression, protein synthesis, and secretion by human vascular smooth muscle cells.[18]

Free radicals are involved in BBB opening and cell death, and cyclooxygenase (COX) inhibition limits BBB disruption following ischemic stroke and bacterial meningitis. Latent MMP-9 is activated by free radicals, nitric oxide synthase-dependent *S*-nitrosylation, and proteolysis by MMP-3.[3,19] Direct injection of TNF-α into the brain results in expression of MMP-9.[20] In a mouse model, LPS intracerebral injection caused induction of MMP-3 and -9 with disruption of the BBB. Induced MMP-3 colocalized primarily with macrophages and microglia cells. In MMP-3 knockout mice, significantly reduced BBB opening and neutrophil infiltration was observed.[21] In addition to MMP-3 and MMP-9, intracerebral injection of TNF-α induced a significant upregulation of COX-2; TNF-α significantly depleted glutathione as compared to saline.[22] Indomethacin, an inhibitor of COX-1 and -2, reduced BBB damage at 24 h and significantly attenuated MMP-9 and -3 expression and activation, preventing the loss of endogenous radical scavenging capacity and indicating that MMP-mediated BBB disruption during neuroinflammation can be significantly reduced by administration of COX inhibitors.

Plasminogen activators and plasmin activate the MMPs (Fig. 2). When tPA is given for the treatment of acute stroke, it converts plasminogen to plasmin and initiates the activation of the MMPs. Exogenous tPA can cross both intact and damaged BBB where it can activate proteolysis by MMPs, thereby enhancing the hemorrhagic conversion of the tissues.[23] Proteases are activators of MMPs and this becomes important in the proteolytic disruption of the BBB where the combination of release of MMP-3 by pericytes and MMP-9 by microglia and ECs results in the activation of MMP-9 by MMP-3 with amplification of damage to the BBB.

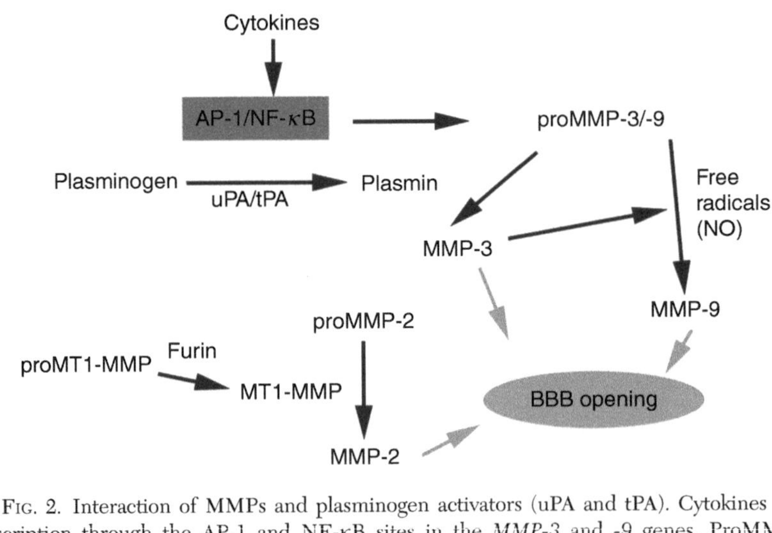

FIG. 2. Interaction of MMPs and plasminogen activators (uPA and tPA). Cytokines induce transcription through the AP-1 and NF-κB sites in the *MMP-3* and *-9* genes. ProMMP-3 is activated by plasmin. ProMMP-9 is activated by MMP-3 and free radicals. The constitutive enzyme, proMT1-MMP, is activated by the convertase Furin and in turn activates proMMP-2. MMPs degrade the basal lamina and tight junctions of endothelial cells, thereby opening the blood–brain barrier (BBB).

III. MMPs in Neuroinflammation

Inflammation occurs in response to injury and initiates pathological responses that potentiate neural injury. The role of MMPs in neuroinflammation has been reported in several neuronal disease models. As part of the neuroinflammatory response in brain hypoxia–ischemia, multiple sclerosis (MS), and CNS infections, MMPs mediate increased permeability of the BBB. In ischemia, the level of MMPs is increased by cytokines and oxidative stress, resulting in degradation of basal lamina surrounding the BBB.[1] MMP-9 is found to be upregulated in the CSF of the patients with MS, an immune disorder characterized by demyelination and axonal loss. Inhibition of MMP expression improved the symptoms of MS.[24] Furthermore, MMP-9 and -12 have been reported to be upregulated after spinal cord injury and contribute to disruption of the blood–spinal cord barrier.[25,26]

Proteolysis is tightly regulated to prevent tissue damage. MMP-2 is constitutively expressed and found normally in the brain and the CSF. The *MMP-2* gene promoter region contains activator protein-2 (AP-2), SP-1, and PEA3-binding sites.[27] MMP-2 is activated at the cell surface by membrane-bound MT1-MMP (MMP-14). By binding the complex to regions close to the membrane, MMP-14 constrains the action of MMP-2.[28] Since MMP-2 is a constitutive enzyme, this constraint controls the extent of damage done to the ECM, whereas MMP-3 and -9, which are secreted into the extracellular space where they can move around freely, cause more extensive damage to the injury site. Several studies suggest that in addition to the constitutive role of MMP-2, it may be upregulated in injury. Hypoxia in cardiac and brain cells induced transcription of MMP-2.[8,29]

MMP gene expression is regulated through the interaction of transcription factors, coactivators, and corepressor proteins with specific elements in the promoter region of MMP genes. A series of inducible genes, including *MMP-3* and -9, are normally expressed at low levels, but respond robustly to inflammatory stimuli. The promoter regions of the inducible MMPs generally contain binding sites for transcription factors, such as activator protein-1 (AP-1) and nuclear factor-κB (NF-κB), which are responsive to oncogenes and cytokines. In addition, the *MMP-14* gene has an NF-κB binding site, suggesting that it can be induced during inflammation. The AP-1 sites bind the transcription factors c-jun and c-fos. Cytokines, such as TNF-α and interleukin-1β (IL-1β), induce MMP-3 and -9 at the transcriptional level, which is important in the neuroinflammatory response under acute and chronic conditions.

Secreted as latent enzymes, MMPs require activation. In the case of MMP-9, several activation mechanisms have been proposed, including other proteases, such as MMP-3, and free radicals, including nitric oxide, which acts through *N*-nitrosylation.[30,31] Activation of MMPs secreted by neutrophils was reported to depend largely on oxidative modification of a cysteine residue in the propeptide domain.[32] Another study revealed that activation of MMP-7 by the proinflammatory oxidant HOCl involves irreversible oxidation of the cysteine residue.[33]

MMPs are regulated upstream by several mechanisms, including oxidative stress, mitogen-activated protein kinase pathways, and nitric oxide. Reactive oxygen species (ROS) stimulate inflammatory cytokine production and protease secretion by microglia, leukocytes, and resident cells of the neurovascular unit.[34] Animal studies have demonstrated a relationship between the generation of ROS and MMP induction in brain injury.[35]

Mechanisms other than proteolysis activate MMPs. During oxidative stress, ROS and peroxynitrite ($ONOO^-$) can activate MMPs via *S*-glutathiolation or *S*-nitrosylation of critical cysteines.[31] $ONOO^-$ causes the *S*-glutathiolation of a critical cysteine residue in the propeptide domains of MMP-2 and -9.[6,32] This activation results in changes of the conformation of the propeptide domain that allows access of the substrate to the catalytic domain.

IV. Tissue Inhibitors of Metalloproteinases and Apoptosis

Activities of metalloproteinases are regulated by transcription, proenzyme activation, and endogenous inhibitors. There are four TIMPs.[36] The TIMPs are relatively small proteins with molecular weights between 21 and 28 kDa. All human TIMPs are composed of N- and C-terminal domains and share significant sequence identity (Table II). A highly conserved region in the N-terminal

TABLE II
NOMENCLATURE, MOLECULAR WEIGHTS, FUNCTIONS, AND LOCATION OF TISSUE INHIBITORS OF METALLOPROTEINASES

Name	Mol. wt. (kDa)	MMPs inhibited	Other functions	Location
TIMP-1	28	All MMPs, ADAM10	Strong inhibitor of MMP-9	Secreted
TIMP-2	21	All MMPs	Forms trimolecular complex with pro-MMP-2 and MT1-MMP to activate MMP-2	Secreted
TIMP-3	24/27	All MMPs, ADAM10, ADAM17	Apoptosis inhibits angiogenesis	Bound to extracellular matrix
TIMP-4	22	All MMPs,	Inhibits angiogenesis	Secreted

Modified from Ref. 37.

TIMP domain interacts with the metalloproteinase catalytic site and is largely responsible for MMP inhibition. The C-terminal TIMP domain makes stabilizing interactions with MMPs that further enhance TIMP inhibition. Although individual TIMPs have preferences for one or another of the MMPs, they can inhibit all of the MMPs. TIMP-2 inhibits MMP-2, but may also participate in the activation of proMMP-2. TIMP-1 inhibits MMP-9, and TIMP-3 acts on MMPs and TACE. TIMP-3 is unique in that it binds to the ECM after secretion.[38] The main substrates inhibited by TIMP-3 (MT1-MMP, MMP-3, and TACE) play central roles in several important processes involved in growth, cell death, and tissue repair.[1] By reducing the activation of MT-MMP, TIMP-3 in turn affects the activation of MMP-2.

In response to injury, TIMP-1 is increased and inhibits MMP-9. TIMP-3 is expressed early in ischemia and contributes to apoptosis of neurons in the middle cerebral artery suture occlusion model.[39,40] TIMP-3 mRNA was found to be overexpressed in developing brain and after injury in rats.[41,42] LPS stimulation of cultured astrocytes leads to the formation of TIMP-3, which inhibits activation of MMP-2 in neurons and may protect the BBB.[43] Although a fourth TIMP has been identified, its function in the brain is not known.

After injuries, molecular cascades are initiated in neurological diseases with the purpose of removing damaged cells and preparing the tissues for repair.[44] This dual role necessitates a full understanding of the timing of the phases involved in injury and repair. Early after injury, constitutive enzymes are activated and begin the process of disassembling the ECM, opening the BBB, and initiating cell death by apoptosis.[45–47] The second stage of injury involves MMPs in the processes of angiogenesis and neurogenesis.[48–50] In this second phase, treatment may interfere with repair. Remodeling of the ECM characterizes the third phase when gliosis results in impenetrable scar tissues that block the regrowth of axons. The action of the MMPs on the basal lamina and tight junction proteins (TJPs) of ECs is the final common pathway for opening of the BBB, which allows cells to enter the central nervous system and attack invading organisms. This step is probably protective in preantibiotic infections, but in noninfectious diseases, it contributes to tissue damage by injury to the matrix around neurons. When matrix proteins around neurons are degraded, there is loss of contact and cell death by anoikis.[31] Treatment with a selective inhibitor of MMP-2 and -9 blocks neuronal cell death induced by ECM degradation.[51]

Pharmaceutical companies have identified MMP inhibitors for the treatment of cancer.[27] Several agents entered clinical trials.[52] The results were disappointing because long-term use of these agents resulted in overgrowth of ECM in joints, thereby limiting their use. Subsequently, a shift has occurred with the realization that short-term use of MMP inhibitors may be possible in neurological disorders, particularly for treatment of cerebrovascular and cardiovascular diseases.[53]

V. The Neurovascular Unit, Tight Junction Proteins, and MMPs

The structural and functional integrity of the brain depends on the delicate balance between substrate delivery through blood flow and energy demands imposed by neural activity. Complex cerebrovascular control mechanisms ensure that active brain regions receive an adequate amount of blood. Brain capillaries, which are the primary interface between the blood and the brain tissues, play a major role in controlling the neuronal microenvironment. ECs, interendothelial tight junctions, the basal lamina, and perivascular astrocytes are jointly referred to as the BBB or, more recently, as the neurovascular unit. Around the cerebral ECs is a basal lamina composed primarily of laminin, fibronectin, and heparan sulfate. The basal lamina provides a structural barrier to extravasation of cellular blood elements and anchors ECs and astrocytes. Barriers limiting transport into and out of the brain are found at the TJP and at the basal lamina.

Neurons and astrocytes, which are in close proximity, are functionally coupled to smooth muscle cells and ECs. Recent advances in cerebrovascular neurobiology have revealed that activity-induced hemodynamic responses require complex signaling mechanisms that involve not only ECs but also neurons, astrocytes, pericytes, and the ECM. These cells, which constitute a functional unit, act together to generate, coordinate, and transduce the molecular signals that underlie the changes in cerebral blood flow.

Cerebrovascular endothelial tight junctions restrict molecules from moving between the blood and the brain. TJPs join together with ECs to form an interface between the blood and the brain.[54] Transmembrane TJPs consist of three integral proteins: claudins, occludin, and junctional adhesion molecules (JAM).[55] Zona occludens (ZO)-1 and -3, and cingulin are considered to be cytoplasmic tight junction accessory proteins which connect tight junctions to the actin cytoskeleton. The extracellular loops of occludin, claudins, and JAM originating from neighboring cells form the paracellular barrier of the TJ, which selectively prevents most blood-borne substances from entering the brain. In rodents and adult human brains, claudin-1, claudin-5, and occludin have been found to be present in brain endothelial tight junctions forming the BBB. TJ are dynamic structures. Proteins of TJ are subject to changes in expression, subcellular location, posttranslational modification, and protein–protein interactions under both physiological and pathophysiological conditions. Occludin, claudin-5, and ZO-1, which are the main structural barrier proteins, are considered sensitive indicators of normal and disturbed functional states of the BBB. Taken together, the regulation of TJPs is essential for the maintenance of BBB permeability. Disruption of BBB TJ by disease or drugs can lead to impaired BBB function and thus compromise the CNS.

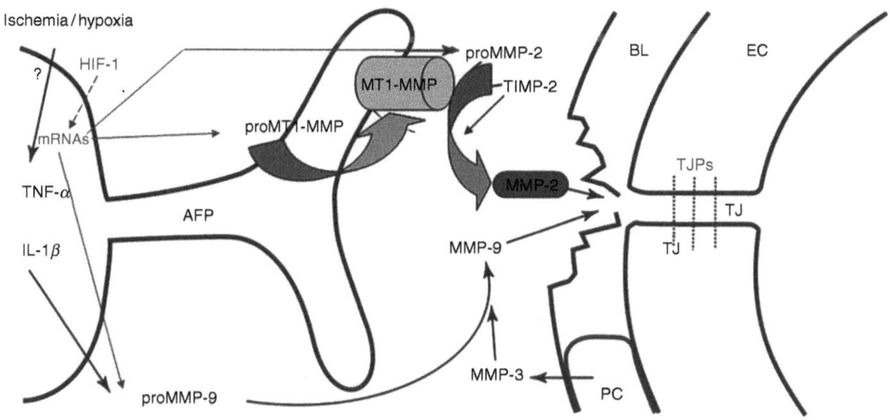

FIG. 3. Schematic drawing showing that activation of MMP-2 occurs through the action of the trimolecular complex during the early opening of the BBB in a 3 h reperfusion after 90 min MCAO. In astrocytic foot processes (AFP), MT1-MMP joins with TIMP-2 to activate proMMP-2 in a spatially constrained manner close to the basal lamina (BL). Pericytes (PC) are in the BL. The ECs have tight junctions (TJs). Activated MMP-2 has direct access to the portion of the BL beneath the AFP, and components of the BL are degraded. The manner in which this disruption of the BL leads to increased permeability is unclear since the role of the BL in maintaining the integrity of the blood vessel is uncertain. Secondary neuroinflammation with the formation of cytokines (TNF and IL-1) and induction of MMP-9 and -3 occurs. ProMMP-9 is activated by MMP-3 and degrades the TJs and BL. Modified from Ref. 91.

MMPs are found in all of the elements of the neurovascular unit, but different MMPs have a predilection for certain cell types. ECs produce mainly MMP-9, pericytes express MMP-3 and -9, while astrocytes express MMP-2 and MT-MMP in the endfeet that surround the ECs (Fig. 3). This pattern of MMPs facilitates the opening of the BBB in inflammation, but also allows for the gradual changes in the ECM that are most likely ongoing and involve the action of the MMP-2/MT-MMP complex in remodeling the matrix to prevent excessive buildup.

VI. MMPs and Cell Death

In addition to their role in the proteolysis of extracellular protein targets, recent studies have revealed that MMPs also localize to various intracellular sites. It has been recognized that MMPs rapidly act on intracellular substrates in response to brain and heart injury during hypoxia. Recent studies have demonstrated increased activity of MMPs including MMP-2, -9, and -13, in neuronal nuclei of brains at an early stage of ischemic stroke and reperfusion injury.[8,56,57] The nucleus has a matrix that resembles the ECM and provides structural and organizational support for various nuclear processes. Biological processes such as apoptosis, cell cycle regulation, and nuclear matrix degradation involve proteolytic processing of nuclear proteins. The intranuclear gelatinase activity in ischemic neurons suggests a possible role for gelatinases in nuclear matrix proteolysis. Our most recent study showed that intranuclear MMP-2 and -9 activity facilitates oxidative injury in neurons during early ischemic insult by cleaving nuclear proteins poly-ADP-ribose polymerase 1 (PARP-1) and X-ray cross-complementary factor 1 (XRCC1).[8] Both PARP-1 and XRCC1 are DNA repair enzymes that are important in DNA fragmentation and cell death following the induction of DNA damage. DNA base excision repair (BER) machinery is the main mechanism in mammalian neuronal nuclei to repair various types of oxidative DNA damage. DNA strand breaks induce PARP-1 activity, which triggers DNA repair by binding to oxidatively damaged DNA and recruiting BER factors, including XRCC1. XRCC1 plays a central role in DNA repair by interacting with major DNA repair enzymes in the BER pathway.[58] Overactivation of PARP-1 caused by massive DNA damage may result in cell necrosis by ATP depletion, and DNA fragmentation and cell apoptosis induced by translocation of apoptosis-inducing factor. However, PARP-1 activation may play a neuroprotective role at an early stage during ischemia/reperfusion injury when tissue has not yet become excessively damaged and PARP-1 has not yet become overactivated. The cleavage of DNA repair proteins by MMP activity in neuronal nuclei interferes with oxidative DNA damage repair, which could contribute to neuronal apoptosis in ischemic injuries (Fig. 4). Mice overexpressing superoxide dismutase (SOD) are protected from ischemic injury and have reduced MMP production.[59] Activation of intracellular MMPs could be one of the earliest pathological events triggered downstream of oxidative stress. MMP inhibition may prove to be a novel therapeutic strategy to prevent neuronal apoptotic injury in the brain.

Microglial activation and inflammation are associated with progressive neuronal apoptosis in neurodegenerative human brain disorders. Dopaminergic neurons that are under cellular stress and undergoing apoptosis release active MMP-3, which in turn triggers microglial activation and production of proinflammatory biomolecules. This action indicates that MMP-3 signaling may trigger inflammatory reactions by activating microglia as well as exacerbation of neuronal apoptosis that leads to rapid phagocytosis of apoptotic neurons.[7] MMP-3 may diminish the death-promoting activity of death receptors by shedding membrane-bound Fas-ligand to soluble form. *In vitro* studies demonstrated a neuroprotective action of MMP-3 against Fas-mediated neuronal apoptotic death during oxygen–glucose deprivation (OGD).[40]

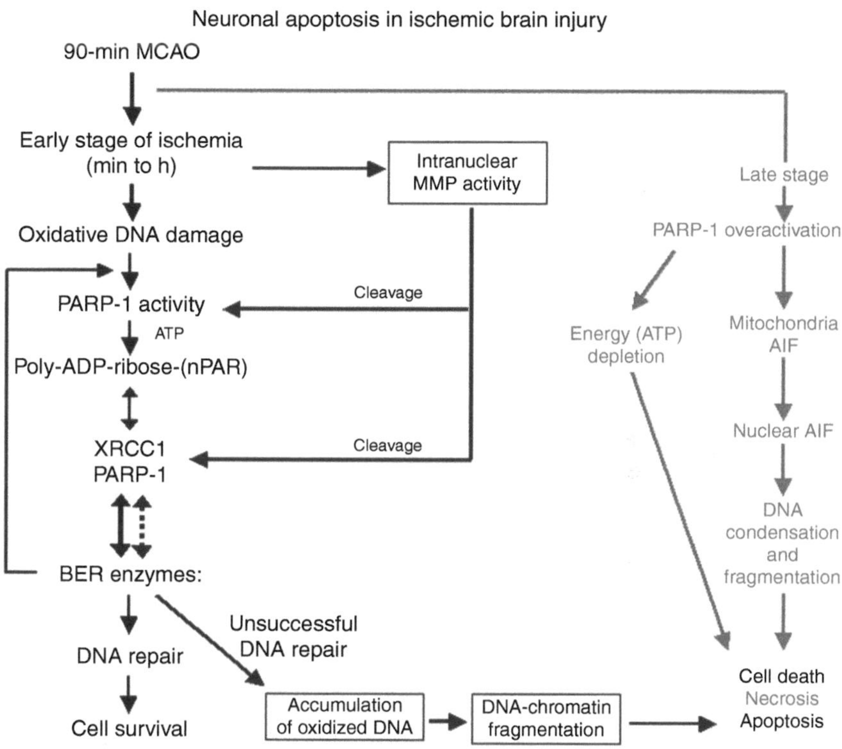

FIG. 4. Schematic drawing of the proposed mechanism for intranuclear MMPs to facilitate oxidative DNA damage in neurons after an ischemic insult. Cerebral ischemia/reperfusion triggers oxidative DNA damage, which leads to PARP-1 activation and initiation of the DNA BER (base excision repair) pathway for repair of oxidized DNA. When repair is successful, the neuron may be protected from apoptosis. However, excessive nuclear gelatinase (MMP-2/-9) proteolysis in neurons impairs the protective action of the DNA BER pathway by cleaving PARP-1 and XRCC1, leading to neuronal DNA fragmentation and apoptosis. Modified from Ref. 8.

VII. Neurological Disorders

A. Multiple Sclerosis

MMPs are important in MS where they disrupt the BBB and contribute to demyelination. CSF of MS patients in an acute exacerbation has elevated levels of MMP-9.[60] *In vitro* studies demonstrate that MMPs attack myelin and break myelin into myelin basic protein fragments, but the role of MMPs *in vivo* is less clear.[61] High-dose steroids, which are used to treat acute MS exacerbations, dramatically reduce MMP-9 in the CSF.[24] Steroids combine with molecules in the cytosol before moving into the nucleus to block the AP-1 site in the MMP-9 gene promoter region. The MMP-9 detected in CSF in MS comes from endogenous production by brain cells; this was demonstrated in studies that indexed MMP-9 with levels of albumin, which is passively transported across the BBB. A greater increase in MMP-9 in the CSF compartment occurred than expected from the movement of albumin across the BBB.[62]

Treatment of animals with experimental allergic encephalomyelitis with MMP inhibitors reduces the severity of the illness.[63] A tetracycline derivative, minocycline, was found to improve symptoms in MS patients in a small study.[64,65] Minocycline suppresses production of MMP-9, but has other effects on the inflammatory response, including reduction in the infiltration of microglia. Further studies with larger numbers of patients and controls are needed to corroborate these results.

B. Bacterial Meningitis

Acute inflammation of the meninges triggers the release of MMPs, which can be detected in the CSF, opening the BBB at the brain surface. Elevated levels of MMPs are found in viral, bacterial, and fungal meningitis.[66] MMPs and ADAM-17, also known as TNF-α converting enzyme (TACE), contribute synergistically to the pathophysiology of bacterial meningitis; TACE proteolytically releases several cell-surface proteins, including the proinflammatory cytokine TNF-α and its receptors, which in turn stimulate cells to produce active MMPs, facilitating leukocyte extravasation and brain edema by degradation of ECM components.[67] Treatment with BB-1101, a hydroxamate inhibitor of MMP and TACE, downregulated the CSF concentration of TNF-α and decreased the incidences of seizures and mortality. Several types of meningitis cause an increase of MMPs in the CSF, including lyme disease, viral infections, and tuberculosis.[66] A water-soluble MMP inhibitor, TNF484, with actions against both MMPs and TACE, was shown to be effective in an experimental model of bacterial meningitis.[68]

C. Hypoxia/Ischemia

A complex pattern of MMP expression in hypoxic/ischemic injury is emerging from many laboratories.[45,69–71] In the reperfusion model, there is an early increase in MMP-2, which is transient, but results in the early reversible opening of the BBB. Elevation of MMP-2 in the early stages of the injury has been observed in rodents and nonhuman primates.[47,72] TJPs are degraded by MMP-2, but remain within the vessels after 3 h of reperfusion. However, by 24 h, the TJPs are no longer seen in the vessels.[47] Following the initial opening of the BBB, there is a second opening between 24 and 48 h, depending on the time of occlusion. During this phase, there is marked increase in MMP-9, which leads to more extensive damage to the blood vessels.

Treatment of stroke with tPA within 3 h of onset improves outcome. However, tPA increases the risk of hemorrhage ~10-fold.[73] MMPs are released during the infarct and aggravate the hemorrhage. Some bleeding after stroke is common in embolic strokes, but less common in thrombotic infarcts. Since treatment with tPA removes thrombi and promotes reperfusion of an infracted region, there is an enhanced danger of bleeding. In a study of the normal history of embolic strokes, magnetic resonance imaging (MRI) showed hemorrhagic transformation in 69% of infarcts, suggesting that hemorrhagic transformation is a regular finding in medium and large cardioembolic infarcts.[74] Plasma levels of MMP-9 correlate with hemorrhagic transformation and intracerebral hemorrhage.[75] Treatment of rats with tPA increases the mortality and increases the opening of the BBB and hemorrhage; when a broad-spectrum MMP inhibitor, BB-94, is given with the tPA, it blocks the opening of the BBB and dramatically reduces hemorrhage.[76,77] Opening of the BBB after tPA leads to increased mortality; when the BBB is closed with an MMPs inhibitor, the death rate is dramatically reduced (Fig. 5). tPA increased the expression and activation of MMP-9. At 12 h, tPA-treated rats showed significantly higher levels of proMMP-9 and cleaved MMP-9 than untreated controls. By 24 h, all rats showed evidence of hemorrhagic transformation in the ischemic territory. Rats treated with BB-94 and tPA showed significantly reduced hemorrhage volumes compared with those that received tPA alone.[78] When the BBB remains intact, the fibrinolytic agent acts on fibrin within the blood vessels; however, if the BBB is compromised, the tPA escapes into the brain and acts on the MMPs. Agents that maintain the integrity of the BBB may therefore extend the therapeutic window for treatment.

D. Vascular Cognitive Impairment and MMP-Mediated Inflammation

Vascular disease is an important cause of dementia and accelerates the course of Alzheimer's disease.[79] Vascular cognitive impairment (VCI) is a heterogeneous disease with involvement of the small and large blood vessels. Binswanger's disease is a disease of the small vessels of the brain that is related to hypertension, diabetes, vasculitis, and other causes of blood vessel damage. Clinically, patients with Binswanger's disease and white matter hyperintensities (WMH) on MRI scans have trouble walking, incontinence, and executive dysfunction on neuropsychological testing.[80] Arteriolosclerosis of the penetrating vessels of the brain leads to hypoperfusion of the deep white matter. Several blood vessel territories converge on the deep white matter, making it a border zone or watershed region that can be made hypoxic when the cerebral blood flow is reduced by damage to the heart or the flow restricted by narrowing of hypertensive vessels. Hypoxia also can cause damage to the

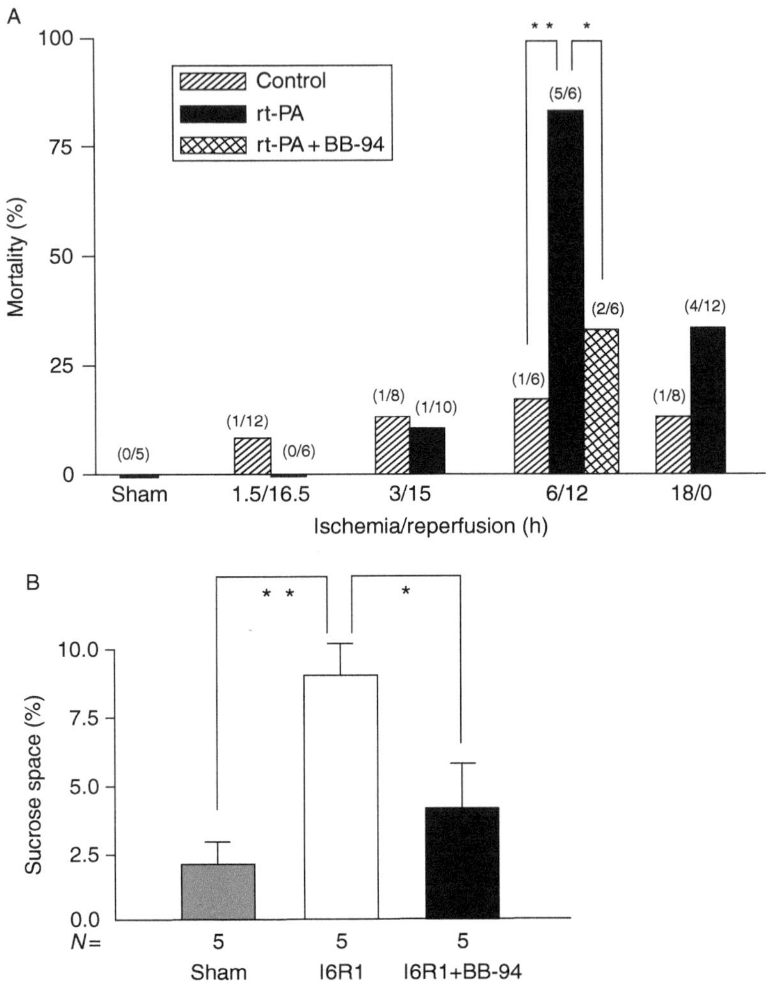

FIG. 5. (A) Effect of rt-PA on mortality in animals with different intervals of ischemia and reperfusion. Mortality is shown in percent with the numbers of animals dying and the number of animals studied shown in parentheses above the bars. When reperfusion was delayed to 6 h, mortality was increased markedly in rt-PA-treated animals (**$P < 0.01$). Treatment with BB94 reduced rt-PA-associated mortality significantly (*$P < 0.05$). Control animals had MCAO without rt-PA treatment. (B) Opening of the BBB as measured by sucrose space in rats with 6 h of ischemia and 1 h of reperfusion (I6R1). Compared with sham-operated animals, BBB permeability was markedly increased in untreated rats (**$P < 0.01$). BB94 given 2 and 5 h after MCAO markedly decreased BBB opening (*$P < 0.01$). Adapted from Ref. 77.

deep white matter. Either hypoperfusion of blood or hypoxia results in an elevation of HIF-1α. HIF induces the *fur* gene, which encodes a convertase that activates MMP-14 and initiates the activation of MMP-2.

Animal models of VCI based on hypoxic hypoperfusion are induced in rats by permanent bilateral carotid artery occlusion (BCAO),[81] and stenosis is induced by a coil placed around both carotids in the mouse.[82] Rats with BCAO show progressive damage to the white matter of the corpus callosum and the optic tract. White matter hypoperfusion leads to hypoxic injury, which is characterized by astrocytosis and proliferation of microglial cells. Damage to the BBB then occurs and leads to vasogenic edema. There is loss of myelin, which can continue for several months and result in behavioral changes.[83] The reactive cells in rats with BCAO produce MMP-2.[84] Elevated levels of MMPs in the white matter lead to breakdown of the BBB; MMP-2 knockout mice have reduced opening of the BBB and an inhibitor to MMP-2 reduced the injury to the BBB.[82]

MRI of rats with BCAO shows increased apparent diffusion coefficients (ADC) in the regions of the white matter and hippocampus. The lesions are often unilateral and correlate with leakage of Evans blue, as observed by fluorescent microscopy.[85] MMPs are seen in the blood vessels with Evans blue leakage. The conclusions from the animal studies and the human CSF and autopsy studies provide strong evidence implicating MMPs in the pathogenesis of VCI. Whether the MMPs attack the myelin directly or create vasogenic edema with secondary demyelination will require further study.

CSF of patients with VCI contains elevated levels of MMP-9, which is absent in patients with Alzheimer's disease.[86] When corrections are made for the amount of protein entering the CSF from the blood by indexing the MMPs to albumin in the blood and the CSF, MMP-2 appears to be involved in Binswanger's disease (Rosenberg *et al.*, unpublished data). Autopsy tissues from patients with VCI show evidence of vascular damage in the white matter consistent with disruption of the BBB.[87] White matter lesions show increased MMP-2 in astrocytes and MMP-3 in macrophages/microglia.[88] While the connection between the expression of MMPs, vascular disruption, and the loss of myelin remains uncertain, *in vitro* studies show that MMPs can degrade myelin basic protein.[89] Thus, the expression and activation of MMPs in the white matter of VCI patients, possibly related to hypoxic hypoperfusion episodes, may explain both the injury to the BBB and demyelination. A possible scheme to explain the damage secondary to chronic hypoxic hypoperfusion is shown in Fig. 6.

VIII. Conclusions

MMPs have dual roles in brain diseases. Initially, they contribute to the molecular cascade of injury by forming a final common pathway for BBB opening and cell death. Later, however, they participate in angiogenesis and

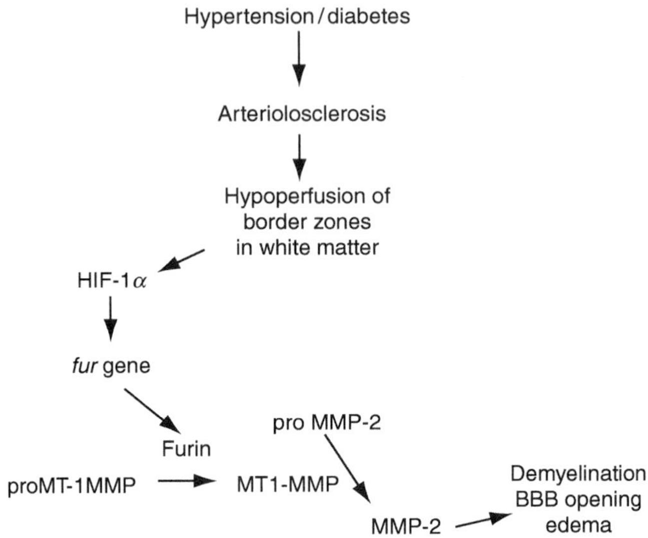

FIG. 6. Possible mechanism for MMP-mediated neuroinflammation in vascular cognitive impairment. Hypertensive blood vessels with fibrosis and arteriolosclerosis reduce perfusion to the deep white matter leading to hypoxia. Induction of hypoxia inducible factor (HIF) leads to activation of MMP-2. Opening of the blood–brain barrier leads to vasogenic edema. MMPs degrade myelin, causing damage to the white matter.

neurogenesis as part of the repair process. This dual role makes the use of MMP inhibitors time dependent. MMP inhibitors have been used in a number of animal studies to block BBB injury, decrease infarct size, and prevent cell death. However, the same inhibitors blocked angiogenesis and neurogenesis. The major effects of the MMP inhibitors have been in the early stages of stroke when closing of the BBB is desirable in order to extend the treatment window for tPA. In some conditions, where inflammation is a greater concern than hypoxia, such as in MS and meningitis, the use of MMP inhibitors may be feasible for 1–2 weeks without incurring unwanted side effects.

While long-term use of MMP inhibitors may lead to side effects, short-term use in stroke, acute MS, and bacterial meningitis seems reasonable to test clinically, since the preclinical studies are promising.[53] Chronic conditions such as VCI will be more challenging because of the need for treatment over months and years. Some agents, such as minocycline and doxycycline, are being tested in preventing the changes in vascular diseases, such as aortic aneurysm, and these agents may be worth testing.[90]

References

1. Cunningham LA, Wetzel M, Rosenberg GA. Multiple roles for MMPs and TIMPs in cerebral ischemia. *Glia* 2005;**50**:329–39.
2. Yong VW. Metalloproteinases: mediators of pathology and regeneration in the CNS. *Nat Rev Neurosci* 2005;**6**:931–44.
3. Jian LK, Rosenberg GA. Matrix metalloproteinases and free radicals in cerebral ischemia. *Free Radic Biol Med* 2005;**39**:71–80.
4. Lo EH, Dalkara T, Moskowitz MA. Mechanisms, challenges and opportunities in stroke. *Nat Rev Neurosci* 2003;**4**:399–415.
5. Page-McCaw A, Ewald AJ, Werb Z. Matrix metalloproteinases and the regulation of tissue remodelling. *Nat Rev Mol Cell Biol* 2007;**8**:221–33.
6. Okamoto T, Akuta T, Tamura F, van Der Vliet A, Akaike T. Molecular mechanism for activation and regulation of matrix metalloproteinases during bacterial infections and respiratory inflammation. *Biol Chem* 2004;**385**:997–1006.
7. Kim YS, Kim SS, Cho JJ, Choi DH, Hwang O, Shin DH, et al. Matrix metalloproteinase-3: a novel signaling proteinase from apoptotic neuronal cells that activates microglia. *J Neurosci* 2005;**25**:3701–11.
8. Yang Y, Candelario-Jalil E, Thompson JF, Cuadrado E, Estrada EY, Rosell A, et al. Increased intranuclear matrix metalloproteinase activity in neurons interferes with oxidative DNA repair in focal cerebral ischemia. *J Neurochem* 2010;**112**:134–49.
9. Jin R, Yang G, Li G. Molecular insights and therapeutic targets for blood-brain barrier disruption in ischemic stroke: critical role of matrix metalloproteinases and tissue-type plasminogen activator. *Neurobiol Dis* 2010;**38**:376–85.
10. Shipley JM, Wesselschmidt RL, Kobayashi DK, Ley TJ, Shapiro SD. Metalloelastase is required for macrophage-mediated proteolysis and matrix invasion in mice. *Proc Natl Acad Sci USA* 1996;**93**:3942–6.
11. Choi DH, Kim EM, Son HJ, Joh TH, Kim YS, Kim D, et al. A novel intracellular role of matrix metalloproteinase-3 during apoptosis of dopaminergic cells. *J Neurochem* 2008;**106**:405–15.
12. Zucker S, Pei D, Cao J, Lopez-Otin C. Membrane type-matrix metalloproteinases (MT-MMP). *Curr Top Dev Biol* 2003;**54**:1–74.
13. McMahon S, Grondin F, McDonald PP, Richard DE, Dubois CM. Hypoxia-enhanced expression of the proprotein convertase furin is mediated by hypoxia-inducible factor-1: impact on the bioactivation of proproteins. *J Biol Chem* 2005;**280**:6561–9.
14. Gottschall PE, Deb S. Regulation of matrix metalloproteinase expressions in astrocytes, microglia and neurons. *Neuroimmunomodulation* 1996;**3**:69–75 (Review).
15. Mun-Bryce S, Rosenberg GA. Gelatinase B modulates selective opening of the blood-brain barrier during inflammation. *Am J Physiol* 1998;**274**:R1203–11.
16. Mun-Bryce S, Lukes A, Wallace J, Lukes-Marx M, Rosenberg GA. Stromelysin-1 and gelatinase A are upregulated before TNF-alpha in LPS-stimulated neuroinflammation. *Brain Res* 2002;**933**:42–9.
17. Rosell A, Foerch C, Murata Y, Lo EH. Mechanisms and markers for hemorrhagic transformation after stroke. *Acta Neurochir Suppl* 2008;**105**:173–8.
18. Doronzo G, Russo I, Mattiello L, Trovati M, Anfossi G. C-reactive protein increases matrix metalloproteinase-2 expression and activity in cultured human vascular smooth muscle cells. *J Lab Clin Med* 2005;**146**:287–98.
19. Hahn-Dantona E, Ramos-DeSimone N, Sipley J, Nagase H, French DL, Quigley JP. Activation of proMMP-9 by a plasmin/MMP-3 cascade in a tumor cell model. Regulation by tissue inhibitors of metalloproteinases. *Ann N Y Acad Sci* 1999;**878**:372–87.

20. Rosenberg GA, Estrada EY, Dencoff JE, Stetler-Stevenson WG. Tumor necrosis factor-alpha-induced gelatinase B causes delayed opening of the blood-brain barrier: an expanded therapeutic window. *Brain Res* 1995;**703**:151–5.
21. Gurney KJ, Estrada EY, Rosenberg GA. Blood-brain barrier disruption by stromelysin-1 facilitates neutrophil infiltration in neuroinflammation. *Neurobiol Dis* 2006;**23**:87–96.
22. Candelario-Jalil E, Taheri S, Yang Y, Sood R, Grossetete M, Estrada EY, et al. Cyclooxygenase inhibition limits blood-brain barrier disruption following intracerebral injection of tumor necrosis factor-alpha in the rat. *J Pharmacol Exp Ther* 2007;**323**:488–98.
23. Adibhatla RM, Hatcher JF. Tissue plasminogen activator (tPA) and matrix metalloproteinases in the pathogenesis of stroke: therapeutic strategies. *CNS Neurol Disord Drug Targets* 2008;**7**:243–53.
24. Rosenberg GA, Dencoff JE, Correa Jr. N, Reiners M, Ford CC. Effect of steroids on CSF matrix metalloproteinases in multiple sclerosis: relation to blood-brain barrier injury. *Neurology* 1996;**46**:1626–32.
25. Noble LJ, Donovan F, Igarashi T, Goussev S, Werb Z. Matrix metalloproteinases limit functional recovery after spinal cord injury by modulation of early vascular events. *J Neurosci* 2002;**22**:7526–35.
26. Wells JE, Hurlbert RJ, Fehlings MG, Yong VW. Neuroprotection by minocycline facilitates significant recovery from spinal cord injury in mice. *Brain* 2003;**126**:1628–37.
27. Overall CM, Lopez-Otin C. Strategies for MMP inhibition in cancer: innovations for the post-trial era. *Nat Rev Cancer* 2002;**2**:657–72.
28. Strongin AY, Collier I, Bannikov G, Marmer BL, Grant GA, Goldberg GI. Mechanism of cell surface activation of 72-kDa type IV collagenase. Isolation of the activated form of the membrane metalloprotease. *J Biol Chem* 1995;**270**:5331–8.
29. Alfonso-Jaume MA, Bergman MR, Mahimkar R, Cheng S, Jin ZQ, Karliner JS, et al. Cardiac ischemia-reperfusion injury induces matrix metalloproteinase-2 expression through the AP-1 components FosB and JunB. *Am J Physiol* 2006;**291**:H1838–46.
30. Nagase H. Activation mechanisms of matrix metalloproteinases. *Biol Chem* 1997;**378**:151–60 (Review).
31. Gu Z, Kaul M, Yan B, Kridel SJ, Cui J, Strongin A, et al. S-nitrosylation of matrix metalloproteinases: signaling pathway to neuronal cell death. *Science* 2002;**297**:1186–90.
32. Ali MA, Schulz R. Activation of MMP-2 as a key event in oxidative stress injury to the heart. *Front Biosci* 2009;**14**:699–716.
33. Fu X, Kassim SY, Parks WC, Heinecke JW. Hypochlorous acid oxygenates the cysteine switch domain of pro-matrilysin (MMP-7). A mechanism for matrix metalloproteinase activation and atherosclerotic plaque rupture by myeloperoxidase. *J Biol Chem* 2001;**276**:41279–87.
34. Mukhopadhyay P, Rajesh M, Batkai S, Kashiwaya Y, Hasko G, Liaudet L, et al. Role of superoxide, nitric oxide, and peroxynitrite in doxorubicin-induced cell death in vivo and in vitro. *Am J Physiol* 2009;**296**:H1466–83.
35. Wakisaka Y, Chu Y, Miller JD, Rosenberg GA, Heistad DD. Spontaneous intracerebral hemorrhage during acute and chronic hypertension in mice. *J Cereb Blood Flow Metab* 2010;**30**:56–69.
36. Brew K, Nagase H. The tissue inhibitors of metalloproteinases (TIMPs): an ancient family with structural and functional diversity. *Biochim Biophys Acta* 2010;**1803**:55–71.
37. Baker AH, Edwards DR, Murphy G. Metalloproteinase inhibitors: biological actions and therapeutic opportunities. *J Cell Sci* 2002;**115**:3719–27.
38. Leco KJ, Khokha R, Pavloff N, Hawkes SP, Edwards DR. Tissue inhibitor of metalloproteinases-3 (TIMP-3) is an extracellular matrix-associated protein with a distinctive pattern of expression in mouse cells and tissues. *J Biol Chem* 1994;**269**:9352–60.

39. Wallace JA, Alexander S, Estrada EY, Hines C, Cunningham LA, Rosenberg GA. Tissue inhibitor of metalloproteinase-3 is associated with neuronal death in reperfusion injury. *J Cereb Blood Flow Metab* 2002;**22**:1303–10.
40. Wetzel M, Li L, Harms KM, Roitbak T, Ventura PB, Rosenberg GA, et al. Tissue inhibitor of metalloproteinases-3 facilitates Fas-mediated neuronal cell death following mild ischemia. *Cell Death Differ* 2008;**15**:143–51.
41. Jaworski DM. Differential regulation of tissue inhibitor of metalloproteinase mRNA expression in response to intracranial injury. *Glia* 2000;**30**:199–208.
42. Jaworski DM, Fager N. Regulation of tissue inhibitor of metalloproteinase-3 (Timp-3) mRNA expression during rat CNS development. *J Neurosci Res* 2000;**61**:396–408.
43. Liu W, Furuichi T, Miyake M, Rosenberg GA, Liu KJ. Differential expression of tissue inhibitor of metalloproteinases-3 in cultured astrocytes and neurons regulates the activation of matrix metalloproteinase-2. *J Neurosci Res* 2007;**85**:829–36.
44. Dirnagl U, Iadecola C, Moskowitz MA. Pathobiology of ischaemic stroke: an integrated view. *Trends Neurosci* 1999;**22**:391–7.
45. Heo JH, Lucero J, Abumiya T, Koziol JA, Copeland BR, del Zoppo GJ. Matrix metalloproteinases increase very early during experimental focal cerebral ischemia. *J Cereb Blood Flow Metab* 1999;**19**:624–33.
46. Rosenberg GA, Estrada EY, Dencoff JE. Matrix metalloproteinases and TIMPs are associated with blood-brain barrier opening after reperfusion in rat brain. *Stroke* 1998;**29**:2189–95.
47. Yang Y, Estrada EY, Thompson JF, Liu W, Rosenberg GA. Matrix metalloproteinase-mediated disruption of tight junction proteins in cerebral vessels is reversed by synthetic matrix metalloproteinase inhibitor in focal ischemia in rat. *J Cereb Blood Flow Metab* 2007;**27**:697–709.
48. Lee SR, Kim HY, Rogowska J, Zhao BQ, Bhide P, Parent JM, et al. Involvement of matrix metalloproteinase in neuroblast cell migration from the subventricular zone after stroke. *J Neurosci* 2006;**26**:3491–5.
49. Zhao BQ, Wang S, Kim HY, Storrie H, Rosen BR, Mooney DJ, et al. Role of matrix metalloproteinases in delayed cortical responses after stroke. *Nat Med* 2006;**12**:441–5.
50. Wang L, Zhang ZG, Zhang RL, Gregg SR, Hozeska-Solgot A, LeTourneau Y, et al. Matrix metalloproteinase 2 (MMP2) and MMP9 secreted by erythropoietin-activated endothelial cells promote neural progenitor cell migration. *J Neurosci* 2006;**26**:5996–6003.
51. Gu Z, Cui J, Brown S, Fridman R, Mobashery S, Strongin AY, et al. A highly specific inhibitor of matrix metalloproteinase-9 rescues laminin from proteolysis and neurons from apoptosis in transient focal cerebral ischemia. *J Neurosci* 2005;**25**:6401–8.
52. Coussens LM, Fingleton B, Matrisian LM. Matrix metalloproteinase inhibitors and cancer: trials and tribulations. *Science* 2002;**295**:2387–92.
53. Hu J, Van den Steen PE, Sang QX, Opdenakker G. Matrix metalloproteinase inhibitors as therapy for inflammatory and vascular diseases. *Nat Rev Drug Discov* 2007;**6**:480–98.
54. Hawkins BT, Davis TP. The blood-brain barrier/neurovascular unit in health and disease. *Pharmacol Rev* 2005;**57**:173–85.
55. Furuse M, Fujita K, Hiiragi T, Fujimoto K, Tsukita S. Claudin-1 and -2: novel integral membrane proteins localizing at tight junctions with no sequence similarity to occludin. *J Cell Biol* 1998;**141**:1539–50.
56. Amantea D, Corasaniti MT, Mercuri NB, Bernardi G, Bagetta G. Brain regional and cellular localization of gelatinase activity in rat that have undergone transient middle cerebral artery occlusion. *Neuroscience* 2008;**152**:8–17.
57. Cuadrado E, Rosell A, Borrell-Pages M, Garcia-Bonilla L, Hernandez-Guillamon M, Ortega-Aznar A, et al. Matrix metalloproteinase-13 is activated and is found in the nucleus of neural cells after cerebral ischemia. *J Cereb Blood Flow Metab* 2009;**29**:398–410.

58. Vidal AE, Boiteux S, Hickson ID, Radicella JP. XRCC1 coordinates the initial and late stages of DNA abasic site repair through protein-protein interactions. *EMBO J* 2001;**20**:6530–9.

59. Morita-Fujimura Y, Fujimura M, Gasche Y, Copin JC, Chan PH. Overexpression of copper and zinc superoxide dismutase in transgenic mice prevents the induction and activation of matrix metalloproteinases after cold injury-induced brain trauma. *J Cereb Blood Flow Metab* 2000;**20**:130–8.

60. Gijbels K, Masure S, Carton H, Opdenakker G. Gelatinase in the cerebrospinal fluid of patients with multiple sclerosis and other inflammatory neurological disorders. *J Neuroimmunol* 1992;**41**:29–34.

61. Chandler S, Cossins J, Lury J, Wells G. Macrophage metalloelastase degrades matrix and myelin proteins and processes a tumour necrosis factor-alpha fusion protein. *Biochem Biophys Res Commun* 1996;**228**:421–9.

62. Liuzzi GM, Trojano M, Fanelli M, Avolio C, Fasano A, Livrea P, et al. Intrathecal synthesis of matrix metalloproteinase-9 in patients with multiple sclerosis: implication for pathogenesis. *Mult Scler* 2002;**8**:222–8.

63. Hewson AK, Smith T, Leonard JP, Cuzner ML. Suppression of experimental allergic encephalomyelitis in the Lewis rat by the matrix metalloproteinase inhibitor Ro31-9790. *Inflamm Res* 1995;**44**:345–9.

64. Metz LM, Zhang Y, Yeung M, Patry DG, Bell RB, Stoian CA, et al. Minocycline reduces gadolinium-enhancing magnetic resonance imaging lesions in multiple sclerosis. *Ann Neurol* 2004;**55**:756.

65. Yong VW, Giuliani F, Xue M, Bar-Or A, Metz LM. Experimental models of neuroprotection relevant to multiple sclerosis. *Neurology* 2007;**68**:S32–7, discussion S43–S54.

66. Leppert D, Lindberg RL, Kappos L, Leib SL. Matrix metalloproteinases: multifunctional effectors of inflammation in multiple sclerosis and bacterial meningitis. *Brain Res Brain Res Rev* 2001;**36**:249–57.

67. Leib SL, Clements JM, Lindberg RL, Heimgartner C, Loeffler JM, Pfister LA, et al. Inhibition of matrix metalloproteinases and tumour necrosis factor alpha converting enzyme as adjuvant therapy in pneumococcal meningitis. *Brain* 2001;**124**:1734–42.

68. Echchannaoui H, Leib SL, Neumann U, Landmann RM. Adjuvant TACE inhibitor treatment improves the outcome of TLR2-/- mice with experimental pneumococcal meningitis. *BMC Infect Dis* 2007;**7**:25.

69. Rosenberg GA, Navratil M, Barone F, Feuerstein G. Proteolytic cascade enzymes increase in focal cerebral ischemia in rat. *J Cereb Blood Flow Metab* 1996;**16**:360–6.

70. Gasche Y, Fujimura M, Morita-Fujimura Y, Copin JC, Kawase M, Massengale J, et al. Early appearance of activated matrix metalloproteinase-9 after focal cerebral ischemia in mice: a possible role in blood-brain barrier dysfunction. *J Cereb Blood Flow Metab* 1999;**19**:1020–8.

71. Wang X, Jung J, Asahi M, Chwang W, Russo L, Moskowitz MA, et al. Effects of matrix metalloproteinase-9 gene knock-out on morphological and motor outcomes after traumatic brain injury. *J Neurosci* 2000;**20**:7037–42.

72. Chang DI, Hosomi N, Lucero J, Heo JH, Abumiya T, Mazar AP, et al. Activation systems for latent matrix metalloproteinase-2 are upregulated immediately after focal cerebral ischemia. *J Cereb Blood Flow Metab* 2003;**23**:1408–19.

73. Anonymous . Tissue plasminogen activator for acute ischemic stroke. The National Institute of Neurological Disorders and Stroke rt-PA Stroke Study Group. *N Engl J Med* 1995;**333**:1581–7 (see comments).

74. Hornig CR, Bauer T, Simon C, Trittmacher S, Dorndorf W. Hemorrhagic transformation in cardioembolic cerebral infarction. *Stroke* 1993;**24**:465–8.

75. Rosell A, Ortega-Aznar A, varez-Sabin J, Fernandez-Cadenas I, Ribo M, Molina CA, et al. Increased brain expression of matrix metalloproteinase-9 after ischemic and hemorrhagic human stroke. *Stroke* 2006;**37**:1399–406.
76. Lapchak PA, Chapman DF, Zivin JA. Metalloproteinase inhibition reduces thrombolytic (Tissue plasminogen Activator)-induced hemorrhage after thromboembolic stroke [In Process Citation). *Stroke* 2000;**31**:3034–40.
77. Pfefferkorn T, Rosenberg GA. Closure of the blood-brain barrier by matrix metalloproteinase inhibition reduces rtPA-mediated mortality in cerebral ischemia with delayed reperfusion. *Stroke* 2003;**34**:2025–30.
78. Sumii T, Lo EH. Involvement of matrix metalloproteinase in thrombolysis-associated hemorrhagic transformation after embolic focal ischemia in rats. *Stroke* 2002;**33**:831–6.
79. Bowler JV. Modern concept of vascular cognitive impairment. *Br Med Bull* 2007;**83**:291–305.
80. Caplan LR. Binswanger's disease—revisited. *Neurology* 1995;**45**:626–33, (Review).
81. Wakita H, Tomimoto H, Akiguchi I, Kimura J. Glial activation and white matter changes in the rat brain induced by chronic cerebral hypoperfusion: an immunohistochemical study. *Acta Neuropathol (Berl)* 1994;**87**:484–92.
82. Nakaji K, Ihara M, Takahashi C, Itohara S, Noda M, Takahashi R, et al. Matrix metalloproteinase-2 plays a critical role in the pathogenesis of white matter lesions after chronic cerebral hypoperfusion in rodents. *Stroke* 2006;**37**:2816–23.
83. Farkas E, Luiten PG, Bari F. Permanent, bilateral common carotid artery occlusion in the rat: a model for chronic cerebral hypoperfusion-related neurodegenerative diseases. *Brain Res Brain Res Rev* 2007;**54**:162–80.
84. Ihara M, Tomimoto H, Kinoshita M, Oh J, Noda M, Wakita H, et al. Chronic cerebral hypoperfusion induces MMP-2 but not MMP-9 expression in the microglia and vascular endothelium of white matter. *J Cereb Blood Flow Metab* 2001;**21**:828–34.
85. Sood R, Yang Y, Taheri S, Candelario-Jalil E, Estrada EY, Walker EJ, et al. Increased apparent diffusion coefficients on MRI linked with matrix metalloproteinases and edema in white matter after bilateral carotid artery occlusion in rats. *J Cereb Blood Flow Metab* 2009;**29**:308–16.
86. Adair JC, Charlie J, Dencoff JE, Kaye JA, Quinn JF, Camicioli RM, et al. Measurement of gelatinase B (MMP-9) in the cerebrospinal fluid of patients with vascular dementia and Alzheimer disease. *Stroke* 2004;**35**:e159–62.
87. Lin JX, Tomimoto H, Akiguchi I, Matsuo A, Wakita H, Shibasaki H, et al. Vascular cell components of the medullary arteries in Binswanger's disease brains: a morphometric and immunoelectron microscopic study. *Stroke* 2000;**31**:1838–42.
88. Rosenberg GA, Sullivan N, Esiri MM. White matter damage is assoiciated with matrix metalloproteinases in vascular dementia. *Stroke* 2001;**32**:1162–8.
89. Chandler S, Coates R, Gearing A, Lury J, Wells G, Bone E. Matrix metalloproteinases degrade myelin basic protein. *Neurosci Lett* 1995;**201**:223–6.
90. Chung AW, Yang HH, Radomski MW, van Breemen C. Long-term doxycycline is more effective than atenolol to prevent thoracic aortic aneurysm in marfan syndrome through the inhibition of matrix metalloproteinase-2 and -9. *Circ Res* 2008;**102**:e73–85.
91. Candelario-Jalil E, Yang Y, Rosenberg GA. Diverse roles of matrix metalloproteinases and tissue inhibitors of metalloproteinases in neuroinflammation and cerebral ischemia. *Neuroscience* 2009;**159**:983–94.

Regulation of Caspases in the Nervous System: Implications for Functions in Health and Disease

CAROL M. TROY,[*,†] NSIKAN AKPAN,[*,†] AND YING Y. JEAN[*,†]

[*]*Department of Pathology & Cell Biology, Taub Center for the Study of Alzheimer's Disease and the Aging Brain, Columbia University College of Physicians and Surgeons, New York, USA*

[†]*Department of Neurology, Taub Center for the Study of Alzheimer's Disease and the Aging Brain, Columbia University College of Physicians and Surgeons, New York, USA*

I. The Caspase Family of Death Proteases 266
A. Caspase Activation 268
B. Endogenous Inhibitors of Caspases, the IAP Family 272
C. Endogenous Inhibitors of IAPs 272
D. Regulation of Mitochondria in Apoptosis 272
E. Measurement of Caspase Activation/Activity 273
F. Targets of Caspases 275
II. Caspases and Developmental Neuronal Death 275
III. Nonapoptotic Caspase Function in the Nervous System 276
IV. Caspase Function in Neurodegenerative Diseases 277
A. Alzheimer's Disease 277
B. Cerebral Ischemia 284
C. Linking Stroke and AD 290
V. Future Directions 292
References 292

Caspases, initially identified as a family of proteases regulating cell death, have been found to have nonapoptotic functions as well. Some family members are critical for mediating programmed cell death in development. After development, caspases are downregulated in the nervous system, but continue to perform important nonapoptotic functions relevant for neurogenesis and synaptic plasticity. In neurodegenerative diseases, where aberrant neuronal death is an outstanding feature, there is an increase in caspase activity. The specific caspase death pathways leading to dysfunction and death have still not been fully clarified, despite the plethora of scientific literature addressing these issues. In this chapter, we will present the current knowledge of caspase

Progress in Molecular Biology
and Translational Science, Vol. 99
DOI: 10.1016/S1877-1173(11)99007-3

1877-1173/11 $35.00

activation and activity pathways, the current tools for examining caspases, and functions of caspases in the nervous system in health and in disease. Alzheimer's Disease, the most common neurodegenerative disorder, and cerebral ischemia, the most common cause of neurologic death, are used to illustrate our current understanding of death signaling in neurodegenerative diseases. A better understanding of how caspases function in health and disease would provide appropriate specific targets for the development of therapeutic interventions for these diseases.

Life and death are exquisitely regulated at the cellular level from development through maturity. During development, neuronal death is the major factor shaping the nervous system. This death is mainly caspase-mediated apoptosis. Once the waves of developmental death have passed (death occurs at different times in different parts of the nervous system), there is downregulation of the death machinery, as the postmitotic neurons should live for the life of the organism. Aberrant neuronal death is a major part of neurodegenerative disorders, but there is still no clear understanding of the processes leading to the phenotypes of the various diseases. Even the type of death that occurs continues to be debated, whether it is apoptotic, necrotic, or autophagic, or some combination of these death mechanisms. Here, we will discuss the role that the caspases play in neuronal function, dysfunction, and death. First, we will discuss the regulation of caspase activation and activity. We will examine the current understanding of caspase function in developmental neuronal death and then illustrate the role of caspases in neuronal death in disease employing two diseases of neuronal loss, Alzheimer's Disease (AD), which is the most common chronic neurodegenerative disorder, and cerebral ischemia/stroke, the third most common cause of death in Western society, which is an acute neuronal disorder with chronic sequelae.

I. The Caspase Family of Death Proteases

Since the seminal work on death pathways in the model organism *Caenorhabditis elegans*, studies of apoptosis have increased exponentially. While it is agreed that caspases are critical mediators of apoptosis, many details of the death pathways remain to be explicated. Caspases are a highly conserved family, from worms to humans. One major difference, however, is the number of caspases: in worms, there is one caspase, while in mammals, there are 13 (Fig. 1). This difference raises the question of whether individual mammalian caspases have unique functions. Despite the plethora of studies, this is still an open question. In terms of disease intervention, specific nonredundant functions of individual caspases would provide therapeutic targets. As we increase our understanding of how caspase activation and activity are regulated, we can more critically evaluate current studies and reevaluate former studies.

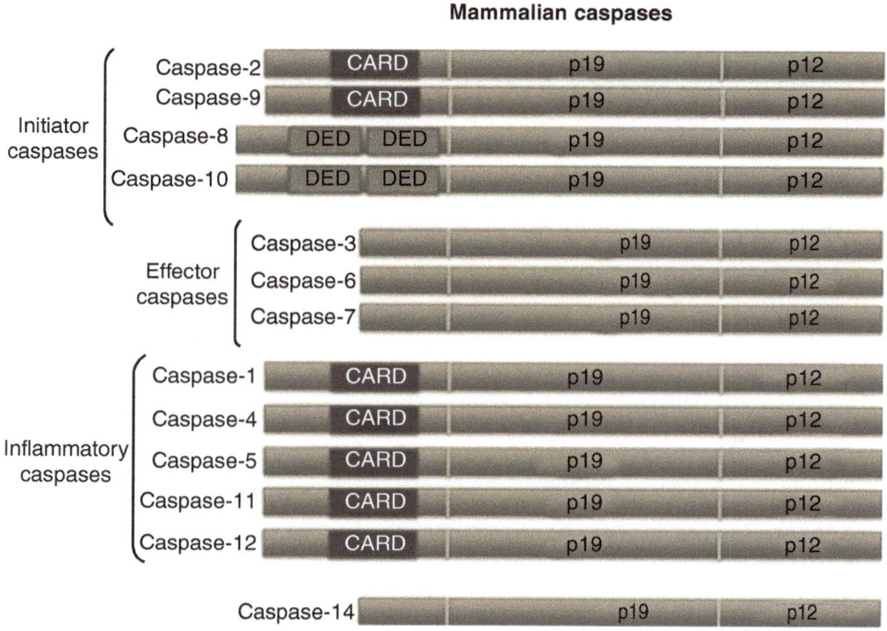

FIG. 1. *Mammalian caspases.* Mammalian caspases are schematically represented and grouped by activity. Yellow lines indicate cleavage sites. (For interpretation of the references to color in this figure legend, the reader is referred to the Web version of this chapter.)

In 1992, the interleukin-1β cleaving enzyme (ICE),[1,2] now known as caspase-1, was identified and within a few months, ced-3, an enzyme with significant homology to ICE, was identified in *C. elegans* and found to execute apoptosis.[3] Many more mammalian homologs were identified in the ensuing years, and the term "caspase," for *c*ysteine dependent, *asp*artate-specific pro*tease*, was agreed upon for the mammalian family. The 13 different mammalian caspases can be divided either by structure or by putative actions. From a structural perspective, there are two general groups, caspases with short prodomains (caspases-3, -6, -7, and -14) and with long prodomains (caspases-1, -2, -4, -5, -8, -9, -10, -11, and -12). Caspase-14 is involved in keratinocyte differentiation and does not appear to have a function in the nervous system[4] and will not be discussed further. The effectors of apoptosis are the caspases with short prodomains, caspases-3, -6, and -7. The caspases with long prodomains are further subdivided. Caspases-2, -8, -9, and -10 are initiators of apoptosis, although caspase-2 may act as both an initiator and effector. Caspases-1, -4, -5, and -11 process cytokines and contribute to inflammation: caspases-4 and -5 are found only in humans, while caspase-11 is present only in rodents.

The inflammatory caspases can also lead to cell death by autocrine mechanisms. Thus, although these caspases may not be direct apoptosis initiators, their activity can lead to death. Caspase-12 is somewhat of an enigma. It may be involved in cytokine processing or in death, particularly in endoplasmic reticulum-associated death. Additionally, while the gene for caspase-12 is expressed in rodents and humans, the protein is expressed in rodents but only in a small number of humans.[5] There are studies of caspase-12 that suggest that it acts as a dominant-negative regulator of caspase-1 activity and that the caspase-12 proteolytic activity results only in autoprocessing.[6] Work also supports a role for caspase-12 in ER stress as an activator of matrix metalloprotease 3.[7]

A. Caspase Activation

Caspases are also grouped based on the mechanism of activation (Fig. 2). While the activation of effector caspases is dependent on cleavage at an aspartate residue in the intersubunit linker,[8,9] the activation of initiator

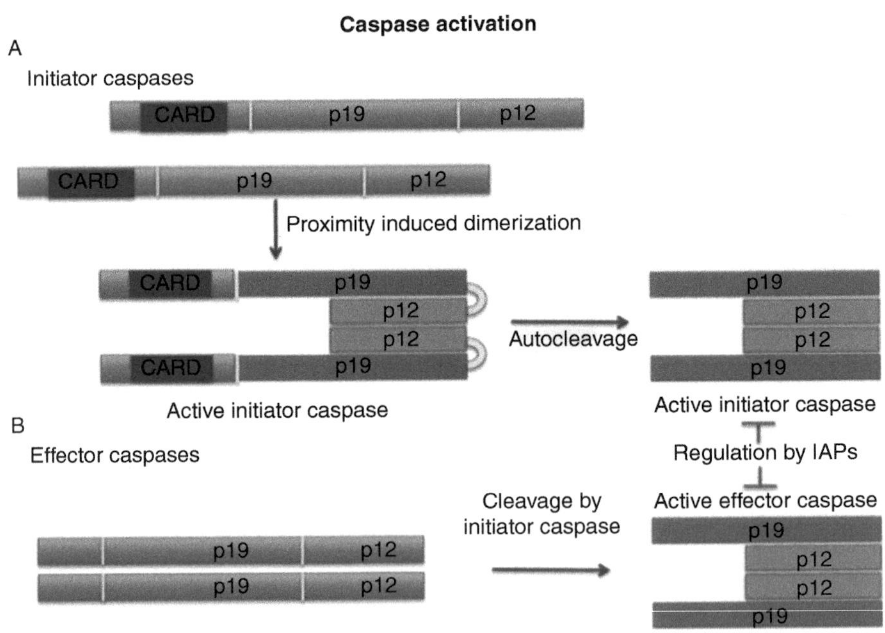

FIG. 2. *Activation mechanisms of effector and initiator caspases.* Activation models for initiator and effector caspases are shown. Blue indicates inactive caspase while orange denotes activated caspase. Some of the activated caspases are subject to regulation by IAPs. (See Color Insert.)

caspases is more complex and new studies are refining the activation mechanisms. As a generalization, caspases with long prodomains undergo dimerization with a structural change that provides for the formation of an active site.[9–11] The longer intersubunit linker allows flexibility, leading to the formation of an active site. Most of the initiators exist as monomers in the zymogen form; caspase-2 zymogen is a dimer.[12] Interaction of the specific adaptor protein with the prodomain leads to dimerization and activation. Once the zymogen is activated, there is limited proteolysis of the interdomain. For caspase-9, proteolysis of the interdomain is not required: uncleavable mutants retain activity.[13] However, recent work on caspase-8 shows that both dimerization and cleavage are required for optimal activity.[14] This finding suggests that there are subtle differences in activation requirements that have implications for measurements of activation/activity of each caspase.

The long prodomain caspases have unique activation platforms. The intrinsic pathway of apoptosis centers on the mitochondrial regulation of death (Fig. 3). The apoptosome,[15] the caspase-9 activation platform, was the first mammalian activation platform described. Release of cytochrome *c* from mitochondria leads to an ATP-dependent recruitment of caspase-9 by the adaptor,

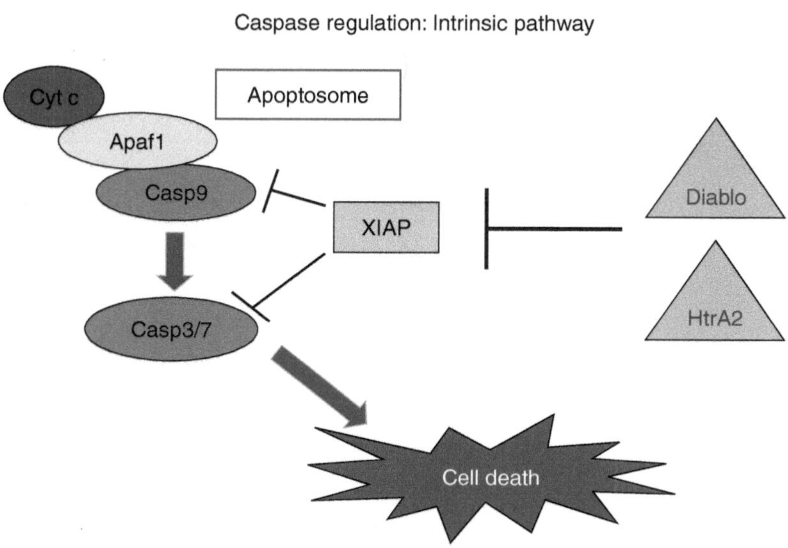

FIG. 3. *Caspase regulation: intrinsic pathway.* The intrinsic death pathway is activated by permeabilization of the mitochondria, leading to the release of cytochrome *c* and formation of the apoptosome, the caspase-9 activation platform. Once activated, caspase-9 can cleave and activate caspases-3 and -7. Caspases-3, -7, and -9 are subject to inhibition by XIAP. XIAP can be inhibited by Diablo/SMAC or HtrA2/Omi which are released from permeabilized mitochondria.

Apaf-1. Caspase-9 then undergoes limited autocleavage to a p35 fragment, which can be inhibited by XIAP, X-linked inhibitor of apoptosis protein. As noted above, cleavage is not required for activation of caspase-9 and, since the autocleavage allows inhibition of caspase-9 activity, it can actually lead to decreased activity. Caspase-3 can also cleave caspase-9 to a p37 fragment; this cleavage is not an activation step but potentiates activation by preventing XIAP inhibition of caspase-9,[16] as described below.

The extrinsic death pathway is activated by ligand binding to death receptors (Fig. 4). The caspase-8 activation platform is the death-inducing signaling complex (DISC). Ligand binding to a cell surface death receptor leads to the recruitment of an adaptor protein, which recruits caspase-8. As an example, Fas ligand binds to the fas receptor to recruit FADD and subsequently caspase-8. Binding to FADD dimerizes caspase-8 and then autocleavage provides for complete activation.[14,17] Cleavage, without dimerization, by another enzyme,

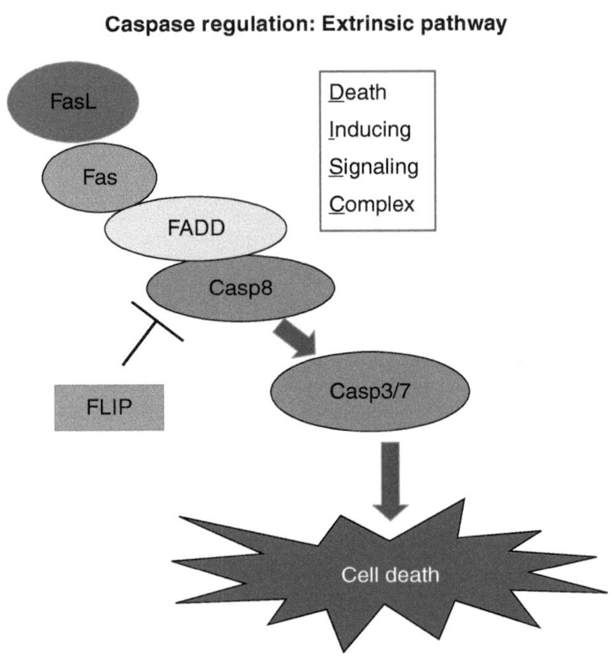

FIG. 4. *Caspase regulation: extrinsic pathway.* The extrinsic pathway is activated when a ligand binds to a death receptor; binding of FasL to the Fas receptor is shown. This leads to recruitment of an adaptor protein (FADD), which recruits caspase-8, forming the DISC, leading to dimerization and activation of caspase-8. Caspase-8 then cleaves and activates effector caspases. FLIP inhibits the formation of active caspase-8 dimers.

an effector caspase or granzyme B, does not produce activation of caspase-8.[14] Recent work suggests that TNF-α induces two different caspase-8 activation pathways, a RIPK1-dependent and a RIPK1-independent pathway.[18] There is also phosphorylation-mediated regulation of caspase-8 activity.[19]

The activation platform for caspase-2 is proposed to be the PIDDosome, a complex of PIDD, RAIDD, and caspase-2.[20,21] However, two different lines of PIDD null mice appear to undergo caspase-2-dependent death, although caspase-2 activity was not measured in those studies.[22,23] Studies of the PIDD complex in tumor cell lines showed that overexpression of PIDD induced RAIDD-dependent cell death[24] and that PIDD function in those cells depends on the cleavage state of PIDD.[25] PIDD autocleaves to PIDD-C and PIDD-CC. Partially cleaved PIDD, PIDD-C, complexes with NEMO and leads to NFκB activation and survival signaling. Further cleavage of PIDD produces a fragment, PIDD-CC, that, in the overexpression system, complexes with RAIDD and leads to cell death. There are three isoforms of PIDD[26]: isoform 1 contains full-length PIDD and can be processed to PIDD-C and PIDD-CC; isoform 2 has a deletion of 146 amino acids at the N-terminus and an 11 amino acid deletion at position aa580; isoform 3 has a 17 amino acid deletion at aa705. Isoform 2 cannot form the PIDD-CC fragment. Expression of these isoforms is tissue- and cell-type specific: isoform 2 is present mainly in transformed cell lines and in normal liver, pancreas, and leukocytes, while isoforms 1 and 3 are found in normal tissue with the exception of the skeletal muscle. Isoform 1 is also expressed in transformed cell lines and isoforms 1 and 2 are found in several tumors. Since isoform 2 cannot form PIDD-CC, it should not interact with RAIDD. It is possible that PIDD function is different in cells with isoform 2 rather than the isoforms that can form PIDD-CC. This hypothesis would have interesting implications for the regulation of caspase-2 function in the different cell types. Caspase-2 activity is also regulated by phosphorylation.[27–29]

Caspase-1 is activated in the inflammasome.[30–32] Several inflammasomes have been identified, the IPAF (ICE-protease activating factor), the NLRP1 (NACHT-, LRR-, and pyrin-domain containing proteins), and the NLRP2/3 inflammasomes.[31–34] The basic assembly involves the aforementioned proteins (IPAF or NLRP), which contain pyrin domains, caspase-1, which contains a CARD domain, and an adaptor protein with a CARD domain, such as ASC (apoptosis-associated speck-like protein containing a CARD) or CARDINAL (CARD inhibitor of NFkB-activating ligands). Caspase-1 is incorporated by ASC into the inflammasome through its CARD domain.[35,36] The inflammasomes are activated by receptor-mediated signaling through TLRs (toll-like receptors) or in the cytosol by the NLR family in response to stress signals.

B. Endogenous Inhibitors of Caspases, the IAP Family

The *I*nhibitor of *A*poptosis *P*rotein family is classified by the presence of BIR (*b*aculovirus *I*AP *r*epeat) domains. In mammals, there are eight IAPs (see Ref. 37 for a review). Of these, cIAP1, cIAP2, and XIAP have been shown to bind to caspases, but XIAP is the only one which inactivates caspases.[38] Other IAPs do inhibit cell death but not by directly inhibiting caspase activity. XIAP binds to and inactivates active caspases-3 and -7 via the BIR2-linker region and binds to and inactivates caspase-9 via the BIR3 domain. cIAP1 and cIAP2 can also bind to active caspases-3 and -7 but lack the region necessary for inactivating the caspases.[39] XIAP-BIR3 binds to the neoepitope of caspase-9 that is generated by autocleavage of caspase-9 at aa315. XIAP cannot bind to the capase-9 neoepitope generated by caspase-3 cleavage of caspase-9 at aa330. Thus, caspase-3 cleavage promotes activity of caspase-9 by preventing the inhibition of caspase-9 by XIAP. In a similar fashion, the p20 fragment of caspase-3 can be inhibited by XIAP, but the p17 fragment cannot be inhibited. The BIR3 domains of cIAP1 and cIAP2 contain only one of the four residues required to inactivate caspase-9.[38] Caspase-8 also has an endogenous inhibitor, c-FLIP, which is caspase-8 without the active site.[40] FLIP dimerizes with caspase-8, preventing its activation rather than inhibiting active caspase-8.

C. Endogenous Inhibitors of IAPs

There are several endogenous inhibitors of IAPs. Smac/Diablo[41,42] and HtrA2/Omi[43] are found in the mitochondria of healthy cells, and the N-terminal amino acids of these proteins form the IAP-binding motif (IBM). Mitochondrial permeabilization leads to release of these proteins, which then interfere with the IAP-inhibition of caspases, resulting in potentiation of cell death. Another endogenous inhibitor of IAPs is XAF1.[44] In nonneuronal tissues, XAF1 functions as a tumor suppressor. Originally identified as a specific inhibitor of XIAP, it is now reported that XAF1 can also bind to cIAP1 and cIAP2.[45] As the function of cIAP1/2 is not quite clear, it is also not clear what the function of XAF1 binding to these molecules would be.

D. Regulation of Mitochondria in Apoptosis

Mitochondrial outer membrane permeabilization (MOMP) allows for selective release of molecules that participate in death. These molecules include cytochrome *c*, Smac/Diablo, HtrA2/Omi, and AIF. The Bcl2 family of proteins participates in the regulation of MOMP.[11] This multimembered family is identified by the presence of BH domains and has both antiapoptotic members, which contain 4 BH domains, and proapoptotic members, which contain either 3 BH domains or only a BH3 domain. Bcl-2 family proteins reside either in the mitochondrial outer membrane or in the cytoplasm of healthy cells.

E. Measurement of Caspase Activation/Activity

Much of the literature regarding caspases should be reevaluated with regard to what the measures of caspases used in a particular study actually showed. Multiple methods have been used to measure caspase activity and activation, including Western blotting, immunocytochemistry, synthetic peptide substrate/inhibitor assays, and fluorescent indicators. Western blotting with specific antibodies can be used to quantify changes in zymogen expression and reveal the appearance of cleavage fragments. The relevance of cleavage to activation state varies among the caspases, as noted above. Proteolytic processing of the zymogen is required for effector caspases (3, 6, and 7) and appearance of the p17/19 and p10/12 fragments indicates that these caspases have been activated. For the initiator caspases, dimerization is required and there is a variable requirement for autocleavage (see above), but appearance of cleaved fragments is not a reliable read-out for activation since it will not distinguish between a cleavage event in the process of activation and a cleavage event that occurred without prior dimerization.

An additional subtle point that is often missed is that caspase activation and activity are not equivalent: IAPs can inhibit the activity of activated caspases. This distinction is critically relevant from a functional perspective. Only an active caspase will enact death. Caspases-3, -7, or -9 can be inhibited by XIAP *in situ*. Lysis of the cells for Western blotting disrupts IAP-caspase complexes and the Western blot shows cleaved caspases. Thus, activated but inhibited caspases cannot be distinguished from active caspases using Western blotting of cell lysates. For initiator caspases, interpretation of Western blotting data is even more problematic. Both caspase-8 and caspase-9 undergo autoproteolysis when incorporated into their respective activation complexes, although the full-length caspases have activity as well.[16,46,47] As stated before, caspase-8 requires further processing for optimal activity, while additional processing of caspase-9 does not substantially increase activity.[14] Caspase-2 also undergoes autoproteolysis,[12] but it has not yet been clear how much this proteolysis contributes to total caspase-2 activity.

Another frequently used technique is immunocytochemistry, using antibodies which detect either the zymogen and cleaved fragments or are specific for neoepitopes generated during cleavage of zymogens. This technique is excellent for studying changes in the subcellular localization of a caspase, and for appearance of neoepitopes of effector caspases, with the same caveats noted above for Western blotting: appearance of the neoepitope does not necessarily indicate that there is an active caspase. Many studies inaccurately use the appearance of cleaved caspase-3 as a measure of apoptosis but, without another measure, cleaved caspase-3 alone does not mean that the cell is dying, as the caspase can be inhibited by XIAP.

Synthetic peptide pseudosubstrate caspase inhibitors/substrates, such as DEVD or IETD, have been and continue to be used by many groups as a method of distinguishing the role of individual caspases, despite studies published over a decade ago showing that these reagents are not specific for individual caspases.[48,49] More recently, a comprehensive study evaluated the specificity of these reagents and concluded that all of the so-called specific caspase inhibitors/substrates are best at detecting caspase-3 activity, and some do not even detect their putative targeted caspase.[50] To evaluate data, it is also important to understand how these peptides inhibit caspases: the peptide binds to the active caspase and inhibits downstream events. For *in vivo* measures of caspase activity, fluorescent versions of these peptides are being utilized as substrates that fluoresce when bound to the active caspase. These peptides bind irreversibly to the active site and cause inhibition of downstream events. Thus, once the fluorescent peptide is added, caspase activity is inhibited. Any cellular events that occur subsequent to the addition of the fluorescent peptide are not the result of caspase activation and could be the result of caspase inhibition. Thus, data acquired using these reagents should be reevaluated.

There are a few methods available to show that a specific caspase is active. One of these utilizes a broad-spectrum caspase substrate/irreversible inhibitor linked to biotin, bVADfmk. This active site affinity ligand binds irreversibly to the active caspase in the cell and inhibits the caspase. Preloading cells with bVADfmk before initiation of death captures initiator caspases and addition of bVADfmk at the time of harvesting cells undergoing death detects active effector caspases.[51] The biotinylated affinity ligand–caspase complex is isolated with streptavidin, run on SDS-PAGE, and the specific caspase bound is detected with Western blotting for individual caspases. Studies using this method show that the active initiator caspases isolated by the affinity ligand are uncleaved, supporting the proximity-induced dimerization model of activation (no cleavage necessary initially) for initiator caspases. Adaptation of this method to animal studies could enable the identification of which caspases are active *in vivo*. Groups are also developing specific fluorescent read-outs of caspase activity. For caspase-2, such a system has been used in cultured cells to show dimerization and activation of caspase-2 after heat shock.[52] This system could be adapted for other initiator caspases.

Functional relevance for individual caspases can be obtained using molecular knockdown/knockout of individual caspases. Abrogation of death shows that specific caspases are critical for execution of the death pathway under study. Studies have used antisense and siRNA knockdown of individual caspases. The ability to use siRNA to knockdown caspase-2 has been questioned because of the lack of reproducibility of published work using siRNA.[53] However, it appears that it may be the sequence used in the studies in question that is at fault, rather than the ability of siRNA to ablate expression of caspase-2, as siRNA have been successfully used to knockdown other caspases.

Caspase null mice have been valuable tools for studying caspases with a role in development, as discussed below. However, caspase null mice without profound phenotypes suggest either that these caspases do not have a major role in development or that there are compensatory changes in other caspases, which obscure the function of the targeted caspase. Use of conditional knockouts could address these issues.

F. Targets of Caspases

Caspases can cleave many proteins; cleavage is limited (cleavage at one or two sites) as illustrated by the activation of effector caspases by initiator caspases. There have been multiple reports of substrate cleavage changing the activity of the substrate: that is, converting a proapoptotic protein to an antiapoptotic one, or vice versa. Such events serve to amplify the signaling pathways leading to death or survival. Although predictions have been made, careful time courses of substrate cleavage events, combined with specific blockade of these events, will be required to determine the function of such cleavage in individual death pathways.

II. Caspases and Developmental Neuronal Death

During development, programmed cell death, which is mediated by caspases, is the major force that shapes the organism. This process is essential to the formation of a normal nervous system. Two general populations die during neuronal development: neuronal precursors and postmitotic neurons. Caspase-9 and the caspase-9 adapter, Apaf1, are essential for death of the neuronal precursors. Mice lacking either of these proteins have severe malformations of the nervous system.[54–57] The initial report of the caspase-3 null mice also showed a severe neuronal developmental phenotype, very similar to the caspase-9 null mice, with death of the mice soon after birth.[58] Care must be taken when drawing conclusions, because changing the genetic background of the caspase-3 null mice from a mixed background to a C57/Black6 background modified the phenotype drastically[59]—on this background, neuronal development is grossly normal, and the mice are able to grow to adulthood and to breed. Thus, the dependence of neuronal developmental death on caspase-3 is unclear. It is also not clear that all neuronal developmental death is dependent only on the caspase-9 pathway. The timing of death varies among regions of the brain, from the early embryonic to the early postnatal periods.

Mice lacking many of the other caspases, such as caspase-1, -2, -6, -7, -11, or -12, do not have an obvious neuronal phenotype.[59–63] These caspases may play a role in regional pruning of neurons or in the plasticity of the nervous system, and be important in the maintenance of the mature nervous system.

There is also an issue of compensatory changes in other caspases or caspase regulators, which can confound the simple interpretation of results from mouse knockout studies. This issue is a significant problem with apoptotic genes, where phenotypic selection in developing embryos can severely alter normal expression patterns, as has been shown in caspase-2, -3, and -9 null mice.[63,64] Compensations have been shown at the level of expression of other caspases[63] or of regulators of caspase activity.[64] The relative expression of the caspases and their regulators is a major factor in the choice of execution pathway. It would be wise to keep in mind this potential complexity when evaluating data on caspase involvement in neurodegenerative diseases.

III. Nonapoptotic Caspase Function in the Nervous System

There is increasing evidence that there are nonapoptotic functions of caspases in the nervous system. These data are most compelling for a function for caspases in neurogenesis and synaptic activity.[65] Studies of neurogenesis indicate that caspase-3 activity in neuronal progenitors facilitates neurogenesis.[66] Inhibition of caspase activity also blocked neurite extension, suggesting a role for caspases in the morphologic changes that occur during neurogenesis. Differentiation of mouse neural stem cells was inhibited by inhibition of caspase activity, and caspase inhibition also blocked p53 phosphorylation and transcriptional activation.[67] These data suggest a critical, nonapoptotic function of caspase-3 in neurogenesis.

Neuronal circuits are regulated by selective pruning of synapses, axons, and dendrites. In *Drosophila*, the initiator caspase Dronc is required for pruning.[68] Studies of trophic factor deprivation-mediated death of cultured dorsal root ganglion neurons suggest that there is selective activation of caspase-6 in axons, which leads to axonal pruning but not to death.[69] Caspases have also been implicated in neuronal plasticity. *In vivo* studies in the zebra finch showed that caspase-3 activity is required for the birds to learn a new song.[70] The cleaved caspase-3 is in the synapses and is associated with XIAP, suggesting that XIAP modulates the activity of caspase-3, so that synapse remodeling occurs but not death. Studies using DEVDfmk in rats support a role for caspase activity in long-term potentiation[71] and in active avoidance learning.[72] It has been recently shown that caspase-3 activation is required for long-term depression. LTD was inhibited in caspase-3 null hippocampal slices and XIAP-BIR3, a specific inhibitor of caspase-9, also inhibited LTD.[73] These nonapoptotic caspase functions also indicate why it is not appropriate to connote caspase-3 cleavage with apoptosis.

These nonapoptotic functions of caspases need to be better understood for the design of appropriate therapeutic interventions to treat diseases with aberrant caspase activation and neuronal death.

IV. Caspase Function in Neurodegenerative Diseases

The study of diseases requires the use of models of the disease. We will present two diseases that are among the most highly prevalent human neurological disorders of aging. We will consider what the appropriate models are and what the current understanding of the caspase pathways is in each disease.

A. Alzheimer's Disease

Over the past century, the life span in industrialized countries has increased an average of 25 years. This increased life expectancy has shifted the average age of the population and more people are living beyond 80 years of age. With an increase in the aged population, there is an increase in the prevalence of diseases associated with aging. At age 65, about 2–3% of people have AD, but the incidence of AD doubles for every 5 years of age afterward, and approximately 50% of people over the age of 85 have AD. There are about 5.3 million cases of AD in the United States.[74] As a result of increasing life expectancy and expanding population, the number of AD cases should triple over the next 40 years.[74] AD is the most common form of dementia among the elderly population and represents the fourth leading cause of death in industrialized countries.[75,76]

AD was characterized by Alois Alzheimer in 1907 as a progressive impairment of cognitive functions.[77] The most frequent symptom is gradual loss of episodic memory, which is memory of recent events. With disease progression, impairment is found in language, cognition, reason, and temporal-spatial orientation, together with other changes in mood and personality. After the initial clinical symptoms, there is continuous and progressive decline leading to mutism, vegetative state, inanition, and finally death. Survival ranges from 5 to 20 years from detection of the first symptoms of the disease to death. The definitive diagnosis of AD is pathologic but biopsies are rarely done and tissue is usually obtained postmortem. Thus, the clinical diagnosis is of possible or probable AD. The two main histopathological hallmarks required for the definitive diagnosis of AD are senile plaques, extracellular congophyllic deposits composed of Aβ peptide, and neurofibrillary tangles (NFTs), intraneuronal filamentous aggregates composed of hyperphosphorylated tau proteins and paired helical filaments (PHF).[78]

Multiple factors have contributed to the lack of full understanding of the order of events in the development and progression of AD; and these factors include the insidious onset of clinical symptoms, the lack of a definitive clinical diagnosis, and the chronic nature of the disease. The progressive loss of synapses and neurons in limbic and cortical areas, translated functionally as a disconnection between different brain areas, is postulated to lead to the clinical symptoms.[79,80] This initial disconnection does not seem to have repercussions on behavior for several decades, since the gradual loss of synapses and neurons seems to start 20–40 years before the manifestation of the first symptoms.[81,82] Distribution of NFT correlates better than amyloid plaque burden with the severity of dementia and neuronal death.[83–87] But, as no studies of progressive pathologic analysis of single patients over time exist, all of the human data should be considered as correlative rather than definitive with regards to molecular causality. The use of imaging of amyloid plaques in AD patients may provide information about the progression of the disease, when correlated with symptoms.

Three genes have been linked to familial AD: amyloid precursor protein (APP),[88] located on chromosome 21, and presenilin-1 (PS1)[89] and presenilin-2 (PS2),[90] located on chromosomes 14 and 1, respectively. About 85% of EOAD cases correspond to mutations in the PS1 gene, and mutations in PS2 or APP are much less frequent.[91,92] There have been 157 mutations in PS1 and 10 in PS2 described worldwide. There are families with a clear genetic component that have not been associated with mutations in any of these three genes, which suggests that other genes may be involved.

1. Mouse Models in AD

Knowledge of the genetic mutations in familial AD enabled the development of mouse models expressing these human mutations in APP and PSEN1/2 as tools to study AD. The mouse models do exhibit some of the pathological hallmarks of AD, including amyloid deposits, abnormal tau hyperphosphorylation, and gliosis, and also display impairments of memory and learning.[93–99] The mice do not have NFTs or neuronal loss. Additional models were generated containing mutant APP and PS1 or both of these and mutant tau, in an attempt to provide a phenocopy for AD.[100–110] Since some of these models overexpress mutant genes that are not even found in AD cases, the results extracted from these models should be taken with caution.

Although these mice do not exhibit every single hallmark of the disease, they are a valuable tool for understanding the interaction between genetic factors able to modify the *in vivo* production and deposition of Aβ, identifying molecular targets, and evaluating new therapies designed to stop or slow down the pathological and clinical manifestations of this devastating disease. Some of the mouse models may represent good models of the early, preclinical disease. The models have been exploited to show how plaques develop in real time using 2-photon microscopy on live animals. Surprisingly, these studies revealed that individual plaques develop in several hours and then remain constant over months.[111] This technique promises to be useful for examining pathways and structural changes.

2. *In Vitro* Models of AD

In vitro models utilizing primary neuronal cultures provide an opportunity to decipher molecular mechanisms of AD. These models have been instrumental in our understanding of how familial AD gene mutations alter Aβ metabolism, secretion, and/or degradation.[112] Work using these models has shown that alterations in the γ-secretase complex produce an enhancement of APP processing[113,114] and lead to an increase in the intracellular concentration of Aβ species.[115] These modifications in the processing of APP alter the $A\beta_{40}/A\beta_{42}$ ratio, increasing the relative amount of $A\beta_{42}$.[116–118] *In vitro* models have provided critical insights into where in the cell APP is processed and how Aβ is generated and transported within the cell.[119]

Arguably, the most relevant use of *in vitro* models is to identify the neurotoxic species triggering the progression of the disease. The lack of correlation between the amyloid plaque burden and the clinical progression of the disease has led to the ongoing debate about the role of amyoid in the etiology of AD. However, studies of soluble oligomeric Aβ species show that early memory impairment and synaptic dysfunction correlate better with the levels of oligomeric Aβ than with the fibrillar species.[120] It is important to keep in mind that Aβ assemblies are highly dynamic and are in active equilibrium with the environment. This property makes it very difficult to ascribe specific toxic characteristics to a certain type of aggregation state.

However, several studies have examined the relationship between toxicity and aggregation. In 1998, the characterization of soluble, nonfibrillar oligomeric species (ADDL: Aβ derived diffusible ligands) and demonstration of their neurotoxic properties[121] suggested that oligomers, not fibrils, were the toxic species. ADDLs caused neuronal death in primary hippocampal neuron cultures and inhibited long-term potentiation in cultured slices from rat brains.[121,122] Another modification of the *in vitro* systems employs naturally produced, cell-derived Aβ oligomers[123,124] that are toxic in cultures and inhibit LTP when injected into rat brain.[124] The relevant size of the oligomers is a topic of debate, and oligomers as large as dodecamers have been isolated from human AD brains[125] and from transgenic mice.[126] These species are also toxic in primary neuronal cultures and impair LTP in hippocampal slices.[127]

3. Caspases in AD

a. Caspases in AD brains. Neuronal loss is an outstanding feature of AD. A role for apoptosis in this neuronal loss is suggested by the increased expression of caspases and cleaved caspase substrates in postmortem AD brains.[128,129] mRNA expression of caspases-1, -2, -3, -5, -6, -7, -8, and -9 is increased in the brain of AD patients compared to controls.[130] Pyramidal neurons from vulnerable regions involved in the disease showed an increase in activated caspase-3 and -6.[131,132] Negative correlations have been found between caspase-8 levels and age at disease onset and age at patient death, suggesting a role in disease regulation.[133] Postmortem AD brain tissue showed extensive expression of cleaved caspase-8 (p18) in neurons and reactive astrocytes in the hippocampus and the entorhinal cortex[134]; the immunoreactivity of cleaved caspase-8 is colocalized in tangle-bearing and plaque-associated neurons. As noted above, it is not clear whether this is active caspase-8 or effector-processed (nonactive) caspase-8.

Synaptosomes prepared from AD brain frontal cortices showed an enrichment in caspase-9 compared to nondemented controls.[135] While there is a great deal of correlative data showing that caspases are somehow involved in AD, there is not a great deal of specific data for individual caspases. Much of the data either measures changes in expression, which does not necessarily translate into a functional role for proteins that require activation, or utilizes the nonspecific peptide inhibitor/substrates to imply the involvement of specific caspases. The peptide inhibitor/substrate data can be viewed as indications of caspase function, but these data certainly cannot be construed as implicating individual caspases in AD dysfunction pathways.

b. Caspase cleavage of APP. APP contains caspase cleavage sites and it has been described that caspases-3, -6, -7, and -8 can cleave APP.[129,131,136] Cultures of neuronal cells deprived of serum showed activation of caspase-6 and subsequent processing of APP by caspase-6, generating a 6.5-kDa fragment that contains Aβ.[131] In this context, caspases may play an active role in Aβ-induced neurotoxicity. Amyloid plaques are enriched in caspase-cleaved APP.[129]

In vitro systems have been instrumental in defining the cytotoxic properties of APP cleavage fragments, particularly the C-terminal fragments (CTF) derived from γ- and β-secretase cleavage of APP. C99 is toxic to neurons[137] and expression of C99 in rodent brains induces Aβ deposition, neurodegeneration, alterations in behavior, and synaptic deficits.[138] C99 can be further cleaved by caspases to generate C31 and it may be the C31 that leads to the CTF toxicity.[138,139] The specific caspase that cleaves C99 has not been adequately identified, although caspases-8 and -9 have been implicated.[140] Expression of C31 alone is cytotoxic, suggesting that the cleavage of CTFs to generate C31 may potentiate susceptibility to apoptosis.[135] Mutations in APP to abolish the caspase cleavage site significantly attenuate the cytotoxic effects of C99.[139] PDAPP mice carrying the cleavage mutation [PDAPP(D664A) mice] deposit Aβ but do not develop behavioral deficits, supporting a role for C31 in the potentiation of AD deficits.[141] However, a recent study using these mice showed that there is no effect of expression of APPD664A on the development of behavioral defects.[142] Further work is needed to clarify the functional relevance of C31.

c. Caspase processing of tau. Caspases do not only process APP; caspase cleavage of tau enhances tau filament polymerization *in vitro.*[143,144] Cleaved caspase-6 in the cortex and hippocampus of human AD brains is colocalized with plaques and NFTs.[145,146] There is also an increase in caspase-6 cleaved tau fragments, detected with an antibody to the neoepitope generated when caspase-6 cleaves tau.[146] A study of cultured dorsal-root ganglion neurons suggests that NGF-withdrawal leads to APP cleavage, which then induces caspase-6 activation.[69] Taken together, these studies suggest that cleaved APP activates caspase-6, which may induce NFT formation. In support of a role for caspases in NFT formation are 2-photon studies. Live imaging from mice expressing mutant tau shows that there is activation of caspases prior to development of tangles. However, this study used a fluorescent caspase substrate/inhibitor (FLICA), which binds irreversibly to active sites, to follow caspase activation, so the causal link between caspase activation and tangle development cannot be inferred, despite the conclusions drawn from this study.[147] This work illustrates that caspase activation may be an early event in the death pathway. This is supported by studies discussed below showing that Aβ induces caspase activation in primary neurons within 2 h of addition to the cultures.[148]

d. Caspases and inflammation in AD. Inflammation is believed to contribute to the progression of AD.[149] AD brains have increased mRNA and protein expression of the inflammatory cytokines IL-1β and IL-18 in both neurons and glia that are colocalized with Aβ and tau.[150,151] Microglia isolated from autopsies of AD patients showed stronger IL-1β response to Aβ than did microglia isolated from control brains.[152] In the Tg2576 AD mouse model, astrocytes and microglia surround amyloid plaques, and IL-1β expression is upregulated in these astrocytes, suggesting that Aβ induces reactive astrocytes, leading to an increase in the inflammatory cytokine.[153]

While caspases may not be directly involved in regulating the expression of cytokines, they do regulate their release. Caspase-1 can regulate Aβ-mediated release of cytokines. Aβ treatment of mouse microglia causes increased IL-1β release.[154] Caspase-1 activation was assessed in this study by measuring its cleavage profile, which does not indicate whether the cleaved caspase-1 is active (see above), but the increase in IL-1β is strong support for activation of caspase-1.

What is the role of the NLRP caspase-1 inflammasome in the context of AD? No direct study has been done linking AD and caspase-1 activation in the human brain, but NLRP1 protein is found in neurons and oligodendrocytes.[155] It has been shown in cell lines and primary mouse macrophages that NLRP3 is activated following intracellular K+ withdrawal.[156] In spinal cord neurons treated with valinomycin, a K+ depleting agent, there is activation of the

NLRP1 caspase-1 inflammasome.[157] Moreover, cerebellar neurons grown in K+ free medium exhibited higher NLRP1 protein levels.[158] A link between AD and K+ channel dysfunction has been reported. Aβ has been implicated in inducing a decrease in intracellular K+ levels.[159,160] Based on these studies, it can be inferred that increased Aβ in the brain may activate the NLRP1 caspase-1 inflammasome through intracellular K+ depletion.

e. Caspases and Aβ toxicity. Caspase-2 has been shown to be required for Aβ-mediated death of primary neurons using antisense knockdown; neurons from caspase-2 null mice are resistant to Aβ toxicity.[161] Most recently, caspase-2 activity has been measured using the bVAD affinity ligand method in neurons treated with Aβ; measurable activity is induced within 2 h of treatment.[148] Other studies of Aβ induction of caspase-2 have employed peptide substrates to measure activity; these measures are not specific for caspase-2 (see above).

Cortical neurons from caspase-12 null mice are resistant to Aβ.[161,162] Caspase-12 has been shown to be involved in the signaling pathways associated with ER stress-induced apoptosis in rodent AD models.[162,163] However, due to a frame shift mutation in the human caspase-12 gene in most humans, the full-length protein is expressed in only a subset of people and there is no increase of AD in this population.[5,164] Hence, caspase-12 is not believed to play any role in human AD. Human caspase-4 shares a high homology with caspase-12, and it is also induced by ER stress.[165] Aβ increases caspase-4 cleavage in the human neuroblastoma cell line, SK-N-SH, within 24 h.[165]

It has also been proposed that Aβ cross-links the cell surface death receptors Fas/TNFR. In neuronal PC12 cells, Aβ induces downregulation of FLIPs, the endogenous caspase-8 inhibitor, leading to cleavage of caspase-8.[166,167] Another study utilizing SH-SY5Y neuroblastoma cells revealed that caspase-8 is recruited to the APP complex following Aβ stimulation. Once in the complex, caspase-8 is cleaved to a p10 fragment. It is argued that the cleaved/activated caspase-8 cleaves APP to induce cell death.[168] These studies have not shown that removal of caspase-8 blocks Aβ effects.

Calcium and the calmodulin-dependent protein kinase II (CaMKII) could play a partial role in regulating Aβ-induced caspase activity.[169] This work points to a potential calcium dysregulation induced by Aβ to activate CaMKII, which is sufficient to elicit caspase activity, measured with peptide substrates. However, CamKII may also phosphorylate caspase-2 to inactivate the caspase.[27] The function of CamKII with regard to Aβ and caspase-2 needs to be studied further. The role of calcium in regulating caspase actions is further demonstrated in a study that utilized a human neuronal cell line (hNT), revealing that Aβ induces accumulation of a lipid metabolite, 1-*o*-hexadecyl-2-acetyl-*sn*-glycero-3-phosphocholine (C16:0PAF). This lipid product causes elevated intracellular calcium levels as a result of ER stress that leads to caspase activation.

Peptide substrates were used to measure the caspase activity.[170] Furthermore, m-calpain was activated before detectable caspase activity using the peptide substrates, implicating this enzyme in the regulation of caspase activation.

f. Caspases and Aβ generation. In addition to responding to Aβ toxicity, caspase-2 or other caspases may play a role in generating Aβ. BACE, the β-secretase that generates Aβ from APP, has been shown to be stabilized by caspase-3 cleavage of GGA3, an adaptor protein involved in BACE trafficking.[171] As discussed below, it was shown that in a model of cerebral ischemia, this mechanism leads to an increase in Aβ.

Taken together, caspases are very closely linked to the degeneration process found in AD brains, but the specific caspases involved and the downstream effect of these caspases has not been adequately defined. Figure 5 proposes how caspases may act in AD pathogenesis. Table I summarizes the current evidence of a role for each caspase in AD.

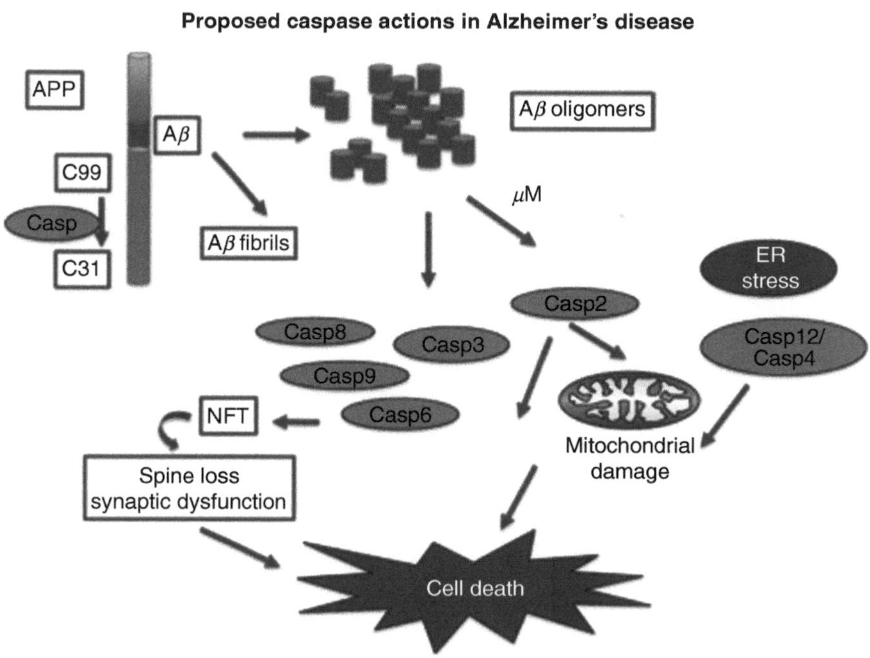

FIG. 5. *Proposed caspase actions in Alzheimer's Disease.* The amyloid precursor protein (APP) can be proteolytically processed to release multiple products, including Aβ, C99, and C31. C99 and C31 are toxic to neurons. Aβ can form fibrils, which deposit as amyloid plaques and soluble oligomers and can lead to ER stress, caspase activation, tau hyperphosphorylation, neuronal dysfunction, and neuronal death.

TABLE I
CASPASES IMPLICATED IN ALZHEIMER'S DISEASE

Caspase	Paradigm	Measure	References
1	AD brain Aβ treated microglia	mRNA Casp1 cleavage *In vitro* Tau cleavage	128,143,154
2	AD brain Aβ treated hippocampal neurons Aβ production from H4-C99 cells	mRNA Casp2 null neurons siRNA depletion Casp2 cleavage z-VDVAD-fmk	128,161,235
3	AD brain Aβ production from H4-C99 cells Tg4519 mice	mRNA z-DEVD-fmk *In vitro* Tau cleavage cleaved Casp3 IHC Casp3 cleavage	128,143,235–238
4	Aβ treatment of SK-N-SH cells	Casp4 cleavage siRNA depletion	165,239
5	AD brain	mRNA	128
6	AD brain Tg4510 mice	mRNA *In vitro* Tau cleavage Casp6 cleavage	128,131,143,236
7	AD brain Tg4510 mice	mRNA *In vitro* Tau cleavage	128,143,236
8	AD brain Aβ production from H4-C99 cells	mRNA siRNA depletion z-IETD-fmk *In vitro* Tau cleavage cleaved Casp8 IHC	128,134,143,235,240
9	AD brain	mRNA cleaved Casp9 IHC	128,240
12	Aβ treated cortical neurons Hippocampal organotypic slices treated with Aβ	Antisense to Casp12 Casp12 null neurons Casp12 cleavage	162,241

B. Cerebral Ischemia

Stroke, an acute neurodegenerative disorder, is the third most common cause of death and a leading cause of disability in major industrialized countries.[172] Approximately 85% of strokes are ischemic, resulting from a thrombus that occludes a major cerebral artery and leads to a focal loss of blood flow. Unless normal blood flow is recovered within a short period of time, under 3 h from onset of occlusion, there is massive cellular death within the ischemic territory. Strokes can also be a consequence of either cardiac arrest or intraparenchymal hemorrhage.

The series of events instigated by ischemia develops over many days. Inflammation and altered microvascular permeability produce tissue edema. Direct effects on cells trigger glutamatergic excitotoxicity, ionic imbalance, and free-radical reactions, among others. There are also dynamic interactions of the neurovascular unit (the complex of neurons, the microvessels that supply them, the supportive cells: astroglia and microglia, and other resident inflammatory cells).[173,174]

Over the past decade, the mode of cell death induced by ischemia has been shown to be combined necrotic and apoptotic cell death. The center of the lesion, the core, has complete energy depletion, and there is a gradient of energy depletion from the core toward the bordering zone, the penumbra. From a morphologic perspective, the core has been considered mainly necrotic,[175,176] while the penumbra has been considered apoptotic.[177–179] However, it has been suggested that the initial events in the core are apoptotic, and secondary necrosis results from a rapid failure to fully develop the apoptotic program because of the maintained depletion of apoptosis-requiring energy stores in the core.[180]

Neuronal death in the adult brain does not necessarily exhibit the classic morphological manifestations of apoptosis, as initially described by Kerr *et al.*,[181] while in the neonatal brain, hypoxic-ischemic insults can cause typical morphological changes of apoptosis. In the adult brain, neurons are equipped with antiapoptotic molecules that raise the apoptotic threshold.[182] There is a large body of evidence that suggests there is caspase-mediated cell death after ischemia,[183] and multiple studies implicate apoptotic pathways as critical mediators of ischemia.[184] For more than a decade, it has been suggested that cell death in the CNS following injury can coexist as apoptosis, necrosis, and hybrid forms along an apoptosis–necrosis continuum.[185]

The damage associated with the sudden return of blood flow, or reperfusion, to the oxygen-starved, energetically compromised tissue is mechanistically and therapeutically important.[186] With timely reperfusion, either spontaneous or therapeutic, territory in the penumbra may be salvaged, but a delay in reperfusion will cause cell death. This phenomenon is termed "reperfusion injury." It is held that much of this injury is produced during transient ischemia by inflammation, excitotoxicity, and apoptotic cell injury,[174] and can result in the disruption of the blood–brain barrier (BBB) and formation of brain edema.[187] An advantage of modeling stroke is that, unlike other neurodegenerative diseases, in stroke, the primary insult that triggers the pathology is known, and the models provide a highly reproducible, clinically relevant approach for studying this acute insult.

1. Animal Models of Stroke

Animal models are widely used to mimic human stroke. In contrast to other neurodegenerative diseases, there are excellent animal models of stroke that provide evidence for mechanisms, as well as prevention and treatment efficacy *in vivo*, that allow for control of collateral factors.

While cerebral ischemia can be modeled in large animals (pigs, baboons, etc.), rodent models are the most widely used. Although there are differences in anatomy and functionality between the rodent and the human brain, these models permit a careful dissection of mechanisms of injury and neuroprotection and allow examination of events that occur from onset of the insult to weeks or months after the ischemic event [for a review, see Ref. 188].

There are a variety of models that reflect the assortment of insults in humans. These models can be divided into three subgroups: global ischemia, focal ischemia, and hemorrhagic infarct. Global models, where the major blood vessels to the forebrain are occluded, are models of the consequences of cardiac arrest rather than stroke.[189] In focal models, a smaller arterial branch is occluded. Most models occlude the middle cerebral artery (MCA), as occlusion of this vessel is found in the majority of human ischemic strokes.[189] Global and focal occlusion models are either permanent or transient, allowing examination of the damage that results from both ischemia and reperfusion. Hemorrhagic stroke models involve the stereotactic injection of collagenase into the cerebral parenchyma.[189]

An overriding question of much animal model research in the last decade has been the translational value of the models. The lack of efficacy of neuroprotectants against stroke in clinical trials raised the question of the validity of the animal models for preclinical research. A proof of confidence in the value of animal models is that changing several physiological parameters, like reperfusion, hyperglycemia, hyperthermia, or blood pressure, has similar effects on the outcome of stroke in humans and in animal models [for a review, see Ref. 189]. Nevertheless, although animal models mimic the human stroke pathology, small mammalian brains are quite different from human brains in function and morphology, and this may influence the efficacy of the therapies. This phenomenon was recently represented in an *in silico* model, which considered the differences in the percentage of white matter and glia between rodents and humans.[190] If this hypothesis is correct, the efficacy of the therapies should be apparent when animals with a more similar brain composition to humans are used for models.

From a practical, logistical perspective, it is difficult to treat patients at early time points; therefore, clinical investigators should focus on mechanisms that occur later in the ischemic cascade. However, there are also clinical settings where the risk of stroke is increased, such as cardiac surgery, where preventative treatments would be useful. Additionally, it is important to consider how much protection is achieved in preclinical trials, where a reduction of at least 50% of the ischemic damage has been the goal, and also to develop appropriate behavioral tests. There are multiple mechanisms occurring simultaneously and sequentially during stroke. Therefore, it is important to know the chronology and relevance of the different mechanisms that are occurring as a consequence of ischemia and reperfusion. The best treatment strategy might rely on a combination of neuroprotective agents, each geared toward a different death mechanism, administered at the appropriate time to interfere in the multiple ischemia-induced death pathways.

2. *In Vitro* Models of Ischemia

In vitro models have been utilized for over two decades in the study of cerebral ischemia.[191] Much of our understanding of the pathogenesis of stroke derives from studies of cell lines, and primary and organotypic cultures exposed to hypoxia, anoxia, oxygen, and glucose deprivation. These models allow the dissection of pathways in a controlled setting. Of the models under study, the oxygen–glucose deprivation (OGD) model is believed to most resemble the *in vivo* ischemia setting.[192–194] Typical insults last for 15–90 min and are immediately followed by reperfusion, that is, addition of glucose to media and return to normoxic conditions.[192,193] Acute ischemic insults (10–20 min) are used for electrophysiological analysis in hippocampal neurons[195,196] as well as spiny/aspiny striatal neurons.[197,198] The discussion below gives a brief overview of work in hippocampal primary cultures or brain slices, unless otherwise stated.

Early evidence with the OGD model suggested that ischemic injury was mainly necrotic and mediated by glutamate excitotoxicity,[199–201] but subsequent work showed that apoptosis is masked by glutamate receptor activation in OGD in primary cultures.[193] Apoptosis in primary cultures is a delayed event (~24–36 h postinsult), while glutamate-mediated necrosis is observed as early as 2 h after the insult.[192] As with some other insults, the extent of the insult, in this case OGD, determines how much necrosis occurs. Shorter periods of OGD (~15 min) result in most neurons dying by apoptosis, while longer periods of OGD (~75 min) do not alter apoptotic death, but increase the amount of necrosis.[202] The relative timing of apoptosis and necrosis are different in hippocampal brain slices exposed to OGD compared to primary cultures. In slices, apoptosis and glutamate-dependent necrosis both occur within 3 h after the ischemic episode.[203–205] These discrepancies may signify the importance of the neuron–glia context in studying ischemia *in vitro*.

The role of caspases has been studied in limited detail in OGD.[202,204–207] These studies show that caspase-3 is activated by OGD and can be regulated by pan-caspase inhibitors. OGD cells treated with NMDA receptor antagonists undergo apoptotic death that can be inhibited by a pan-caspase inhibitor, z-VAD-fmk.[204,206] Another caspase inhibitor (Ac-YVAD-cmk) exhibited similar neuroprotection in OGD brain slices.[205] These inhibitors are not specific for individual caspases,[50] and therefore, these studies do not clarify which caspases are important in promoting death in this ischemia model. Knockout of caspase-3 is also neuroprotective against OGD.[208] Cleavage of caspases-3, -8, and -9 has been observed in microglia exposed to OGD,[209] although the causality of the different caspases has not been determined. It has also been shown that cleaved caspase-7 and caspase-3 are found exclusively in microglia and neurons, and not astrocytes in OGD in mixed primary cultures,[202] again suggesting that the neuron–glia interactions are important in modeling ischemia.

As will be discussed below, caspase-1 and generation of interleukin-1β are found in animal models of ischemia. Activation of these pathways and subsequent neuronal death are found when SOD-1, the cytoplasmic superoxide dismutase, is downregulated.[210,211] This death is mediated by activation of caspase-1, secretion of IL-1β, and generation of peroxynitrite, all molecules that have been shown to be important in the progression of ischemia *in vivo*.[212–215] Thus, this model may provide a cellular model for investigation of the mechanism of neuronal death in ischemia.

3. Caspases in Stroke

In 1997, a function for caspases in mammalian neurodegeneration was first shown with the demonstration that a dominant-negative mutant of caspase-1 provided protection against tMCAo.[216] During the ensuing years, multiple studies have been published, and it is clear that caspases are key molecules in the death mechanisms induced during ischemia.[183,186] We will present data for caspases that are conserved in human and rodents.

a. Caspase-1. As noted above, the first studies of caspase in neuronal disease showed that ablation of caspase-1 activity provided protection against tMCAo, but those studies used a dominant-negative caspase-1 that could block activity of other caspases in addition to caspase-1.[216] This work was followed with studies using caspase-1 null mice which exhibited reduced ischemic damage.[217] Indeed, several lines of evidence point to an important role for this caspase. The pro-inflammatory cytokine IL-1β, which is the main target of caspase-1 processing, is rapidly induced by focal/global ischemia.[218] Ablation of IL-1β expression provides more than 80% protection from ischemic damage.[214] Caspase-1 has been reported to be increased in the core of the stroke[180] and in the penumbra.[219] A recent review discusses the role of inflammation in acute neurodegeneration.[30]

b. Caspase-2. The initial characterization of the caspase-2 null mice showed that the caspase-2 null mice were not protected from tMCAo.[61] A more recent study of caspase-2 and PIDD in global ischemia shows a role for PIDD in this model of ischemia.[220] However, this study does not directly measure caspase-2 activity and, since the PIDD null mice appear to undergo caspase-2 dependent death[22,23] (although neurons have not been studied), it is not clear that caspase-2 is part of the ischemic death mechanism. But it must be considered that, since the caspase-2 null mice have been shown to have an increased expression of caspase-9 and Smac/DIABLO in the brain,[64] the compensations in death molecules might mask a role for caspase-2 in ischemia. Thus, the role of caspase-2 and of PIDD in ischemia requires further study.

c. Caspase-3. Caspase-3 is present in the ischemic penumbra[59,180] and its deletion renders mice more resistant to ischemic injury.[208] However, although caspase activation is generally described in the penumbra of focal infarcts, immunohistochemical analysis in MCAo has detected neurons containing caspase-3 in the infarct core.[221,222] In humans, although there is limited information available, there is a procaspase-3 increase in expression within hours resulting from permanent arterial occlusion,[223] and activated caspase-3 and cleaved PARP have been detected in some neurons several days after cardiac arrest with reperfusion.[183] Cleaved caspase-3 has also been detected in glial cells 24 h postinfarction.[224]

d. Caspase-6. Little data exists with regard to a role for caspase-6 in stroke. Studies of mRNA (see below) showed an increase in caspase-6 expression in rats subjected to pMCAo.

e. Caspase-7. Much of the data with regard to caspase-7 is overlapping with the data about caspase-3. A clear, distinct role for caspase-7 in stroke has not yet been described. Studies of mRNA (see below) showed an increase in caspase-7 expression in rats subjected to pMCAo.

f. Caspase-8. Increased expression of caspase-8 has been seen in both the core and in the penumbra,[180] as has increased expression of FasL.[219]

g. Caspase-9. A limited study of human brains obtained from 4 h to 5 days postinfarction suggests an increase in caspase-9 protein expression peaking at 24 h postinfarct.[225] In rodent models, caspase-9 is increased in the penumbra.[180]

h. mRNA expression. Harrison *et al.*[226] reported a strict transcriptional regulation of caspase expression following pMCAo in rats. There is a pattern of expression with an increase of caspase-1, -3, -6, -7, -8, and -11 mRNA, and a decrease of caspase-9 was detected at different time points. Caspase-2 mRNA shows no changes. But, in this report, the cellular types expressing these proteases were not studied.

Figure 6 presents an overview of what is presently known about the localization of different caspases during tMCAo. The timing of appearance and cell type location for the different caspases still needs further study. Actual

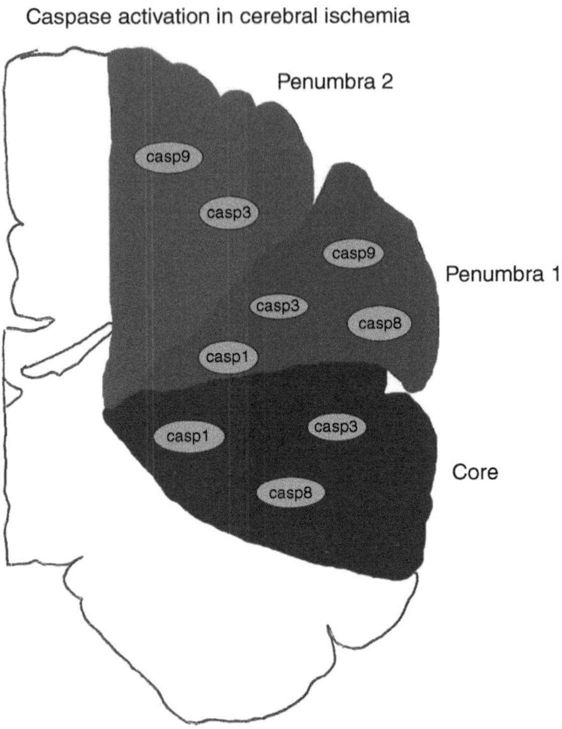

FIG. 6. *Caspase activation in cerebral ischemia.* Caspases are activated in the ischemic core early in stroke and then in the penumbra as the stroke progresses.

measures of specific caspase activity induced by ischemia are also needed. The knowledge of the pathways and molecules that interconnect these caspases is still elusive, and more research has to be done in this area. Table II summarizes the evidence for a role for each caspase in stroke.

C. Linking Stroke and AD

In this chapter, we have summarized the current knowledge of the roles of caspases in AD and stroke. Caspases can either participate in the development of the disease or in the progression of the disease. Interestingly, a number of studies have revealed a positive correlation between stroke and AD. Cross-epidemiological studies show that patients who have suffered a stroke have an increased likelihood of developing dementia.[227–229] In a rodent ischemic model, tau hyperphosphorylation is present in the cortices.[230] In terms of the relationship between stroke and amyloid, studies point to caspases as key players in modulating between these two conditions. Rodents subjected to

TABLE II
Caspases Implicated in Ischemia

Caspase	Paradigm	Measure	References
1	tMCAo pMCAo	Dominant-negative Casp1 Casp1 null mice ac-YVAD-AFC mRNA	180,216,217,226,242
2	Global ischemia	Caspase-2 cleavage	220
3	tMCAo pMCAO	Casp3 null mice WB clC3 ac-DEVD-AFC/WB/IHC mRNA	180,208,221,226
6	pMCAo	mRNA levels	226
7	pMCAo	mRNA levels	226
8	pMCAo	mRNA levels ac-IETD-AFC/WB/IHC	180,221,226
9	human stroke tissue tMCAo rodent tMCAo canine	IHC C9 Release of C9 from mitochondria	219,225,243

tMCAo showed increases in APP and Aβ expression in the ischemic area, corpus callosum, thalamus, striatum, caudate-putamen, and hippocampus within days of reperfusion.[231–233] The increased expression levels are observed in both neurons and astrocytes in short-term studies ranging from days to weeks. However, over longer survival time, the levels of APP and Aβ gradually decrease.[231] This change may be explained by a downregulation of APP and Aβ in astrocytes and/or clearance of these peptides by astrocytes and microglia. Although the expressions of APP and Aβ decrease over time, in the thalamus, the expression profile of Aβ changes from diffused to plaque-like deposits.[232] In sum, these rodent studies point to an initial upregulation in APP and Aβ in a diffused pattern in various brain regions following MCAO. Over time, the levels of these proteins are reduced. However, there remains residual damage in the form of plaque-like structures.

The mechanism for elevated APP and Aβ in brain regions following stroke has begun to be addressed by investigators. Administration of zDEVDfmk to rats prior to tMCAo blocked the appearance of cleaved caspase-3 and abrogated the increased expression of BACE1, the enzyme that produces Aβ from APP, and Aβ.[233,234] Similar effects were seen in a model of brain trauma that shows increased Aβ levels.[233,234] While these studies do not specifically examine caspase-3 function, they do implicate activation of caspases in the production of Aβ, providing a mechanism underlying the clinical observations that stroke or head trauma are risk factors for the development of AD.

V. Future Directions

There is an increasing appreciation that caspases have a variety of actions, both death and non-death-related. With this in mind, it is even more important that our understanding of the specific functions of individual caspases in health and disease is increased. In this way, appropriate disease interventions that do not interfere with the non-disease-related actions of caspases can be devised. In order to progress in our understanding of caspase functions, attention must be directed to the tools available to study caspases and the development of better tools for analysis. Multiple studies have now shown that many of the tools used to study caspases are inadequate, yet they continue to be used and to confound the literature. A search of the current literature will show how many studies still use the pseudopeptide inhibitors/substrates as evidence for the function of specific caspases. Many studies still equate caspase cleavage by Western blotting with activity, or caspase-3 cleavage by immunocytochemistry with apoptosis. As of today, the best tools available are molecular manipulation and affinity ligands.

Another issue in the study of human disease is the models available. In the two diseases that we have used to illustrate current knowledge of caspases in neuronal disease, there are different issues. The *in vivo* models of cerebral ischemia mimic the human disease quite accurately and provide excellent systems for studying death mechanisms. The cellular models of ischemia, while providing the advantages of cell-based models in the study of biochemical and molecular events, do not offer the complexity of the various elements clearly involved in the evolution of the ischemic event. For AD, the *in vivo* models available at best model preclinical disease. There is no adequate model to mimic the progression of the disease. The current models have little or no neuronal loss, making the study of death mechanisms in these animals difficult. Thus, at present, the cell models and the hippocampal slice systems offer the best models for studying death mechanisms in AD. Better models and more specific tools are in development and are eagerly awaited by the field.

References

1. Miura M, Zhu H, Rotello R, Hartwieg EA, Yuan J. Induction of apoptosis in fibroblasts by IL-1 beta-converting enzyme, a mammalian homolog of the *C. elegans* cell death gene ced-3. *Cell* 1993;**75**:653–60.
2. Thornberry NA, Bull HG, Calaycay JR, Chapman KT, Howard AD, Kostura MJ, et al. A novel heterodimeric cysteine protease is required for interleukin-1 beta processing in monocytes. *Nature* 1992;**356**:768–74.
3. Yuan J, Shaham S, Ledoux S, Ellis HM, Horvitz HR. The *C. elegans* cell death gene ced-3 encodes a protein similar to mammalian interleukin-1 beta-converting enzyme. *Cell* 1993;**75**:641–52.

4. Eckhart L, Ban J, Fischer H, Tschachler E. Caspase-14: analysis of gene structure and mRNA expression during keratinocyte differentiation. *Biochem Biophys Res Commun* 2000;**277**:655–9.
5. Fischer H, Koenig U, Eckhart L, Tschachler E. Human caspase 12 has acquired deleterious mutations. *Biochem Biophys Res Commun* 2002;**293**:722–6.
6. Roy S, Sharom JR, Houde C, Loisel TP, Vaillancourt JP, Shao W, et al. Confinement of caspase-12 proteolytic activity to autoprocessing. *Proc Natl Acad Sci USA* 2008;**105**:4133–8.
7. Kim EM, Shin EJ, Choi JH, Son HJ, Park IS, Joh TH, et al. Matrix metalloproteinase-3 is increased and participates in neuronal apoptotic signaling downstream of caspase-12 during endoplasmic reticulum stress. *J Biol Chem* 2010;**285**:16444–52.
8. Boatright KM, Salvesen GS. Caspase activation. *Biochem Soc Symp* 2003;**70**:233–42.
9. Pop C, Salvesen GS. Human caspases: activation, specificity, and regulation. *J Biol Chem* 2009;**284**:21777–81.
10. Boatright KM, Renatus M, Scott FL, Sperandio S, Shin H, Pedersen IM, et al. A unified model for apical caspase activation. *Mol Cell* 2003;**11**:529–41.
11. Adams JM, Cory S. The Bcl-2 apoptotic switch in cancer development and therapy. *Oncogene* 2007;**26**:1324–37.
12. Baliga BC, Read SH, Kumar S. The biochemical mechanism of caspase-2 activation. *Cell Death Differ* 2004;**11**:1234–41.
13. Stennicke HR, Deveraux QL, Humke EW, Reed JC, Dixit VM, Salvesen GS. Caspase-9 can be activated without proteolytic processing. *J Biol Chem* 1999;**274**:8359–62.
14. Oberst A, Pop C, Tremblay AG, Blais V, Denault JB, Salvesen GS, et al. Inducible dimerization and inducible cleavage reveal a requirement for both processes in caspase-8 activation. *J Biol Chem* 2010;**285**:16632–42.
15. Riedl SJ, Salvesen GS. The apoptosome: signalling platform of cell death. *Nat Rev* 2007;**8**:405–13.
16. Denault JB, Eckelman BP, Shin H, Pop C, Salvesen GS. Caspase 3 attenuates XIAP (X-linked inhibitor of apoptosis protein)-mediated inhibition of caspase 9. *Biochem J* 2007;**405**:11–9.
17. Kang TB, Oh GS, Scandella E, Bolinger B, Ludewig B, Kovalenko A, et al. Mutation of a self-processing site in caspase-8 compromises its apoptotic but not its nonapoptotic functions in bacterial artificial chromosome-transgenic mice. *J Immunol* 2008;**181**:2522–32.
18. Wang L, Du F, Wang X. TNF-alpha induces two distinct caspase-8 activation pathways. *Cell* 2008;**133**:693–703.
19. Jia SH, Parodo J, Kapus A, Rotstein OD, Marshall JC. Dynamic regulation of neutrophil survival through tyrosine phosphorylation or dephosphorylation of caspase-8. *J Biol Chem* 2008;**283**:5402–13.
20. Tinel A, Tschopp J. The PIDDosome, a protein complex implicated in activation of caspase-2 in response to genotoxic stress. *Science* 2004;**304**:843–6.
21. Troy CM, Ribe EM. Caspase-2: vestigial remnant or master regulator? *Sci Signal* 2008;**1**:pe42.
22. Manzl C, Krumschnabel G, Bock F, Sohm B, Labi V, Baumgartner F, et al. Caspase-2 activation in the absence of PIDDosome formation. *J Cell Biol* 2009;**185**:291–303.
23. Kim IR, Murakami K, Chen NJ, Saibil SD, Matysiak-Zablocki E, Elford AR, et al. DNA damage- and stress-induced apoptosis occurs independently of PIDD. *Apoptosis* 2009;**14**:1039–49.
24. Berube C, Boucher LM, Ma W, Wakeham A, Salmena L, Hakem R, et al. Apoptosis caused by p53-induced protein with death domain (PIDD) depends on the death adapter protein RAIDD. *Proc Natl Acad Sci USA* 2005;**102**:14314–20.
25. Tinel A, Janssens S, Lippens S, Cuenin S, Logette E, Jaccard B, et al. Autoproteolysis of PIDD marks the bifurcation between pro-death caspase-2 and pro-survival NF-kappaB pathway. *EMBO J* 2007;**26**:197–208.

26. Cuenin S, Tinel A, Janssens S, Tschopp J. p53-induced protein with a death domain (PIDD) isoforms differentially activate nuclear factor-kappaB and caspase-2 in response to genotoxic stress. *Oncogene* 2008;**27**:387–96.
27. Nutt LK, Margolis SS, Jensen M, Herman CE, Dunphy WG, Rathmell JC, et al. Metabolic regulation of oocyte cell death through the CaMKII-mediated phosphorylation of caspase-2. *Cell* 2005;**123**:89–103.
28. Shin S, Lee Y, Kim W, Ko H, Choi H, Kim K. Caspase-2 primes cancer cells for TRAIL-mediated apoptosis by processing procaspase-8. *EMBO J* 2005;**24**:3532–42.
29. Nutt LK, Buchakjian MR, Gan E, Darbandi R, Yoon SY, Wu JQ, et al. Metabolic control of oocyte apoptosis mediated by 14-3-3zeta-regulated dephosphorylation of caspase-2. *Dev Cell* 2009;**16**:856–66.
30. Trendelenburg G. Acute neurodegeneration and the inflammasome: central processor for danger signals and the inflammatory response? *J Cereb Blood Flow Metab* 2008;**28**:867–81.
31. Martinon F, Burns K, Tschopp J. The inflammasome: a molecular platform triggering activation of inflammatory caspases and processing of proIL-beta. *Mol Cell* 2002;**10**:417–26.
32. Agostini L, Martinon F, Burns K, McDermott MF, Hawkins PN, Tschopp J. NALP3 forms an IL-1beta-processing inflammasome with increased activity in Muckle-Wells autoinflammatory disorder. *Immunity* 2004;**20**:319–25.
33. Petrilli V, Papin S, Tschopp J. The inflammasome. *Curr Biol* 2005;**15**:R581.
34. Tschopp J, Schroder K. NLRP3 inflammasome activation: the convergence of multiple signalling pathways on ROS production? *Nat Rev Immunol* 2010;**10**:210–5.
35. Martinon F, Tschopp J. Inflammatory caspases and inflammasomes: master switches of inflammation. *Cell Death Differ* 2007;**14**:10–22.
36. Petrilli V, Dostert C, Muruve DA, Tschopp J. The inflammasome: a danger sensing complex triggering innate immunity. *Curr Opin Immunol* 2007;**19**:615–22.
37. Srinivasula SM, Ashwell JD. IAPs: what's in a name? *Mol Cell* 2008;**30**:123–35.
38. Eckelman BP, Salvesen GS, Scott FL. Human inhibitor of apoptosis proteins: why XIAP is the black sheep of the family. *EMBO Rep* 2006;**7**:988–94.
39. Eckelman BP, Salvesen GS. The human anti-apoptotic proteins cIAP1 and cIAP2 bind but do not inhibit caspases. *J Biol Chem* 2006;**281**:3254–60.
40. Irmler M, Thome M, Hahne M, Schneider P, Hofmann K, Steiner V, et al. Inhibition of death receptor signals by cellular FLIP. *Nature* 1997;**388**:190–5.
41. Du C, Fang M, Li Y, Li L, Wang X. Smac, a mitochondrial protein that promotes cytochrome c-dependent caspase activation by eliminating IAP inhibition. *Cell* 2000;**102**:33–42.
42. Verhagen AM, Ekert PG, Pakusch M, Silke J, Connolly LM, Reid GE, et al. Identification of DIABLO, a mammalian protein that promotes apoptosis by binding to and antagonizing IAP proteins. *Cell* 2000;**102**:43–53.
43. Hegde R, Srinivasula SM, Zhang Z, Wassell R, Mukattash R, Cilenti L, et al. Identification of Omi/HtrA2 as a mitochondrial apoptotic serine protease that disrupts inhibitor of apoptosis protein-caspase interaction. *J Biol Chem* 2002;**277**:432–8.
44. Liston P, Fong WG, Kelly NL, Toji S, Miyazaki T, Conte D, et al. Identification of XAF1 as an antagonist of XIAP anti-Caspase activity. *Nat Cell Biol* 2001;**3**:128–33.
45. Arora V, Cheung HH, Plenchette S, Micali OC, Liston P, Korneluk RG. Degradation of survivin by the X-linked inhibitor of apoptosis (XIAP)-XAF1 complex. *J Biol Chem* 2007;**282**:26202–9.
46. Chang DW, Xing Z, Capacio VL, Peter ME, Yang X. Interdimer processing mechanism of procaspase-8 activation. *EMBO J* 2003;**22**:4132–42.
47. Pop C, Fitzgerald P, Green DR, Salvesen GS. Role of proteolysis in caspase-8 activation and stabilization. *Biochemistry* 2007;**46**:4398–407.

48. Thornberry NA, Rano TA, Peterson EP, Rasper DM, Timkey T, Garcia-Calvo M, et al. A combinatorial approach defines specificities of members of the caspase family and granzyme B Functional relationships established for key mediators of apoptosis. *J Biol Chem* 1997;**272**:17907–11.
49. Talanian RV, Quinlan C, Trautz S, Hackett MC, Mankovich JA, Banach D, et al. Substrate specificities of caspase family proteases. *J Biol Chem* 1997;**272**:9677–82.
50. McStay GP, Salvesen GS, Green DR. Overlapping cleavage motif selectivity of caspases: implications for analysis of apoptotic pathways. *Cell Death Differ* 2008;**15**:322–31.
51. Tu S, McStay GP, Boucher LM, Mak T, Beere HM, Green DR. In situ trapping of activated initiator caspases reveals a role for caspase-2 in heat shock-induced apoptosis. *Nat Cell Biol* 2006;**8**:72–7.
52. Bouchier-Hayes L, Oberst A, McStay GP, Connell S, Tait SW, Dillon CP, et al. Characterization of cytoplasmic caspase-2 activation by induced proximity. *Mol Cell* 2009;**35**:830–40.
53. Kitevska T, Spencer DM, Hawkins CJ. Caspase-2: controversial killer or checkpoint controller? *Apoptosis* 2009;**14**:829–48.
54. Kuida K, Haydar TF, Kuan CY, Gu Y, Taya C, Karasuyama H, et al. Reduced apoptosis and cytochrome c-mediated caspase activation in mice lacking caspase 9. *Cell* 1998;**94**:325–37.
55. Cecconi F, Alvarez-Bolado G, Meyer BI, Roth KA, Gruss P. Apaf1 (CED-4 homolog) regulates programmed cell death in mammalian development. *Cell* 1998;**94**:727–37.
56. Hakem R, Hakem A, Duncan GS, Henderson JT, Woo M, Soengas MS, et al. Differential requirement for caspase 9 in apoptotic pathways in vivo. *Cell* 1998;**94**:339–52.
57. Yoshida H, Kong YY, Yoshida R, Elia AJ, Hakem A, Hakem R, et al. Apaf1 is required for mitochondrial pathways of apoptosis and brain development. *Cell* 1998;**94**:739–50.
58. Kuida K, Zheng TS, Na S, Kuan C, Yang D, Karasuyama H, et al. Decreased apoptosis in the brain and premature lethality in CPP32-deficient mice. *Nature* 1996;**384**:368–72.
59. Namura S, Zhu J, Fink K, Endres M, Srinivasan A, Tomaselli KJ, et al. Activation and cleavage of caspase-3 in apoptosis induced by experimental cerebral ischemia. *J Neurosci* 1998;**18**:3659–68.
60. Kuida K, Lippke JA, Ku G, Harding MW, Livingston DJ, Su MS, et al. Altered cytokine export and apoptosis in mice deficient in interleukin-1 beta converting enzyme. *Science* 1995;**267**:2000–3.
61. Bergeron L, Perez GI, Macdonald G, Shi L, Sun Y, Jurisicova A, et al. Defects in regulation of apoptosis in caspase-2-deficient mice. *Genes Dev* 1998;**12**:1304–14.
62. Wang S, Miura M, Jung YK, Zhu H, Li E, Yuan J. Murine caspase-11, an ICE-interacting protease, is essential for the activation of ICE. *Cell* 1998;**92**:501–9.
63. Zheng TS, Hunot S, Kuida K, Momoi T, Srinivasan A, Nicholson DW, et al. Deficiency in caspase-9 or caspase-3 induces compensatory caspase activation. *Nat Med* 2000;**6**:1241–7.
64. Troy CM, Rabacchi SA, Hohl JB, Angelastro JM, Greene LA, Shelanski ML. Death in the balance: alternative participation of the caspase-2 and -9 pathways in neuronal death induced by nerve growth factor deprivation. *J Neurosci* 2001;**21**:5007–16.
65. D'Amelio M, Cavallucci V, Cecconi F. Neuronal caspase-3 signaling: not only cell death. *Cell Death Differ* 2010;**17**:1104–14.
66. Fernando P, Brunette S, Megeney LA. Neural stem cell differentiation is dependent upon endogenous caspase 3 activity. *FASEB J* 2005;**19**:1671–3.
67. Aranha MM, Sola S, Low WC, Steer CJ, Rodrigues CM. Caspases and p53 modulate FOXO3A/Id1 signaling during mouse neural stem cell differentiation. *J Cell Biochem* 2009;**107**:748–58.
68. Williams DW, Kondo S, Krzyzanowska A, Hiromi Y, Truman JW. Local caspase activity directs engulfment of dendrites during pruning. *Nat Neurosci* 2006;**9**:1234–6.

69. Nikolaev A, McLaughlin T, O'Leary DD, Tessier-Lavigne M. APP binds DR6 to trigger axon pruning and neuron death via distinct caspases. *Nature* 2009;**457**:981–9.
70. Huesmann GR, Clayton DF. Dynamic role of postsynaptic caspase-3 and BIRC4 in zebra finch song-response habituation. *Neuron* 2006;**52**:1061–72.
71. Gulyaeva NV, Kudryashov IE, Kudryashova IV. Caspase activity is essential for long-term potentiation. *J Neurosci Res* 2003;**73**:853–64.
72. Stepanichev MY, Kudryashova IV, Yakovlev AA, Onufriev MV, Khaspekov LG, Lyzhin AA, et al. Central administration of a caspase inhibitor impairs shuttle-box performance in rats. *Neuroscience* 2005;**136**:579–91.
73. Li Z, Jo J, Jia JM, Lo SC, Whitcomb DJ, Jiao S, et al. Caspase-3 activation via mitochondria is required for long-term depression and AMPA receptor internalization. *Cell* 2010;**141**:859–71.
74. Gu Y, Nieves JW, Stern Y, Luchsinger JA, Scarmeas N. Food combination and Alzheimer disease risk: a protective diet. *Arch Neurol* 2010;**67**:699–706.
75. Pappolla MA, Chyan YJ, Poeggeler B, Frangione B, Wilson G, Ghiso J, et al. An assessment of the antioxidant and the antiamyloidogenic properties of melatonin: implications for Alzheimer's disease. *J Neural Transm* 2000;**107**:203–31.
76. Blennow K, de Leon MJ, Zetterberg H. Alzheimer's disease. *Lancet* 2006;**368**:387–403.
77. Alzheimer A, Stelzmann RA, Schnitzlein HN, Murtagh FR. An English translation of Alzheimer's 1907 paper, "Uber eine eigenartige Erkankung der Hirnrinde" *Clin Anat* 1995;**8**:429–31.
78. Small GW, Rabins PV, Barry PP, Buckholtz NS, DeKosky ST, Ferris SH, et al. Diagnosis and treatment of Alzheimer disease and related disorders consensus statement of the American Association for Geriatric Psychiatry, the Alzheimer's Association, and the American Geriatrics Society. *JAMA* 1997;**278**:1363–71.
79. Terry RD, Peck A, DeTeresa R, Schechter R, Horoupian DS. Some morphometric aspects of the brain in senile dementia of the Alzheimer type. *Ann Neurol* 1981;**10**:184–92.
80. Scott SA, DeKosky ST, Sparks DL, Knox CA, Scheff SW. Amygdala cell loss and atrophy in Alzheimer's disease. *Ann Neurol* 1992;**32**:555–63.
81. Khachaturian ZS. Toward a comprehensive theory of Alzheimer's disease—challenges, caveats, and parameters. *Ann NY Acad Sci* 2000;**924**:184–93.
82. Khachaturian ZS. Neurobiology of aging: Alzheimer's clinical review series. Bridging bench to bedside: clinical problems in search of basic solution. *Neurobiol Aging* 2000;**21**:843.
83. Arnold SE, Hyman BT, Flory J, Damasio AR, Van Hoesen GW. The topographical and neuroanatomical distribution of neurofibrillary tangles and neuritic plaques in the cerebral cortex of patients with Alzheimer's disease. *Cereb Cortex* 1991;**1**:103–16.
84. Braak H, Braak E. Demonstration of amyloid deposits and neurofibrillary changes in whole brain sections. *Brain Pathol* 1991;**1**:213–6.
85. Braak H, Braak E. Morphological changes in the human cerebral cortex in dementia. *J Hirnforsch* 1991;**32**:277–82.
86. Arriagada PV, Marzloff K, Hyman BT. Distribution of Alzheimer-type pathologic changes in nondemented elderly individuals matches the pattern in Alzheimer's disease. *Neurology* 1992;**42**:1681–8.
87. Arriagada PV, Growdon JH, Hedley-Whyte ET, Hyman BT. Neurofibrillary tangles but not senile plaques parallel duration and severity of Alzheimer's disease. *Neurology* 1992;**42**:631–9.
88. Goate A, Chartier-Harlin MC, Mullan M, Brown J, Crawford F, Fidani L, et al. Segregation of a missense mutation in the amyloid precursor protein gene with familial Alzheimer's disease. *Nature* 1991;**349**:704–6.
89. Sherrington R, Rogaev EI, Liang Y, Rogaeva EA, Levesque G, Ikeda M, et al. Cloning of a gene bearing missense mutations in early-onset familial Alzheimer's disease. *Nature* 1995;**375**:754–60.

90. Levy-Lahad E, Wijsman EM, Nemens E, Anderson L, Goddard KA, Weber JL, et al. A familial Alzheimer's disease locus on chromosome 1. *Science* 1995;**269**:970–3.
91. Kowalska A, Pruchnik-Wolinska D, Florczak J, Modestowicz R, Szczech J, Kozubski W, et al. Genetic study of familial cases of Alzheimer's disease. *Acta Biochim Pol* 2004;**51**:245–52.
92. Kowalska A, Pruchnik-Wolinska D, Florczak J, Szczech J, Kozubski W, Rossa G, et al. Presenilin 1 mutations in Polish families with early-onset Alzheimer's disease. *Folia Neuropathol* 2004;**42**:9–14.
93. Games D, Adams D, Alessandrini R, Barbour R, Berthelette P, Blackwell C, et al. Alzheimer-type neuropathology in transgenic mice overexpressing V717F beta-amyloid precursor protein. *Nature* 1995;**373**:523–7.
94. Hsiao KK, Borchelt DR, Olson K, Johannsdottir R, Kitt C, Yunis W, et al. Age-related CNS disorder and early death in transgenic FVB/N mice overexpressing Alzheimer amyloid precursor proteins. *Neuron* 1995;**15**:1203–18.
95. Hsiao K, Chapman P, Nilsen S, Eckman C, Harigaya Y, Younkin S, et al. Correlative memory deficits, Abeta elevation, and amyloid plaques in transgenic mice. *Science* 1996;**274**:99–102.
96. Johnson-Wood K, Lee M, Motter R, Hu K, Gordon G, Barbour R, et al. Amyloid precursor protein processing and A beta42 deposition in a transgenic mouse model of Alzheimer disease. *Proc Natl Acad Sci USA* 1997;**94**:1550–5.
97. Sturchler-Pierrat C, Abramowski D, Duke M, Wiederhold KH, Mistl C, Rothacher S, et al. Two amyloid precursor protein transgenic mouse models with Alzheimer disease-like pathology. *Proc Natl Acad Sci USA* 1997;**94**:13287–92.
98. Terai K, Iwai A, Kawabata S, Tasaki Y, Watanabe T, Miyata K, et al. beta-amyloid deposits in transgenic mice expressing human beta-amyloid precursor protein have the same characteristics as those in Alzheimer's disease. *Neuroscience* 2001;**104**:299–310.
99. Callahan MJ, Lipinski WJ, Bian F, Durham RA, Pack A, Walker LC. Augmented senile plaque load in aged female beta-amyloid precursor protein-transgenic mice. *Am J Pathol* 2001;**158**:1173–7.
100. Gotz J, Probst A, Spillantini MG, Schafer T, Jakes R, Burki K, et al. Somatodendritic localization and hyperphosphorylation of tau protein in transgenic mice expressing the longest human brain tau isoform. *EMBO J* 1995;**14**:1304–13.
101. Gotz J, Chen F, Barmettler R, Nitsch RM. Tau filament formation in transgenic mice expressing P301L tau. *J Biol Chem* 2001;**276**:529–34.
102. Brion JP, Tremp G, Octave JN. Transgenic expression of the shortest human tau affects its compartmentalization and its phosphorylation as in the pretangle stage of Alzheimer's disease. *Am J Pathol* 1999;**154**:255–70.
103. Spittaels K, Van den Haute C, Van Dorpe J, Bruynseels K, Vandezande K, Laenen I, et al. Prominent axonopathy in the brain and spinal cord of transgenic mice overexpressing four-repeat human tau protein. *Am J Pathol* 1999;**155**:2153–65.
104. Probst A, Gotz J, Wiederhold KH, Tolnay M, Mistl C, Jaton AL, et al. Axonopathy and amyotrophy in mice transgenic for human four-repeat tau protein. *Acta Neuropathol* 2000;**99**:469–81.
105. Lewis J, McGowan E, Rockwood J, Melrose H, Nacharaju P, Van Slegtenhorst M, et al. Neurofibrillary tangles, amyotrophy and progressive motor disturbance in mice expressing mutant (P301L) tau protein. *Nat Genet* 2000;**25**:402–5.
106. Tanemura K, Akagi T, Murayama M, Kikuchi N, Murayama O, Hashikawa T, et al. Formation of filamentous tau aggregations in transgenic mice expressing V337M human tau. *Neurobiol Dis* 2001;**8**:1036–45.
107. Tanemura K, Murayama M, Akagi T, Hashikawa T, Tominaga T, Ichikawa M, et al. Neurodegeneration with tau accumulation in a transgenic mouse expressing V337M human tau. *J Neurosci* 2002;**22**:133–41.

108. Lim F, Hernandez F, Lucas JJ, Gomez-Ramos P, Moran MA, Avila J. FTDP-17 mutations in tau transgenic mice provoke lysosomal abnormalities and Tau filaments in forebrain. *Mol Cell Neurosci* 2001;**18**:702–14.
109. Tatebayashi Y, Miyasaka T, Chui DH, Akagi T, Mishima K, Iwasaki K, et al. Tau filament formation and associative memory deficit in aged mice expressing mutant (R406W) human tau. *Proc Natl Acad Sci USA* 2002;**99**:13896–901.
110. Gotz J, Chen F, van Dorpe J, Nitsch RM. Formation of neurofibrillary tangles in P301l tau transgenic mice induced by Abeta 42 fibrils. *Science* 2001;**293**:1491–5.
111. Meyer-Luehmann M, Spires-Jones TL, Prada C, Garcia-Alloza M, de Calignon A, Rozkalne A, et al. Rapid appearance and local toxicity of amyloid-beta plaques in a mouse model of Alzheimer's disease. *Nature* 2008;**451**:720–4.
112. De Jonghe C, Esselens C, Kumar-Singh S, Craessaerts K, Serneels S, Checler F, et al. Pathogenic APP mutations near the gamma-secretase cleavage site differentially affect Abeta secretion and APP C-terminal fragment stability. *Hum Mol Genet* 2001;**10**:1665–71.
113. Golde TE, Janus C. Homing in on intracellular Abeta? *Neuron* 2005;**45**:639–42.
114. Golde TE. The Abeta hypothesis: leading us to rationally-designed therapeutic strategies for the treatment or prevention of Alzheimer disease. *Brain Pathol* 2005;**15**:84–7.
115. Kowalska A. Genetic aspects of amyloid beta-protein fibrillogenesis in Alzheimer's disease. *Folia Neuropathol* 2004;**42**:235–7.
116. Theuns J, Marjaux E, Vandenbulcke M, Van Laere K, Kumar-Singh S, Bormans G, et al. Alzheimer dementia caused by a novel mutation located in the APP C-terminal intracytosolic fragment. *Hum Mutat* 2006;**27**:888–96.
117. Theuns J, Brouwers N, Engelborghs S, Sleegers K, Bogaerts V, Corsmit E, et al. Promoter mutations that increase amyloid precursor-protein expression are associated with Alzheimer disease. *Am J Hum Genet* 2006;**78**:936–46.
118. Walker ES, Martinez M, Brunkan AL, Goate A. Presenilin 2 familial Alzheimer's disease mutations result in partial loss of function and dramatic changes in Abeta 42/40 ratios. *J Neurochem* 2005;**92**:294–301.
119. Duyckaerts C, Potier MC, Delatour B. Alzheimer disease models and human neuropathology: similarities and differences. *Acta Neuropathol* 2008;**115**:5–38.
120. Koffie RM, Meyer-Luehmann M, Hashimoto T, Adams KW, Mielke ML, Garcia-Alloza M, et al. Oligomeric amyloid beta associates with postsynaptic densities and correlates with excitatory synapse loss near senile plaques. *Proc Natl Acad Sci USA* 2009;**106**:4012–7.
121. Lambert MP, Barlow AK, Chromy BA, Edwards C, Freed R, Liosatos M, et al. Diffusible, nonfibrillar ligands derived from Abeta1-42 are potent central nervous system neurotoxins. *Proc Natl Acad Sci USA* 1998;**95**:6448–53.
122. Wang HW, Pasternak JF, Kuo H, Ristic H, Lambert MP, Chromy B, et al. Soluble oligomers of beta amyloid (1-42) inhibit long-term potentiation but not long-term depression in rat dentate gyrus. *Brain Res* 2002;**924**:133–40.
123. Walsh DM, Tseng BP, Rydel RE, Podlisny MB, Selkoe DJ. The oligomerization of amyloid beta-protein begins intracellularly in cells derived from human brain. *Biochemistry* 2000;**39**:10831–9.
124. Walsh DM, Klyubin I, Fadeeva JV, Rowan MJ, Selkoe DJ. Amyloid-beta oligomers: their production, toxicity and therapeutic inhibition. *Biochem Soc Trans* 2002;**30**:552–7.
125. Barghorn S, Nimmrich V, Striebinger A, Krantz C, Keller P, Janson B, et al. Globular amyloid beta-peptide oligomer—a homogenous and stable neuropathological protein in Alzheimer's disease. *J Neurochem* 2005;**95**:834–47.
126. Lesne S, Koh MT, Kotilinek L, Kayed R, Glabe CG, Yang A, et al. A specific amyloid-beta protein assembly in the brain impairs memory. *Nature* 2006;**440**:352–7.

127. Nimmrich V, Grimm C, Draguhn A, Barghorn S, Lehmann A, Schoemaker H, et al. Amyloid beta oligomers (A beta(1-42) globulomer) suppress spontaneous synaptic activity by inhibition of P/Q-type calcium currents. *J Neurosci* 2008;**28**:788–97.
128. Cotman CW, Anderson AJ. A potential role for apoptosis in neurodegeneration and Alzheimer's disease. *Mol Neurobiol* 1995;**10**:19–45.
129. Gervais FG, Xu D, Robertson GS, Vaillancourt JP, Zhu Y, Huang J, et al. Involvement of caspases in proteolytic cleavage of Alzheimer's amyloid-beta precursor protein and amyloidogenic A beta peptide formation. *Cell* 1999;**97**:395–406.
130. Pompl PN, Yemul S, Xiang Z, Ho L, Haroutunian V, Purohit D, et al. Caspase gene expression in the brain as a function of the clinical progression of Alzheimer disease. *Arch Neurol* 2003;**60**:369–76.
131. LeBlanc A, Liu H, Goodyer C, Bergeron C, Hammond J. Caspase-6 role in apoptosis of human neurons, amyloidogenesis, and Alzheimer's disease. *J Biol Chem* 1999;**274**:23426–36.
132. Stadelmann C, Deckwerth TL, Srinivasan A, Bancher C, Bruck W, Jellinger K, et al. Activation of caspase-3 in single neurons and autophagic granules of granulovacuolar degeneration in Alzheimer's disease. Evidence for apoptotic cell death. *Am J Pathol* 1999;**155**:1459–66.
133. Matsui T, Ramasamy K, Ingelsson M, Fukumoto H, Conrad C, Frosch MP, et al. Coordinated expression of caspase 8, 3 and 7 mRNA in temporal cortex of Alzheimer disease: relationship to formic acid extractable abeta42 levels. *J Neuropathol Exp Neurol* 2006;**65**:508–15.
134. Rohn TT, Head E, Nesse WH, Cotman CW, Cribbs DH. Activation of caspase-8 in the Alzheimer's disease brain. *Neurobiol Dis* 2001;**8**:1006–16.
135. Lu DC, Rabizadeh S, Chandra S, Shayya RF, Ellerby LM, Ye X, et al. A second cytotoxic proteolytic peptide derived from amyloid beta-protein precursor. *Nat Med* 2000;**6**:397–404.
136. Pellegrini L, Passer BJ, Tabaton M, Ganjei JK, D'Adamio L. Alternative, non-secretase processing of Alzheimer's beta-amyloid precursor protein during apoptosis by caspase-6 and -8. *J Biol Chem* 1999;**274**:21011–6.
137. McPhie DL, Golde T, Eckman CB, Yager D, Brant JB, Neve RL. beta-Secretase cleavage of the amyloid precursor protein mediates neuronal apoptosis caused by familial Alzheimer's disease mutations. *Brain Res Mol Brain Res* 2001;**97**:103–13.
138. Koo EH. The beta-amyloid precursor protein (APP) and Alzheimer's disease: does the tail wag the dog? *Traffic* 2002;**3**:763–70.
139. Zheng H, Koo EH. The amyloid precursor protein: beyond amyloid. *Mol Neurodegener* 2006;**1**:5.
140. Lu DC, Soriano S, Bredesen DE, Koo EH. Caspase cleavage of the amyloid precursor protein modulates amyloid beta-protein toxicity. *J Neurochem* 2003;**87**:733–41.
141. Galvan V, Zhang J, Gorostiza OF, Banwait S, Huang W, Ataie M, et al. Long-term prevention of Alzheimer's disease-like behavioral deficits in PDAPP mice carrying a mutation in Asp664. *Behav Brain Res* 2008;**191**:246–55.
142. Harris JA, Devidze N, Halabisky B, Lo I, Thwin MT, Yu GQ, et al. Many neuronal and behavioral impairments in transgenic mouse models of Alzheimer's disease are independent of caspase cleavage of the amyloid precursor protein. *J Neurosci* 2010;**30**:372–81.
143. Gamblin TC, Chen F, Zambrano A, Abraha A, Lagalwar S, Guillozet AL, et al. Caspase cleavage of tau: linking amyloid and neurofibrillary tangles in Alzheimer's disease. *Proc Natl Acad Sci USA* 2003;**100**:10032–7.
144. Rissman RA, Poon WW, Blurton-Jones M, Oddo S, Torp R, Vitek MP, et al. Caspase-cleavage of tau is an early event in Alzheimer disease tangle pathology. *J Clin Invest* 2004;**114**:121–30.
145. Guo H, Albrecht S, Bourdeau M, Petzke T, Bergeron C, LeBlanc AC. Active caspase-6 and caspase-6-cleaved tau in neuropil threads, neuritic plaques, and neurofibrillary tangles of Alzheimer's disease. *Am J Pathol* 2004;**165**:523–31.

146. Albrecht S, Bogdanovic N, Ghetti B, Winblad B, LeBlanc AC. Caspase-6 activation in familial alzheimer disease brains carrying amyloid precursor protein or presenilin i or presenilin II mutations. *J Neuropathol Exp Neurol* 2009;**68**:1282–93.

147. de Calignon A, Fox LM, Pitstick R, Carlson GA, Bacskai BJ, Spires-Jones TL, et al. Caspase activation precedes and leads to tangles. *Nature* 2010;**464**:1201–4.

148. Tizon B, Ribe EM, Mi W, Troy CM, Levy E. Cystatin C protects neuronal cells from amyloid-beta-induced toxicity. *J Alzheimers Dis* 2010;**19**:885–94.

149. Heneka MT, O'Banion MK. Inflammatory processes in Alzheimer's disease. *J Neuroimmunol* 2007;**184**:69–91.

150. Ojala J, Alafuzoff I, Herukka SK, van Groen T, Tanila H, Pirttila T. Expression of interleukin-18 is increased in the brains of Alzheimer's disease patients. *Neurobiol Aging* 2009;**30**:198–209.

151. Zuliani G, Ranzini M, Guerra G, Rossi L, Munari MR, Zurlo A, et al. Plasma cytokines profile in older subjects with late onset Alzheimer's disease or vascular dementia. *J Psychiatr Res* 2007;**41**:686–93.

152. Lue LF, Rydel R, Brigham EF, Yang LB, Hampel H, Murphy Jr. GM, et al. Inflammatory repertoire of Alzheimer's disease and nondemented elderly microglia in vitro. *Glia* 2001;**35**:72–9.

153. Apelt J, Schliebs R. Beta-amyloid-induced glial expression of both pro- and anti-inflammatory cytokines in cerebral cortex of aged transgenic Tg2576 mice with Alzheimer plaque pathology. *Brain Res* 2001;**894**:21–30.

154. Halle A, Hornung V, Petzold GC, Stewart CR, Monks BG, Reinheckel T, et al. The NALP3 inflammasome is involved in the innate immune response to amyloid-beta. *Nat Immunol* 2008;**9**:857–65.

155. Kummer JA, Broekhuizen R, Everett H, Agostini L, Kuijk L, Martinon F, et al. Inflammasome components NALP 1 and 3 show distinct but separate expression profiles in human tissues suggesting a site-specific role in the inflammatory response. *J Histochem Cytochem* 2007;**55**:443–52.

156. Petrilli V, Papin S, Dostert C, Mayor A, Martinon F, Tschopp J. Activation of the NALP3 inflammasome is triggered by low intracellular potassium concentration. *Cell Death Differ* 2007;**14**:1583–9.

157. de Rivero Vaccari JP, Lotocki G, Marcillo AE, Dietrich WD, Keane RW. A molecular platform in neurons regulates inflammation after spinal cord injury. *J Neurosci* 2008;**28**:3404–14.

158. Frederick Lo C, Ning X, Gonzales C, Ozenberger BA. Induced expression of death domain genes NALP1 and NALP5 following neuronal injury. *Biochem Biophys Res Commun* 2008;**366**:664–9.

159. Colom LV, Diaz ME, Beers DR, Neely A, Xie WJ, Appel SH. Role of potassium channels in amyloid-induced cell death. *J Neurochem* 1998;**70**:1925–34.

160. Yu SP, Farhangrazi ZS, Ying HS, Yeh CH, Choi DW. Enhancement of outward potassium current may participate in beta-amyloid peptide-induced cortical neuronal death. *Neurobiol Dis* 1998;**5**:81–8.

161. Troy CM, Rabacchi SA, Friedman WJ, Frappier TF, Brown K, Shelanski ML. Caspase-2 mediates neuronal cell death induced by beta-amyloid. *J Neurosci* 2000;**20**:1386–92.

162. Nakagawa T, Zhu H, Morishima N, Li E, Xu J, Yankner BA, et al. Caspase-12 mediates endoplasmic-reticulum-specific apoptosis and cytotoxicity by amyloid-beta. *Nature* 2000;**403**:98–103.

163. Song S, Lee H, Kam TI, Tai ML, Lee JY, Noh JY, et al. E2-25K/Hip-2 regulates caspase-12 in ER stress-mediated Abeta neurotoxicity. *J Cell Biol* 2008;**182**:675–84.

164. Saleh M, Vaillancourt JP, Graham RK, Huyck M, Srinivasula SM, Alnemri ES, et al. Differential modulation of endotoxin responsiveness by human caspase-12 polymorphisms. *Nature* 2004;**429**:75–9.

165. Hitomi J, Katayama T, Eguchi Y, Kudo T, Taniguchi M, Koyama Y, et al. Involvement of caspase-4 in endoplasmic reticulum stress-induced apoptosis and Abeta-induced cell death. *J Cell Biol* 2004;**165**:347–56.

166. Vaisid T, Barnoy S, Kosower NS. Calpain activates caspase-8 in neuron-like differentiated PC12 cells via the amyloid-beta-peptide and CD95 pathways. *Int J Biochem Cell Biol* 2009;**41**:2450–8.

167. Vaisid T, Kosower NS, Elkind E, Barnoy S. Amyloid beta peptide toxicity in differentiated PC12 cells: calpain-calpastatin, caspase, and membrane damage. *J Neurosci Res* 2008;**86**:2314–25.

168. Lu DC, Shaked GM, Masliah E, Bredesen DE, Koo EH. Amyloid beta protein toxicity mediated by the formation of amyloid-beta protein precursor complexes. *Ann Neurol* 2003;**54**:781–9.

169. Lin KF, Chang RC, Suen KC, So KF, Hugon J. Modulation of calcium/calmodulin kinase-II provides partial neuroprotection against beta-amyloid peptide toxicity. *Eur J Neurosci* 2004;**19**:2047–55.

170. Ryan SD, Whitehead SN, Swayne LA, Moffat TC, Hou W, Ethier M, et al. Amyloid-{beta}42 signals tau hyperphosphorylation and compromises neuronal viability by disrupting alkylacylglycerophosphocholine metabolism. *Proc Natl Acad Sci USA* 2008;**97**:60–78.

171. Tesco G, Koh YH, Kang EL, Cameron AN, Das S, Sena-Esteves M, et al. Depletion of GGA3 stabilizes BACE and enhances beta-secretase activity. *Neuron* 2007;**54**:721–37.

172. Warlow CP. Epidemiology of stroke. *Lancet* 1998;**352**(Suppl. 3):1–4.

173. del Zoppo GJ. Stroke and neurovascular protection. *N Engl J Med* 2006;**354**:553–5.

174. Ribe EM, Serrano-Saiz E, Akpan N, Troy CM. Mechanisms of neuronal death in disease: defining the models and the players. *Biochem J* 2008;**415**:165–82.

175. Garcia JH, Liu KF, Ho KL. Neuronal necrosis after middle cerebral artery occlusion in Wistar rats progresses at different time intervals in the caudoputamen and the cortex. *Stroke* 1995;**26**:636–42 [Discussion: 643].

176. Lipton P. Ischemic cell death in brain neurons. *Physiol Rev* 1999;**79**:1431–568.

177. Linnik MD, Zobrist RH, Hatfield MD. Evidence supporting a role for programmed cell death in focal cerebral ischemia in rats. *Stroke* 1993;**24**:2002–8 [Discussion: 2008–2009].

178. Charriaut-Marlangue C, Margaill I, Represa A, Popovici T, Plotkine M, Ben-Ari Y. Apoptosis and necrosis after reversible focal ischemia: an in situ DNA fragmentation analysis. *J Cereb Blood Flow Metab* 1996;**16**:186–94.

179. Li P, Allen H, Banerjee S, Franklin S, Herzog L, Johnston C, et al. Mice deficient in IL-1 beta-converting enzyme are defective in production of mature IL-1 beta and resistant to endotoxic shock. *Cell* 1995;**80**:401–11.

180. Benchoua A, Guegan C, Couriaud C, Hosseini H, Sampaio N, Morin D, et al. Specific caspase pathways are activated in the two stages of cerebral infarction. *J Neurosci* 2001;**21**:7127–34.

181. Kerr JF, Wyllie AH, Currie AR. Apoptosis: a basic biological phenomenon with wide-ranging implications in tissue kinetics. *Br J Cancer* 1972;**26**:239–57.

182. Mattson MP. Apoptosis in neurodegenerative disorders. *Nat Rev Mol Cell Biol* 2000;**1**:120–9.

183. Love S. Apoptosis and brain ischaemia. *Prog Neuropsychopharmacol Biol Psychiatry* 2003;**27**:267–82.

184. Ferrer I. Apoptosis: future targets for neuroprotective strategies. *Cerebrovasc Dis* 2006;**21** (Suppl 2):9–20.

185. Martin LJ, Al-Abdulla NA, Brambrink AM, Kirsch JR, Sieber FE, Portera-Cailliau C. Neurodegeneration in excitotoxicity, global cerebral ischemia, and target deprivation: a perspective on the contributions of apoptosis and necrosis. *Brain Res Bull* 1998;**46**:281–309.
186. Pan J, Konstas AA, Bateman B, Ortolano GA, Pile-Spellman J. Reperfusion injury following cerebral ischemia: pathophysiology, MR imaging, and potential therapies. *Neuroradiology* 2007;**49**:93–102.
187. Kahles T, Luedike P, Endres M, Galla HJ, Steinmetz H, Busse R, et al. NADPH oxidase plays a central role in blood-brain barrier damage in experimental stroke. *Stroke* 2007;**38**:3000–6.
188. Traystman RJ. Animal models of focal and global cerebral ischemia. *ILAR J* 2003;**44**:85–95.
189. Richard Green A, Odergren T, Ashwood T. Animal models of stroke: do they have value for discovering neuroprotective agents? *Trends Pharmacol Sci* 2003;**24**:402–8.
190. Dronne MA, Grenier E, Chapuisat G, Hommel M, Boissel JP. A modelling approach to explore some hypotheses of the failure of neuroprotective trials in ischemic stroke patients. *Prog Biophys Mol Biol* 2008;**97**:60–78.
191. Rothman SM. Synaptic activity mediates death of hypoxic neurons. *Science* 1983;**220**:536–7.
192. Goldberg MP, Choi DW. Combined oxygen and glucose deprivation in cortical cell culture: calcium-dependent and calcium-independent mechanisms of neuronal injury. *J Neurosci* 1993;**13**:3510–24.
193. Gwag BJ, Lobner D, Koh JY, Wie MB, Choi DW. Blockade of glutamate receptors unmasks neuronal apoptosis after oxygen-glucose deprivation in vitro. *Neuroscience* 1995;**68**:615–9.
194. Laake JH, Haug FM, Wieloch T, Ottersen OP. A simple in vitro model of ischemia based on hippocampal slice cultures and propidium iodide fluorescence. *Brain Res Brain Res Protoc* 1999;**4**:173–84.
195. Schurr A, Payne RS, Miller JJ, Rigor BM. Study of cerebral energy metabolism using the rat hippocampal slice preparation. *Methods* 1999;**18**:117–26.
196. Schurr A, Tseng MT, West CA, Rigor BM. Taurine improves the recovery of neuronal function following cerebral hypoxia: an in vitro study. *Life Sci* 1987;**40**:2059–66.
197. Calabresi P, Centonze D, Bernardi G. Cellular factors controlling neuronal vulnerability in the brain: a lesson from the striatum. *Neurology* 2000;**55**:1249–55.
198. Pedersen JZ, Bernardi G, Centonze D, Pisani A, Rossi L, Rotilio G, et al. Hypoglycemia, hypoxia, and ischemia in a corticostriatal slice preparation: electrophysiologic changes and ascorbyl radical formation. *J Cereb Blood Flow Metab* 1998;**18**:868–75.
199. Benveniste H, Drejer J, Schousboe A, Diemer NH. Elevation of the extracellular concentrations of glutamate and aspartate in rat hippocampus during transient cerebral ischemia monitored by intracerebral microdialysis. *J Neurochem* 1984;**43**:1369–74.
200. Choi DW, Rothman SM. The role of glutamate neurotoxicity in hypoxic-ischemic neuronal death. *Annu Rev Neurosci* 1990;**13**:171–82.
201. Simon RP, Swan JH, Griffiths T, Meldrum BS. Blockade of N-methyl-D-aspartate receptors may protect against ischemic damage in the brain. *Science* 1984;**226**:850–2.
202. Malagelada C, Xifro X, Minano A, Sabria J, Rodriguez-Alvarez J. Contribution of caspase-mediated apoptosis to the cell death caused by oxygen–glucose deprivation in cortical cell cultures. *Neurobiol Dis* 2005;**20**:27–37.
203. Meli E, Pangallo M, Baronti R, Chiarugi A, Cozzi A, Pellegrini-Giampietro DE, et al. Poly (ADP-ribose) polymerase as a key player in excitotoxicity and post-ischemic brain damage. *Toxicol Lett* 2003;**139**:153–62.
204. Moroni F, Meli E, Peruginelli F, Chiarugi A, Cozzi A, Picca R, et al. Poly(ADP-ribose) polymerase inhibitors attenuate necrotic but not apoptotic neuronal death in experimental models of cerebral ischemia. *Cell Death Differ* 2001;**8**:921–32.

205. Ray AM, Owen DE, Evans ML, Davis JB, Benham CD. Caspase inhibitors are functionally neuroprotective against oxygen glucose deprivation induced CA1 death in rat organotypic hippocampal slices. *Brain Res* 2000;**867**:62–9.
206. Gottron FJ, Ying HS, Choi DW. Caspase inhibition selectively reduces the apoptotic component of oxygen-glucose deprivation-induced cortical neuronal cell death. *Mol Cell Neurosci* 1997;**9**:159–69.
207. Taranukhin AG, Taranukhina EY, Saransaari P, Djatchkova IM, Pelto-Huikko M, Oja SS. Taurine reduces caspase-8 and caspase-9 expression induced by ischemia in the mouse hypothalamic nuclei. *Amino Acids* 2008;**34**:169–74.
208. Le DA, Wu Y, Huang Z, Matsushita K, Plesnila N, Augustinack JC, et al. Caspase activation and neuroprotection in caspase-3-deficient mice after in vivo cerebral ischemia and in vitro oxygen glucose deprivation. *Proc Natl Acad Sci USA* 2002;**99**:15188–93.
209. Shang YC, Chong ZZ, Hou J, Maiese K. FoxO3a governs early microglial proliferation and employs mitochondrial depolarization with caspase 3, 8, and 9 cleavage during oxidant induced apoptosis. *Curr Neurovasc Res* 2009;**6**:223–38.
210. Troy CM, Derossi D, Prochiantz A, Greene LA, Shelanski ML. Downregulation of Cu/Zn superoxide dismutase leads to cell death via the nitric oxide-peroxynitrite pathway. *J Neurosci* 1996;**16**:253–61.
211. Davidson TJ, Harel S, Arboleda VA, Prunell GF, Shelanski ML, Greene LA, et al. Highly efficient small interfering RNA delivery to primary mammalian neurons induces MicroRNA-like effects before mRNA degradation. *J Neurosci* 2004;**24**:10040–6.
212. Fujimura M, Morita-Fujimura Y, Noshita N, Sugawara T, Kawase M, Chan PH. The cytosolic antioxidant copper/zinc-superoxide dismutase prevents the early release of mitochondrial cytochrome c in ischemic brain after transient focal cerebral ischemia in mice. *J Neurosci* 2000;**20**:2817–24.
213. Kondo T, Reaume AG, Huang TT, Carlson E, Murakami K, Chen SF, et al. Reduction of CuZn-superoxide dismutase activity exacerbates neuronal cell injury and edema formation after transient focal cerebral ischemia. *J Neurosci* 1997;**17**:4180–9.
214. Boutin H, LeFeuvre RA, Horai R, Asano M, Iwakura Y, Rothwell NJ. Role of IL-1alpha and IL-1beta in ischemic brain damage. *J Neurosci* 2001;**21**:5528–34.
215. Rothwell N. Interleukin-1 and neuronal injury: mechanisms, modification, and therapeutic potential. *Brain Behav Immun* 2003;**17**:152–7.
216. Friedlander RM, Gagliardini V, Hara H, Fink KB, Li W, MacDonald G, et al. Expression of a dominant negative mutant of interleukin-1 beta converting enzyme in transgenic mice prevents neuronal cell death induced by trophic factor withdrawal and ischemic brain injury. *J Exp Med* 1997;**185**:933–40.
217. Schielke GP, Yang GY, Shivers BD, Betz AL. Reduced ischemic brain injury in interleukin-1 beta converting enzyme-deficient mice. *J Cereb Blood Flow Metab* 1998;**18**:180–5.
218. Allan SM, Tyrrell PJ, Rothwell NJ. Interleukin-1 and neuronal injury. *Nat Rev Immunol* 2005;**5**:629–40.
219. Ferrer I, Planas AM. Signaling of cell death and cell survival following focal cerebral ischemia: life and death struggle in the penumbra. *J Neuropathol Exp Neurol* 2003;**62**:329–39.
220. Niizuma K, Endo H, Nito C, Myer DJ, Kim GS, Chan PH. The PIDDosome mediates delayed death of hippocampal CA1 neurons after transient global cerebral ischemia in rats. *Proc Natl Acad Sci USA* 2008;**105**:16368–73.
221. Velier JJ, Ellison JA, Kikly KK, Spera PA, Barone FC, Feuerstein GZ. Caspase-8 and caspase-3 are expressed by different populations of cortical neurons undergoing delayed cell death after focal stroke in the rat. *J Neurosci* 1999;**19**:5932–41.
222. Guegan C, Sola B. Early and sequential recruitment of apoptotic effectors after focal permanent ischemia in mice. *Brain Res* 2000;**856**:93–100.

223. Love S, Barber R, Srinivasan A, Wilcock GK. Activation of caspase-3 in permanent and transient brain ischaemia in man. *NeuroReport* 2000;**11**:2495–9.

224. Giffard RG, Swanson RA. Ischemia-induced programmed cell death in astrocytes. *Glia* 2005;**50**:299–306.

225. Duan SR, Wang JX, Wang J, Xu R, Zhao JK, Wang DS. Ischemia induces endoplasmic reticulum stress and cell apoptosis in human brain. *Neurosci Lett* 2010;**475**:132–5.

226. Harrison DC, Davis RP, Bond BC, Campbell CA, James MF, Parsons AA, et al. Caspase mRNA expression in a rat model of focal cerebral ischemia. *Brain Res Mol Brain Res* 2001;**89**:133–46.

227. Desmond DW, Moroney JT, Sano M, Stern Y. Incidence of dementia after ischemic stroke: results of a longitudinal study. *Stroke* 2002;**33**:2254–60.

228. Tatemichi TK, Desmond DW, Stern Y, Paik M, Sano M, Bagiella E. Cognitive impairment after stroke: frequency, patterns, and relationship to functional abilities. *J Neurol Neurosurg Psychiatry* 1994;**57**:202–7.

229. Tatemichi TK, Paik M, Bagiella E, Desmond DW, Stern Y, Sano M, et al. Risk of dementia after stroke in a hospitalized cohort: results of a longitudinal study. *Neurology* 1994;**44**:1885–91.

230. Wen Y, Yang S, Liu R, Brun-Zinkernagel AM, Koulen P, Simpkins JW. Transient cerebral ischemia induces aberrant neuronal cell cycle re-entry and Alzheimer's disease-like tauopathy in female rats. *J Biol Chem* 2004;**279**:22684–92.

231. Nihashi T, Inao S, Kajita Y, Kawai T, Sugimoto T, Niwa M, et al. Expression and distribution of beta amyloid precursor protein and beta amyloid peptide in reactive astrocytes after transient middle cerebral artery occlusion. *Acta Neurochir (Wien)* 2001;**143**:287–95.

232. van Groen T, Puurunen K, Maki HM, Sivenius J, Jolkkonen J. Transformation of diffuse beta-amyloid precursor protein and beta-amyloid deposits to plaques in the thalamus after transient occlusion of the middle cerebral artery in rats. *Stroke* 2005;**36**:1551–6.

233. Xiong M, Zhang T, Zhang LM, Lu SD, Huang YL, Sun FY. Caspase inhibition attenuates accumulation of beta-amyloid by reducing beta-secretase production and activity in rat brains after stroke. *Neurobiol Dis* 2008;**32**:433–41.

234. Abrahamson EE, Ikonomovic MD, Ciallella JR, Hope CE, Paljug WR, Isanski BA, et al. Caspase inhibition therapy abolishes brain trauma-induced increases in Abeta peptide: implications for clinical outcome. *Exp Neurol* 2006;**197**:437–50.

235. Chae SS, Yoo CB, Jo C, Yun SM, Jo SA, Koh YH. Caspases-2 and -8 are involved in the presenilin1/gamma-secretase-dependent cleavage of amyloid precursor protein after the induction of apoptosis, *J Neurosci Res* 2010;**88**:1926–33.

236. de Calignon A, Fox LM, Pitstick R, Carlson GA, Bacskai BJ, Spires-Jones TL, et al. Caspase activation precedes and leads to tangles, *Nature* 2010;**464**:1201–4.

237. Su JH, Zhao M, Anderson AJ, Srinivasan A, Cotman CW. Activated caspase-3 expression in Alzheimer's and aged control brain: correlation with Alzheimer pathology. *Brain Res* 2001;**898**:350–7.

238. Louneva N, Cohen JW, Han LY, Talbot K, Wilson RS, Bennett DA, et al. Caspase-3 is enriched in postsynaptic densities and increased in Alzheimer's disease. *Am J Pathol* 2008;**173**:1488–95.

239. Matsuzaki S, Hiratsuka T, Kuwahara R, Katayama T, Tohyama M. Caspase-4 is partially cleaved by calpain via the impairment of Ca^{2+} homeostasis under the ER stress, *Neurochem Int* 2010;**56**:352–6.

240. Rohn TT, Rissman RA, Davis MC, Kim YE, Cotman CW, Head E. Caspase-9 activation and caspase cleavage of tau in the Alzheimer's disease brain. *Neurobiol Dis* 2002;**11**:341–54.

241. Ishige K, Takagi N, Imai T, Rausch WD, Kosuge Y, Kihara T, et al. Role of caspase-12 in amyloid beta-peptide-induced toxicity in organotypic hippocampal slices cultured for long periods. *J Pharmacol Sci* 2007;**104**:46–55.
242. Friedlander RM, Brown RH, Gagliardini V, Wang J, Yuan J. Inhibition of ICE slows ALS in mice [letter] [published erratum appears in Nature 1998 Apr 9;392(6676):560]. *Nature* 1997;**388**:31.
243. Krajewska M, Rosenthal RE, Mikolajczyk J, Stennicke HR, Wiesenthal T, Mai J, et al. Early processing of Bid and caspase-6, -8, -10, -14 in the canine brain during cardiac arrest and resuscitation. *Exp Neurol* 2004;**189**:261–79.

Index

A

Activated protein C (APC), 163, 215
ADAMTS13
  activity, regulation of, 118–120
  calcium binding sites, 114–115
  CUB domains, 117
  cysteine-rich domain, 116
  disintegrin-like domain, 115
  gene, biosynthesis and secretion of, 109
  history of, 106–107
  laboratory methods, 133–134
  metalloprotease domain, 112–114
  pathogenesis of, 120–131
  propeptide domain, 111–112
  spacer domain, 116
  structure and function, 109–111
  thrombospondin type 1-like repeat (TSR), 115–116
  thrombotic microangiopathies, 132–133 (*see also* Thrombotic microangiopathies (TMAs), ADAMTS13)
  von Willebrand factor, 107–109
Ala mutants, of thrombin, 158–162
Allosteric enzyme, thrombin
  Botts–Morales scheme, 152–153
  E and E:Na^+ forms, 152
  E^* form, 152
  kinetic scheme, 152
Alzheimer's disease (AD)
  caspases
    and Aβ generation, 283–284
    and Aβ toxicity, 282–283
    APP cleavage, 280
    in brains, 279–280
    and inflammation, 281–282
    tau, 281
  dementia, 277
  in vitro models, 279
  mouse models, 278
  NFT correlation, 278
Amyloid β precursor protein (AβPP), 87–88
Anticoagulant activity, of thrombin mutants
  crystal structures of, 167–169
  ^{D}146–149e, 166
  E^*–E equilibrium, 165
  E217K and WE, 166–167
  surface representation, 168
  thrombomodulin and hirugen, 169
Antihemophilic drug discovery, 90–91
Antithrombin, coagulation inhibitor, 82, 84–85
Antithrombotic drugs, 91–92
Apoptosome, 269
Autosomal recessive ichthyosis with hypotrichosis (ARIH), 19

B

Bacterial meningitis, MMPs, 254
Blood coagulation, factors IX and X assemblies
  domain arrangement and activation pattern of, 53
  in hemostasis, 52
  low-molecular weight cofactors
    calcium and sodium, 64, 66
    ethylene glycol (EG), 61–63
    exosite-active site mutual influence, 68
    heparins, 63–65
    long distance-communication line, 66–68
    sodium-binding sites in, 67
  membranes, nonhumoral cofactors of
    activated platelets and phospholipid membranes, 80–81
    injured vasculature, endothelial cells, 79–80
  pharmaceutical opportunities and challenges
    antihemophilic drug discovery, 90–91
    antithrombotic drugs, 91–92
    plasma-derived factors, 90
  protein cofactors
    factor V, 78–79

Blood coagulation, factors IX and X assemblies (*cont.*)
factor VII, 69–70
factor VIII, 70–74
substrate's role, in Xase function, 74–75
tissue factor (TF)–fVIIa complex, 69
Xase-like activity enhancement, atomic model of, 75–78
protein inhibitors
endogenous inhibitors, 82–88
exogenous inhibitors, 88–89
trypsin-like serine proteases, proteolytic activation of
activation peptide, release, 60–61
N-terminus formation, 54–60
Botts–Morales scheme, 152–153
Bovine pancreatic trypsin inhibitor (BPTI), 85

C

Calcium, blood coagulation enzymes, 64, 66
Canonical inhibitors *vs.* serpins, 219–221
Caspases regulation, nervous system
activation mechanism
apoptosome, 269
effector and initiator, 268
extrinsic death pathway, 270
inflammasome, 271
intrinsic pathway, 269
PIDDosome, 271
and developmental neuronal death, 275–276
IAP family, 272
measurement
bVADfmk, 274
immunocytochemistry, 273
molecular knockdown, 274
null mice, 275
mitochondria regulation, 272
neurodegenerative diseases
Alzheimer's disease (AD), 277–284
cerebral ischemia, 284–290
linking stroke and AD, 290–291
nonapoptotic caspase function, 276
targets, 275
Cerebral ischemia
animal models, 285–287
caspases, 288–290
in vitro models, 287–288
reperfusion injury, 285
stroke, 284
Cofactors, exosites in ternary complexes, 213–217
Corin subfamily, serine proteases, 24–28
crmA variants, 228–229

D

Death proteases, caspases family. *See* Caspases regulation, nervous system
Dermatan sulfate/heparin roe, HCII, 213–215
Differentially expressed in squamous cell carcinoma (DESC1), 29–30

E

Endogenous inhibitors
blood coagulation
AβPP, 87–88
antithrombin, coagulation inhibitor, 82, 84–85
Kunitz-type inhibitors, 85–86
protein Z-dependent protease inhibitor, 85
serpins, 82–83
TFPI, 87
regulation, serine proteases, 7
Endothelial cells, injured vasculature, 79–80
Enteropeptidase
activation, 10
food digestion and absorption, 11
PRSS7 gene, 9
structural elements, 10
substrate specificity, 10–11
Epidermis, matriptase subfamily, 18–19
Ethylene glycol (EG), 61–63
Exogenous inhibitors, blood coagulation
ixolaris, 89
NAP5, 89
nitrophorin 2, 88
TAP, 89
Exosites, serpins
interactions, in binary complexes, 212–213
in ternary complexes
activated protein C (APC), 215
HCII, dermatan sulfate/heparin roe, 213–215

heparin-induced conformational change, 216
ZPI, factor Xa, 213–214

F

Factor V, 78–79
Factor VII, 69–70
Factor VIII, 70–74
Frameshift mutations, *ADAMTS13* gene, 128

G

Glycosylphosphatidylinositol (GPI)-anchored serine proteases
prostasin, 30–32
testisin, 32–33

H

Heparins
binding site mapping, fIXa catalytic domain, 64–65
cofactor II, 213–215
HCII, 213–215
repetitive subunits, molecular structure of, 63
thrombin and antithrombin bridging, 205
Hepsin, 11–13
Human airway trypsin-like (HAT) protease, 28–29
Human serpins
biology of, 191–192
properties, 188–190
Hypoxia/ischemia, MMPs, 254–255

I

Ichthyosis, follicular atrophoderma, and hypotrichosis (IFAH), 19
Interleukin-1β cleaving enzyme (ICE), 267
Ixolaris, 89

K

Kunitz-type inhibitors, 85–86

M

Mammalian caspases, 267
Matriptase subfamily, serine proteases
ARIH/IFAH, 19–20
cancer biology, 21–22
epidermis, development and functions of, 18–19
epithelial tissues, 20
expression, 18
HAI-1 and HAI-2, 21
matriptase-2, 23
matriptase-3, 23–24
Netherton syndrome, 20
polyserase-1/serase-1B, 24
Matrix metalloproteinases (MMPs)
apoptosis, 248–249
biology
gelatinase A and B, 244
matrilysin (MMP-7), 243
zinc-dependent endopeptidases, 243
blood–brain barrier (BBB), free radicals, 245
and cell death
extracellular protein targets, 251
microglial activation and inflammation, 252
oxidative DNA damage, 253
A disintegrin and metalloproteinases (ADAMs), 242
extracellular matrix (ECM) regulation, 242
in central nervous system and endogenous inhibitors, 244
neuroinflammation
gene expression, 247
oxidative stress, 248
proteolysis, 247
neurological disorders
bacterial meningitis, 254
hypoxia/ischemia, 254–255
multiple sclerosis, 253–254
vascular cognitive impairment, 255–257
neurovascular unit
astrocytic foot processes (AFP), 251
cerebrovascular endothelial tight junctions, 250
neurons and astrocytes, 250
zona occludens (ZO), 250
plasminogen activators, 246
serine proteases, 242
TIMPs, 248–249

Membrane-anchored serine proteases
  amino-terminal transmembrane domain, 4
  corin subfamily, 25
    hypertensive patients, 27
    plasma levels, 28
    pro-ANP processing, in heart, 26–27
  DESC1, 29–30
  domain structures, 3
  endogenous inhibitors, regulation, 7
  GPI-anchored
    prostasin, 30–32
    testisin, 32–33
  HAT, 28–29
  HEPSIN/TMPRSS subfamily
    enteropeptidase, 8–11
    Hpn gene, 11
    MSPL, 16–17
    prostate cancer, 13
    spinesin, 16
    substrate-based assays, 12
    TMPRSS2, 13–15
    TMPRSS3, 15
    TMPRSS4, 16
  mammalian development and tissue homeostasis, physiological functions in, 34
  matriptase subfamily
    ARIH/IFAH, 19–20
    cancer biology, 21–22
    epidermis, development and functions of, 18–19
    epithelial tissues, 20
    expression, 18
    HAI-1 and HAI-2, 21
    matriptase-2, 23
    matriptase-3, 23–24
    Netherton syndrome, 20
    polyserase-1/serase-1B, 24
  structural features
    catalytic domains, 5–6
    domains, 6–7
    extracellular stem regions, 6
  TMPRSS11A, 30
  tryptase γ1/transmembrane tryptase (TMT), 7–8
  type II transmembrane serine proteases (TTSPs), 2, 4
Membranes, nonhumoral cofactors
  activated platelets and phospholipid membranes, 80–81
  injured vasculature, endothelial cells, 79–80
Metalloproteinases. *See* Matrix metalloproteinases (MMPs)
Michaelis complexes, serpins
  inferred organization of, 207
  noncovalent and covalent moiety, conformation of, 204
  α_1PI and bovine trypsin, 201
  RCL, 202–203
  thrombin and antithrombin bridging, heparin role, 205
  X-ray structures of, 202
Missense mutations, *ADAMTS13* gene, 129
Mitochondrial outer membrane permeabilization (MOMP), 272
MMPs. *See* Matrix metalloproteinases
Mosaic serine protease long-form (MSPL), 16–17
Multiple sclerosis, MMPs
  acute exacerbation, 253
  minocycline, 254

N

Netherton syndrome, 20
Neurological disorders, MMPs
  bacterial meningitis, 254
  hypoxia/ischemia, 254–255
  multiple sclerosis
    acute exacerbation, 253
    minocycline, 254
  vascular cognitive impairment (VCI)
    apparent diffusion coefficients (ADC), 257
    arteriolosclerosis, 255
    bilateral carotid artery occlusion (BCAO), 257
    Binswanger's disease, 255
    CSF patients, 257
    hypoxia inducible factor (HIF), 258

O

Ovalbumin variants, 228

P

Plasminogen activator inhibitor-1 (PAI-1) variants, 225–228

Platelet activation and phospholipid membranes, 80–81
Polyserase-1, 24
Prostasin, 30–32
Protease inhibitors, serpin superfamily. *See* Protein protease inhibitors, serpin superfamily
Protein cofactors
  factor V, 78–79
  factor VII, 69–70
  factor VIII, 70–74
  substrate's role, in Xase function, 74–75
  tissue factor (TF)–fVIIa complex, 69
  Xase-like activity enhancement, atomic model of, 75–78
Protein protease inhibitors, serpin superfamily
  vs. canonical inhibitors, 219–221
  classification of, 187–190
  cleaved form of, 208
  complexes, clearance of, 210
  complex formation, energetics of, 209–210
  conformational change in, 217–219
  and conformational disease, 195–197
  covalent complex, 203–205
    acyl-enzyme intermediate complex, α_1PI and pancreatic elastase, 206
    inferred organization of, 207
    stability of, 208–209
  crmA variants, 228–229
  discovery of, 186–187
  engineered variants
    PAI-1 variants, 225–228
    α1PI variants, 222–225
    therapeutic potential, 223
  exosites
    activated protein C (APC), 215
    HCII, dermatan sulfate/heparin roe, 213–215
    heparin-induced conformational change, 216
    interactions, in binary complexes, 212–213
    in ternary complexes, cofactors, 213–217
    ZPI, factor Xa, 213–214
  human serpins
    biology of, 191–192
    properties, 188–190
  Michaelis complexes
    inferred organization of, 207
    noncovalent and covalent moiety, conformation of, 204
    α_1PI and bovine trypsin, 201
    RCL, 202–203
    thrombin and antithrombin bridging, heparin role, 205
    X-ray structures of, 202
  ovalbumin variants, 228
  polymerization, models of, 196
  RCL interactions, 211–212
  recombinant serpins, for replacement therapy, 221–222
  structural features of
    conformational states of, 194
    conserved residues in, 194
    metastable fold of, 195
    native structure, 192–193
  suicide substrate inhibition mechanism
    branched pathway, mechanism, 198
    peptide bond cleavage, two-stage mechanism, 200
    stoichiometry of inhibition (SI), 199
Protein Z-dependent protease inhibitor (ZPI), 85
PRSS7 gene, 9
PRSS8 gene, 30

R

Reactive center loop (RCL) interactions, serpin, 211–212
Recombinant serpins, for replacement therapy, 221–222

S

Serase-1B, 24
Serine proteases, 242. *See also* Membrane-anchored serine proteases; Trypsin-like serine proteases, proteolytic activation
Serpins, 7. *See also* Protein protease inhibitors, serpin superfamily
  in coagulation cascade, 83
  reactive center loop (RCL), 82
Single-nucleotide polymorphisms (SNP), ADAMTS13, 129–130
Sodium, blood coagulation enzymes, 64, 66
Spinesin, 16
Splice mutations, *ADAMTS13* gene, 128
Stoichiometry of inhibition (SI), 199

T

Testisin, 32–33
Thrombin
  allosteric enzyme
    Botts–Morales scheme, 152–153
    E and E:Na^+ forms, 152
    E^* form, 152
    kinetic scheme, 152
  amino acid substitutions, activity of, 171
  antithrombotic and anticoagulant therapies, 146–147
  autolysis loop, 173
  E and E:Na^+ and E^* form, structures of
    D189, 155
    differences between, 153
    equilibrium, 157
    R221a, 157
    R187A and D222A mutants, 153
  E:Na^+ form, 148
  functional properties of, 172
  interactions, 147–148
  mutants, anticoagulant activity of
    crystal structures of, 167–169
    D146–149e, 166
    E^*–E equilibrium, 165
    E217K and WE, 166–167
    surface representation, 168
    thrombomodulin and hirugen, 169
  procoagulant and anticoagulant activities, 158–162
    accessible surface area of, 162
    Ala mutants of, 158–162
  structure
    active site and exosites, 151
    A and B chains, 148–151
    Na^+ binding site, 150–151
  W215A/E217A (WE), prototypic anticoagulant thrombin
    activated protein C (APC), 163
    equiefficacious doses of, 164
  W215E/D146–149e, 174
  W215, mutation of, 170
Thrombospondin type 1-like repeat (TSR), 115–116
Thrombotic microangiopathies (TMAs), ADAMTS13
  congenital deficiency
    frameshift mutations, 128
    location, linear map of, 127
    missense mutations, 129
    mutations of, 124–127
    single-nucleotide polymorphisms, 129–130
    splice mutations, 128
  deficiency, pathogenesis of, 120–121
  inflammatory conditions, 132
  inhibitors of, 130–131
  levels, decrease in, 132–133
  occurrence of, 122–123
Tick-derived anticoagulant peptide (TAP), 89
Tight junction proteins, 250–251
TIMPs. *See* Tissue Inhibitors of Metalloproteinases
Tissue factor (TF)–fVIIa complex, 69
Tissue factor pathway inhibitor (TFPI), 87
Tissue inhibitors of metalloproteinases (TIMPs), 248–249
TMPRSS2, 13–15
TMPRSS3, 15
TMPRSS4, 16
TMPRSS5. *See* Spinesin
TMPRSS6, 23
TMPRSS13. *See* Mosaic serine protease long-form (MSPL)
TMPRSS11A, serine proteases, 30
TNF-α converting enzyme (TACE), 254
Transmembrane tryptase (TMT). *See* Tryptase γ1
Trypsin-like serine proteases, proteolytic activation
  activation peptide, release of, 60–61
  N-terminus formation, 56
    chymotrypsinogen to chymotrypsin, 54–55
    fIX, (in)activated species of, 59
    zymogen fIX, activation process of, 57
Tryptase γ1, 7–8
Type II transmembrane serine proteases (TTSPs), 2, 4. *See also* Membrane-anchored serine proteases

V

Vascular cognitive impairment (VCI), MMPs
  apparent diffusion coefficients (ADC), 257
  arteriolosclerosis, 255
  bilateral carotid artery occlusion (BCAO), 257
  Binswanger's disease, 255
  CSF patients, 257
  hypoxia inducible factor (HIF), 258
von Willebrand factor, ADAMTS13, 107–109

X

X-ray structures of serpin-protease complexes, 202

Z

Z-dependent protease inhibitor (ZPI), 213–214

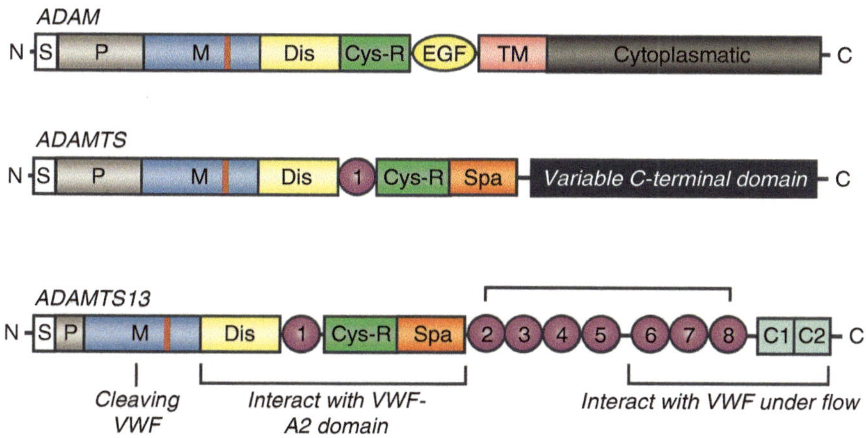

LANCELLOTTI AND DE CRISTOFARO, FIG. 3. Schematic diagram of ADAM, ADAMTS, and ADAMTS13 structure. The structural domains are indicated: signal peptide (S), propeptide (P), metalloprotease (M) (location of zinc-binding motif shown in red), disintegrin domain (Dis), first thrombospondin type 1 (TSP1) repeat (1), cysteine-rich domain (Cys-R), spacer domain (Spa), the second to eighth TSP1 repeats(2) through (8), and two CUB domains (C1 and C2). The metalloprotease domain is the catalytic center that cleaves von Willebrand factor (VWF). The proximal carboxyl-terminal domains from Dis to Spa interact with the A2 domain of VWF. More distal carboxyl-terminal domains (TSP1 2–8) interact with VWF under fluid shear stress. EGF indicates epidermal growth factor-like repeat and TM, transmembrane domain.

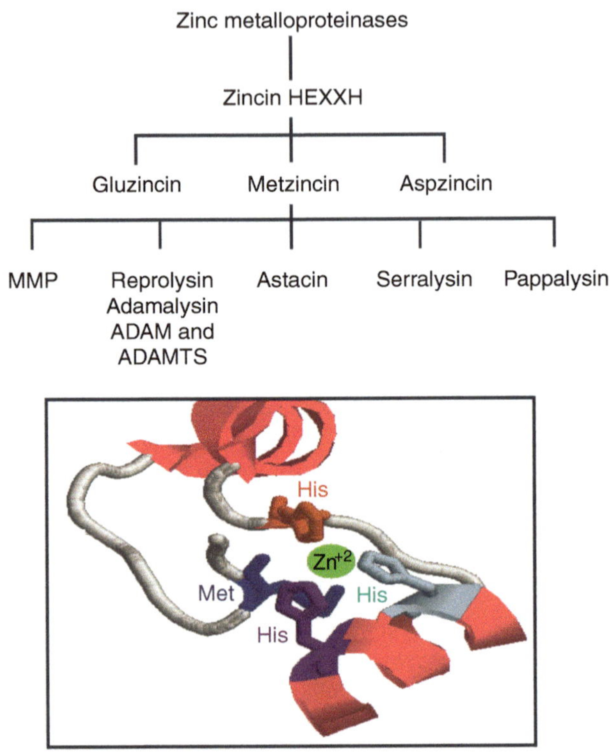

LANCELLOTTI AND DE CRISTOFARO, FIG. 4. The zinc metalloproteinases of the zincin type that have the minimal catalytic zinc-binding motif containing two histidine residues flanking the catalytic glutamate, HEXXH, comprise three superfamilies: the gluzincins, the aspzincins, and the metzincins. Within the metzincins, the major families are the matrixins or matrix metalloproteinases (MMPs), the reprolysins, also known as adamalysins, which include some ADAM (a disintegrin and metalloproteinase) and ADAMTs (ADAMs with thrombospondin repeats proteins), and the astacins. Metzincins have an HEXXHXXGXXHZ. . .M motif with three histidine residues binding the zinc ion and an invariant methionine turn in the active site that generates the name metzincins. X represents any amino acid residue and Z indicates a subfamily specific conserved residue, which is D for both ADAM and ADAMTS members. (Inset) Homology modeling of the metalloproteinase (M) domain of ADAMTS-13. The structure was generated using the program RasMol versus. 2.7.5. The structure of the polypeptide chain 80–290, corresponding to the M-domain of ADAMTS13, was modeled by homology on the crystallographic structure of ADAMTS4 solved at 2.80 Å (PDB entry code: 2RJP). Zinc ion (green) is shown together with the three catalytic His-residues. The "Met-turn" typical of the metzincin family is also indicated.

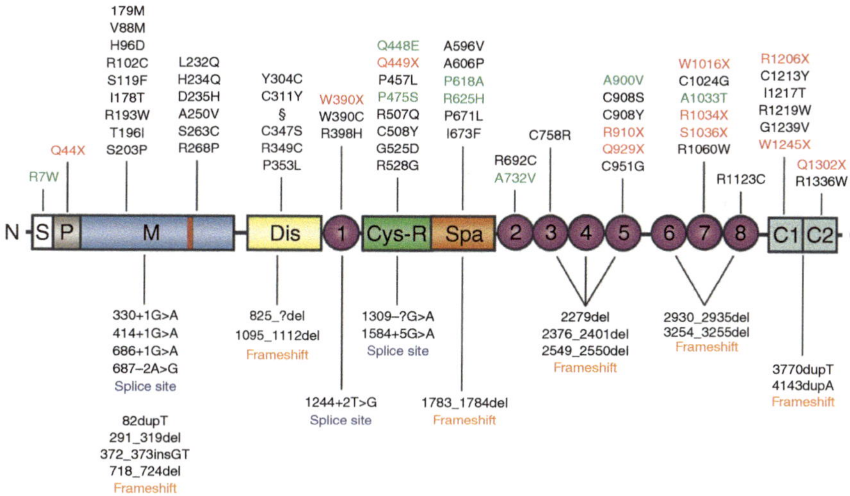

LANCELLOTTI AND DE CRISTOFARO, FIG. 6. Linear map of the location of ADAMTS13 mutations found in patients with congenital thrombotic thrombocytopenic purpura (TTP) (Upshaw–Schulman syndrome (USS)). The missense mutations, nonsense mutations (red), and single nucleotide polymorphisms (SNPs) (green) are shown above the domain structure of ADAMTS13. The mutations that result in alternative splicing of ADAMTS13 mRNA or frameshifts are listed under the domain structure of ADAMTS13. S indicates the signal peptide; P, propeptide; M, metalloprotease (location of zinc-binding motif shown in red); Dis, disintegrin domain; 1, first thrombospondin type 1 (TSP1) repeat; Cys-R, cysteine-rich domain; Spa, spacer domain; 2 through 8, the second to eighth TSP1 repeats; C1 and C2, two CUB domains (for complement C1r/C1s, Uegf, Bmp1 domain) §p.[C322G (+) T323R (+) F324L].

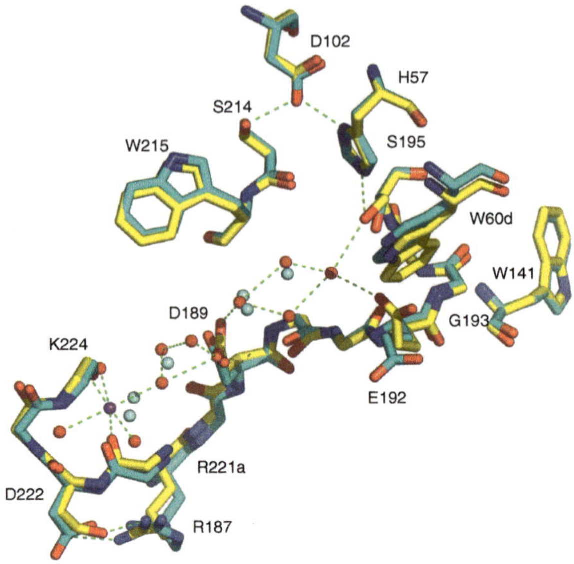

Enrico Di Cera, Fig. 2. Changes induced by Na^+ binding to thrombin revealed by the structures of the E (PDB ID code 1SGI, blue) and E:Na^+ (PDB ID code 1SG8, yellow) forms. The main changes induced by Na^+ (magenta ball) binding are formation of the R187:D222 ion-pair that causes a shift in the backbone O atom of R221a, reorientation of D189 that accounts for the change in substrate binding (lower K_m), shift of the side chain of E192, and shift in the position of the Oγ atom of S195 that engages H57 in an H-bond required for efficient catalysis (higher k_{cat}). A network of water molecules in the E:Na^+ form (red balls) connects Na^+ to the side chain of D189 and continues on to reach the Oγ atom of S195. A critical link in the network is provided by a water molecule that H-bonds to S195 and E192. This water molecule is removed in the E form, causing a reorientation of E192 and S195. The connectivity of water molecules in the E form (cyan balls) is compromised by the lack of Na^+ and proper anchoring of the side chain of D189. H-bonds (broken lines) refer to the E:Na^+ form. No change is observed for W215 and W141, but W60d shifts slightly in the E form.

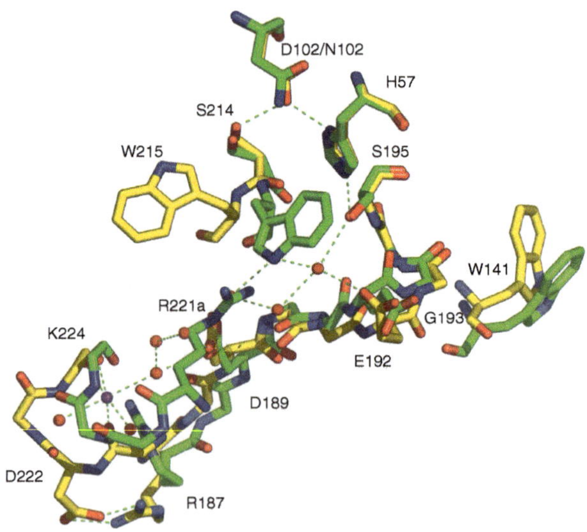

Enrico Di Cera, Fig. 3. The structure of E* (PDB ID code 3BEI, green) compared to E:Na^+ (yellow as in Fig. 2). H-bonds (broken lines) refer to the E:Na^+ form. W215 collapses into the active site, R221a penetrates the protein core and engages the side chain of D189 in a strong ion-pair interaction, and R187 obliterates the Na^+ (magenta ball) site by positioning its guanidinium group within 1 Å from the bound cation. Significant movement is also observed at the level of W141.

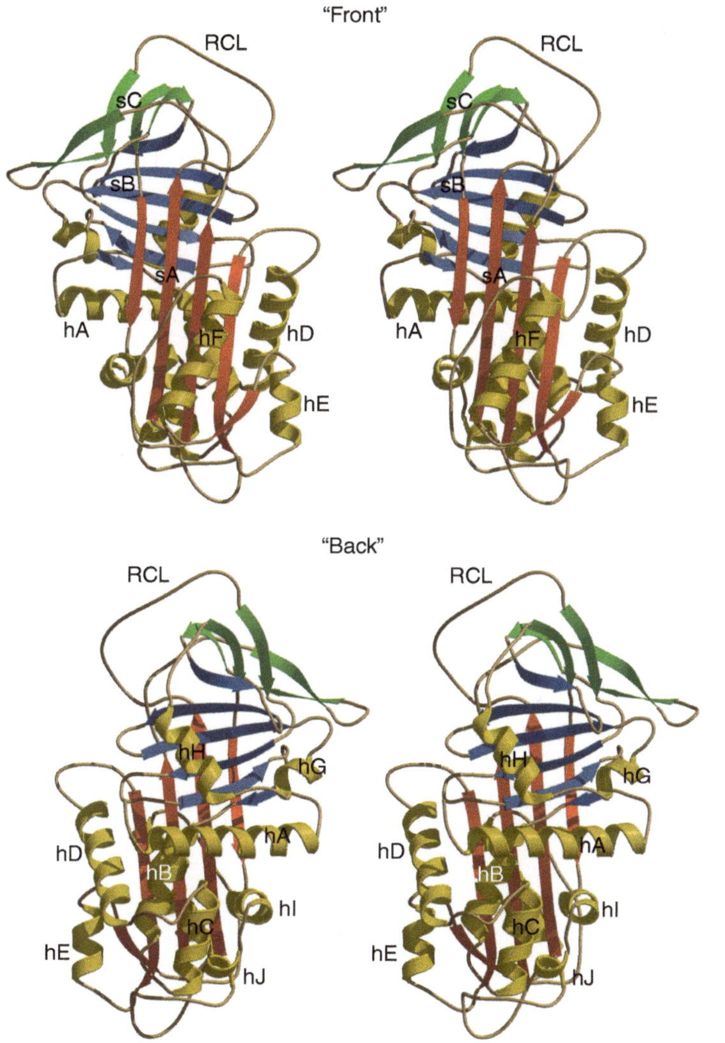

Olson and Gettins, Fig. 1. Native serpin structure. Front and back stereo views of the structure of the serpin, α_1PI in its native state (pdb 1QLP), in ribbon representation. The ~350 amino acid core domain consists of three beta sheets, nine alpha helices (A–J), and a prominent reactive center loop (RCL) that protrudes from the top end of the molecule. Beta sheets are red (sheet A), blue (sheet B), and green (sheet C) and helices are yellow. Reprinted with permission from Ref. 1. Copyright 2002 American Chemical Society.

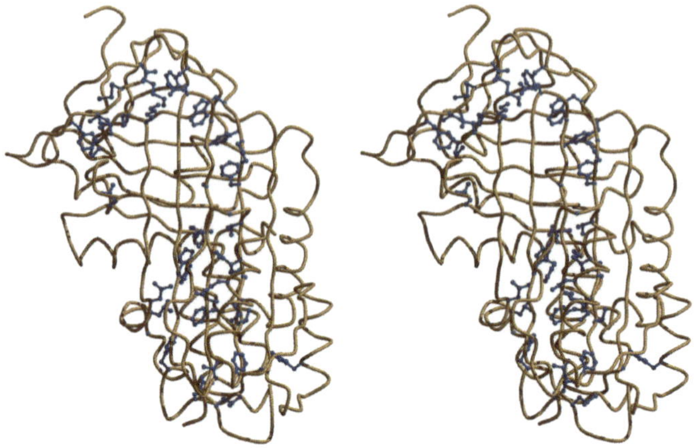

Olson and Gettins, Fig. 2. Conserved residues in serpins. Shown is a stereo view of the RCL-cleaved α_1PI structure in backbone representation (gray, pdb 7API) depicting the location of the 51 highly conserved residues in $>70\%$ of serpins in ball-and-stick representation (blue) that were identified by Irving *et al.*[9] Most of the conserved residues are part of the hydrophobic core. Reprinted with permission from Ref. 1. Copyright 2002 American Chemical Society.

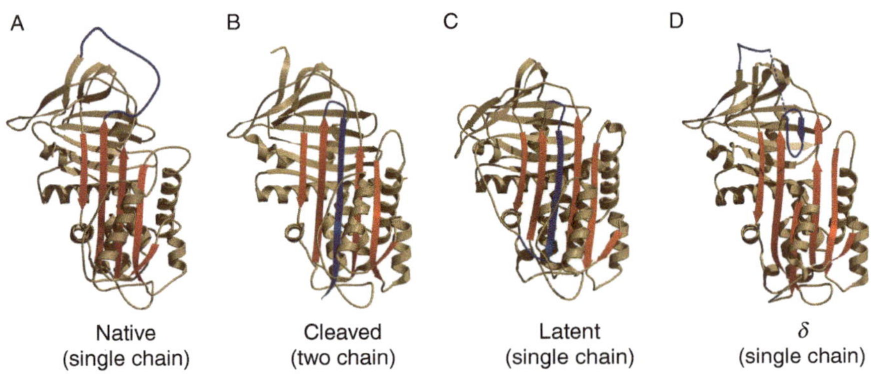

Olson and Gettins, Fig. 3. Conformational states of serpins. Comparison of X-ray structures of the metastable native serpin conformation (A; α_1PI, pdb 1QLP) and the more stable RCL-cleaved (B; α_1PI, pdb 7API), and latent (C; PAI-1, pdb 1C5G) conformations. Also shown is the unusual δ-conformation of an α_1-antichymotrypsin L55P variant with a mutation in the shutter region (D; pdb 1QMN). All structures are shown in ribbon representation with the RCL colored blue, the A β-sheet red, and the remainder of the structure in gray. Reprinted with permission from Ref. 1. Copyright 2002 American Chemical Society.

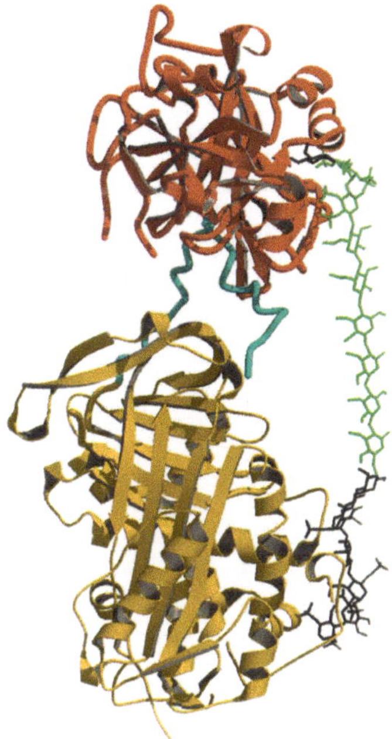

Olson and Gettins, Fig. 9. Role of heparin (green and black) in bridging between thrombin (red) and antithrombin (yellow) to promote formation of the Michaelis complex. The structure (pdb 1SR5)[55] shows specific binding of the high affinity pentasaccharide motif (black) to the heparin-binding site of antithrombin and of the more distal end of the heparin chain to exosite II of thrombin. Reprinted by permission from Macmillan Publishers Ltd,[55] copyright 2004.

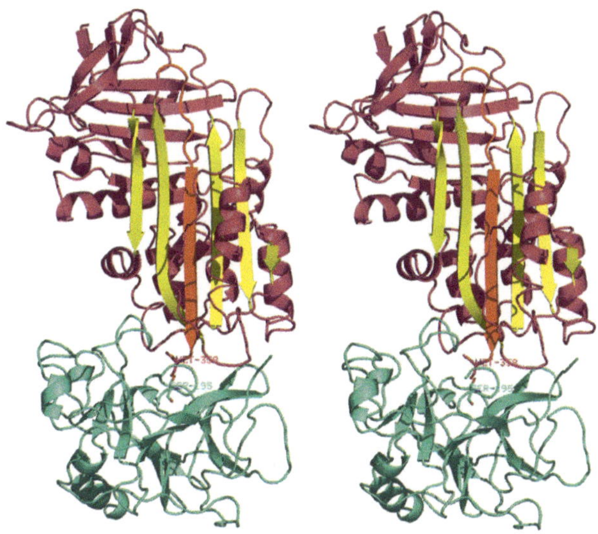

Olson and Gettins, Fig. 10. Stereo view of the covalent, kinetically trapped acyl-enzyme intermediate complex of α_1PI (pink) with pancreatic elastase (cyan) (pdb 2D26).[56] Sheet A of the serpin is highlighted in yellow and the cleaved RCL that has inserted into sheet A is colored red. Within the protease active site, the acyl ester linkage between the catalytic serine and the P1 residue has been displaced relative to the remainder of the catalytic apparatus, rendering the enzyme inactive and trapping the complex kinetically. Other differences in structure between the complex and the constituent free protease and cleaved serpin moieties are minor and limited to the contact interface loops. This research was originally published in The Journal of Biological Chemistry.[56] Copyright the American Society of Biochemistry and Molecular Biology.

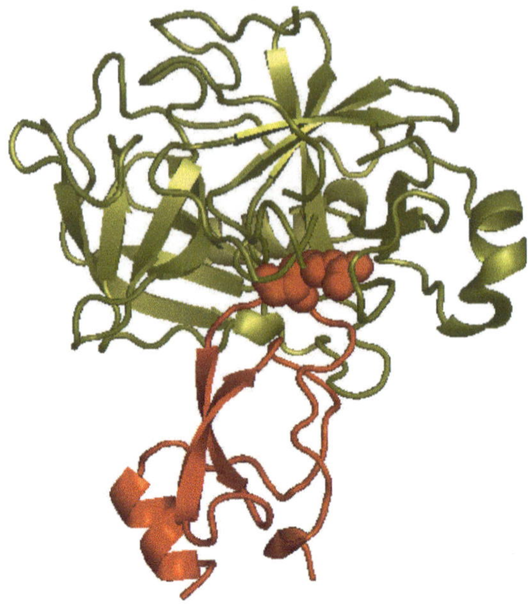

OLSON AND GETTINS, FIG. 14. Structure of a canonical inhibitor–protease complex. X-ray structure of the Kunitz family canonical inhibitor, APPI, in complex with bovine chymotrypsin in ribbon representation (pdb 1CA0). The inhibitor is in red and the protease in yellow. The P1 Arg of the inhibitor is shown in space-filling representation and protrudes into the S1 pocket of the protease. The orientation is the same as in Fig. 7 to illustrate the similar mode of interaction of serpin and canonical inhibitors in the Michaelis complex but a large difference in size of the two inhibitor families.

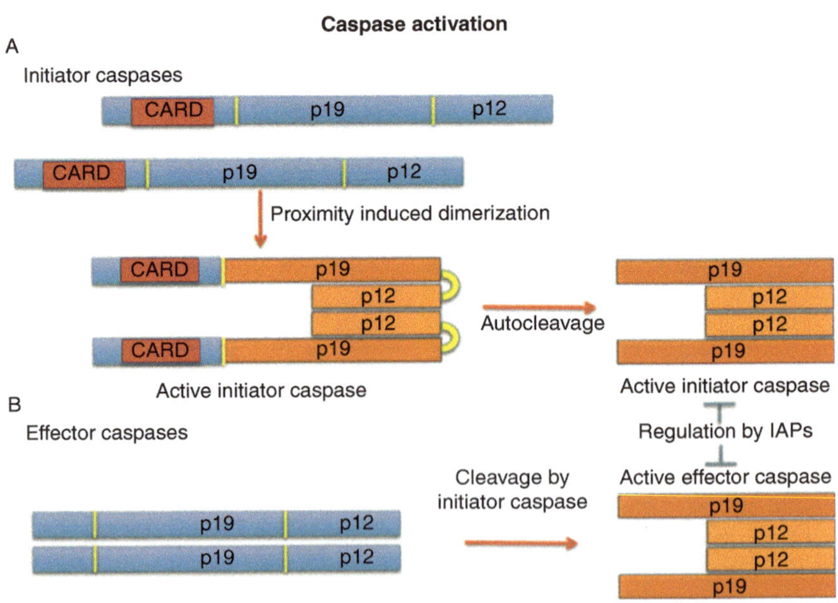

TROY *ET AL.*, FIG. 2. *Activation mechanisms of effector and initiator caspases*. Activation models for initiator and effector caspases are shown. Blue indicates inactive caspase while orange denotes activated caspase. Some of the activated caspases are subject to regulation by IAPs.

Kolibri 360 GmbH
Ferdinand-Jühlke-Straße 7
99095 Erfurt, Deutschland
produktsicherheit@kolibri360.de